TREATISE ON ANALYSIS

Volume VI

This is Volume 10-VI in
PURE AND APPLIED MATHEMATICS

A Series of Monographs and Textbooks

Editors: Samuel Eilenberg and Hyman Bass

A list of recent titles in this series appears at the end of this volume.

Volume 10

TREATISE ON ANALYSIS

10-I. Chapters I–XI, Foundations of Modern Analysis, enlarged ar
corrected printing, 1969

10-II. Chapters XII–XV, enlarged and corrected printing, 1976

10-III. Chapters XVI–XVII, 1972

10-IV. Chapters XVIII–XX, 1974

10-V. Chapter XXI, 1977

10-VI. Chapter XXII, 1978

TREATISE ON
ANALYSIS

J. DIEUDONNÉ
Membre de l'Institut

Volume VI

Translated by

I. G. Macdonald
Queen Mary College
University of London

ACADEMIC PRESS New York San Francisco London 1978
A Subsidiary of Harcourt Brace Jovanovich, Publishers

ACADEMIC PRESS, INC.
111 Fifth Avenue, New York, New York 10003

United Kingdom Edition published by
ACADEMIC PRESS, INC. (LONDON) LTD.
24/28 Oval Road, London NW1 7DX

Library of Congress Cataloging in Publication Data

Dieudonné, Jean Alexandre, Date
Treatise on analysis.

(Pure and applied mathematics, a series of monographs and textbooks : 10)
Except for v. 1, a translation of Eléments d'analyse.
Vol. 2– translated by I. G. MacDonald.
Includes various editions of some volumes.
Includes bibliographies and indexes.
1. Mathematical analysis––Collected works.
I. Title. II. Series.
QA3.P8 vol. 10 1969 510'.8s [515] 75–313532
ISBN 0–12–215506–8 (v. 6)

AMS (MOS) 1970 Subject Classifications: 43–02

PRINTED IN THE UNITED STATES OF AMERICA

"Treatise on Analysis," Volume VI

First published in the French Language under the title "Éléments d'Analyse," tome 6 and copyrighted in 1975 by Gauthier-Villars, Éditeur, Paris, France.

CONTENTS

SCHEMATIC PLAN OF THE WORK

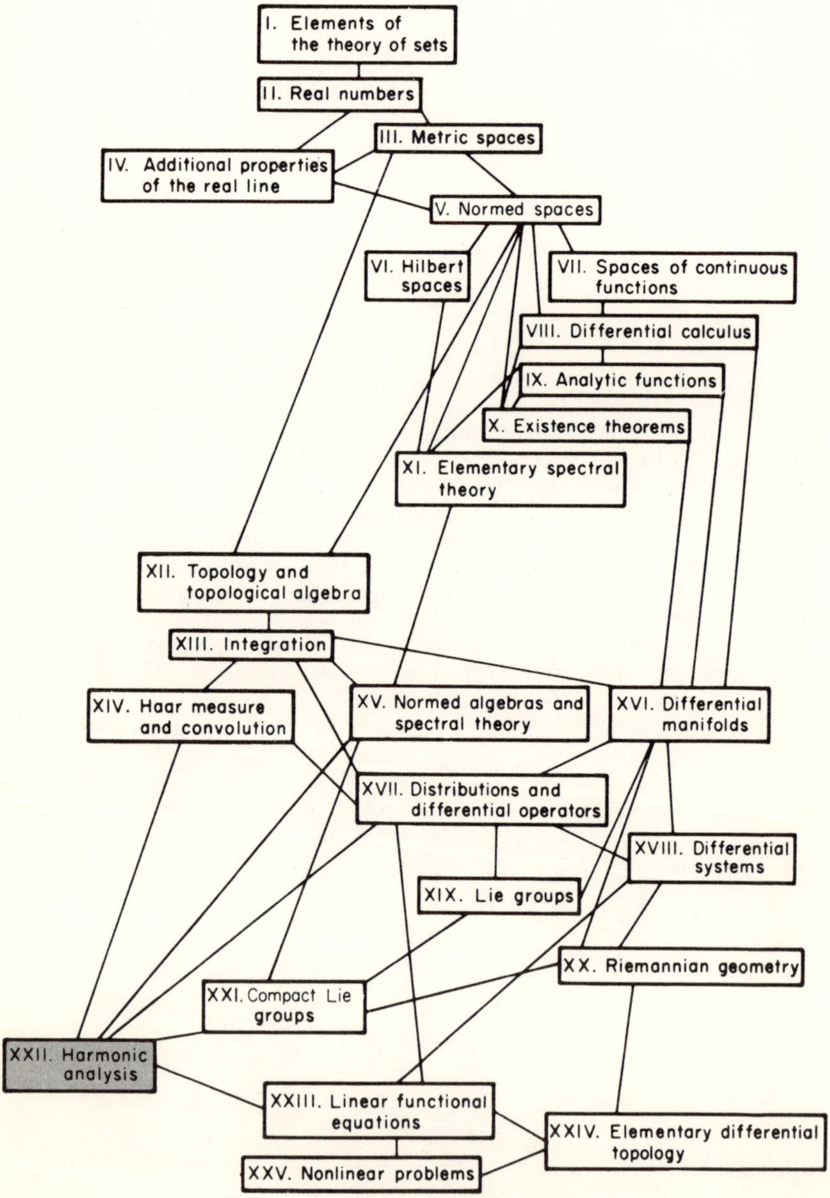

NOTATION

In the following definitions, the first number indicates the chapter and the second number the section of the chapter in which the notation is introduced.

$\mathscr{P}_0$	set of continuous functions p of positive type on G such that $p(e) = 1$: 22.1
$\mathscr{L}^2_{\mathrm{E}}(\mu)$	space of vector-valued functions $\mathbf{f} = \sum_n f_n \mathbf{a}_n$ (with respect to a Hilbert basis $(\mathbf{a}_n)$ of E) such that the f_n are μ-measurable and $\|\mathbf{f}\|^2 = \sum_n \lvert f_n \rvert^2$ is μ-integrable: 22.3
$\mathrm{L}^2_{\mathrm{E}}(\mu)$	Hilbert space of classes of functions in $\mathscr{L}^2_{\mathrm{E}}(\mu)$, with $(\tilde{\mathbf{f}}\|\tilde{\mathbf{g}}) = \int (\mathbf{f}(x)\|\mathbf{g}(x))\, d\mu(x)$, and $\mathrm{N}_2(\mathbf{f}) = (\tilde{\mathbf{f}}\|\tilde{\mathbf{f}})^{1/2}$: 22.3
U^{ind}	representation induced by U: 22.3 and 22.3, prob. 6
$(\boldsymbol{\rho}: \boldsymbol{\sigma})$	multiplicity of $\boldsymbol{\sigma} \in \mathrm{R(H)}$ in the restriction to H of a representation of class $\boldsymbol{\rho}$: 22.5
$\mathscr{C}(\mathrm{G/K})$, $\mathscr{K}(\mathrm{G/K})$, $\mathscr{L}^1_{\mathbf{C}}(\mathrm{G/K})$, $\mathscr{L}^2_{\mathbf{C}}(\mathrm{G/K})$	spaces of functions f belonging, respectively, to $\mathscr{C}(\mathrm{G})$, $\mathscr{K}(\mathrm{G})$, $\mathscr{L}^1_{\mathbf{C}}(\mathrm{G})$, $\mathscr{L}^2_{\mathbf{C}}(\mathrm{G})$ and such that for almost all $s \in \mathrm{G}$, $f(st) = f(s)$ for all $t \in \mathrm{K}$: 22.6
$\mathscr{C}(\mathrm{K\backslash G})$, $\mathscr{K}(\mathrm{K\backslash G})$, $\mathscr{L}^1_{\mathbf{C}}(\mathrm{K\backslash G})$, $\mathscr{L}^2_{\mathbf{C}}(\mathrm{K\backslash G})$	same definition with $f(st) = f(s)$ replaced by $f(st) = f(s)$: 22.6
$\mathscr{C}(\mathrm{K\backslash G/K})$	$\mathscr{C}(\mathrm{G/K}) \cap \mathscr{C}(\mathrm{K\backslash G})$: 22.6

$\mathscr{K}(K\backslash G/K)$	$\mathscr{K}(G/K) \cap \mathscr{K}(K\backslash G)$: 22.6
$\mathscr{L}_c^1(K\backslash G/K)$	$\mathscr{L}_c^1(G/K) \cap \mathscr{L}_c^1(K\backslash G)$: 22.6
$\mathscr{L}_c^2(K\backslash G/K)$	$\mathscr{L}_c^2(G/K) \cap \mathscr{L}_c^2(K\backslash G)$: 22.6
$L_c^1(G/K)$, $L_c^1(K\backslash G)$, $L_c^1(K\backslash G/K)$, $L_c^2(G/K)$, $L_c^2(K\backslash G)$, $L_c^2(K\backslash G/K)$	spaces of classes of functions belonging to the spaces of functions designated by the same notation with L replaced by $\mathscr{L}$: 22.6
$f^\natural$	$f^\natural(s) = \iint_{K\times K} f(tst')\, dm_K(t)\, dm_K(t')$: 22.6
$\mathbf{S}(G/K)$, $\mathbf{S}$	set of spherical functions on G relative to K: 22.6
$\mathscr{F}f$, $\bar{\mathscr{F}}f$	Fourier transform and cotransform of $f \in \mathscr{L}_c^1(G)$: $$\mathscr{F}f(\omega) = \int_G f(x)\omega(x^{-1})\, dm_G(x),$$ $$\bar{\mathscr{F}}f(\omega) = \int_G f(x)\omega(x)\, dm_G(x): 22.7$$
$\mathbf{Z}(G/K)$, $\mathbf{Z}$	set of spherical functions $\omega \in \mathbf{S}(G/K)$ of positive type: 22.7
μ^Δ	Plancherel transform of a linear combination of measures of positive type on G: 22.7 and 22.8
$\mu^\natural$	$\langle \mu^\natural, f\rangle = \langle \mu, f^\natural\rangle$ for $f \in \mathscr{K}(G)$: 22.7
$\bar{\mathscr{F}}'\mu'$	$\bar{\mathscr{F}}'\mu'(x) = \int_{\mathbf{Z}} \omega(x)\, d\mu'(\omega)$, where μ' is a bounded measure on $\mathbf{Z}$: 22.7 and 22.8
$\mathscr{P}_+(K\backslash G/K)$	set of functions of positive type in $\mathscr{C}(K\backslash G/K)$: 22.7
$P_+(G)$	set of measures of positive type on G: 22.8
$P(G)$	set of linear combinations of measures in $P_+(G)$: 22.8
$P'(\mathbf{Z})$	image of P(G) under the Plancherel transformation: 22.8
$\mathscr{P}(G)$	space of linear combinations of continuous functions of positive type on G: 22.8
$\mathscr{P}(K\backslash G/K)$	$\mathscr{P}(G) \cap \mathscr{C}(K\backslash G/K)$: 22.8
$\mathscr{F}\mu$	$(\mathscr{F}\mu)(\omega) = \int_G \omega(x^{-1})\, d\mu(x)$, where μ is a bounded measure on G: 22.8

$\mathscr{P}^1(K\backslash G/K)$	$\mathscr{P}(G) \cap \mathscr{L}^1_{\mathbf{C}}(K\backslash G/K)$: 22.8
$\mathscr{P}^2(K\backslash G/K)$	$\mathscr{P}(G) \cap \mathscr{L}^2_{\mathbf{C}}(K\backslash G/K)$: 22.8
U_ω	unitary representation of G corresponding to the spherical function of positive type ω: 22.9
$\mathrm{Tr}(u)$	trace of an endomorphism of rank 1: 22.9
$\hat{G}$	dual of a locally compact commutative group G: 22.10
$\hat{x}$	element of $\hat{G}$ (i.e., character of G): 22.10
$\langle x, \hat{x} \rangle$	$\hat{x}(x)$: 22.10
$P'(G)$	image of $P(\hat{G})$ under the Plancherel transformation, G being regarded as the dual of $\hat{G}$: 22.10
$\mathscr{P}^1(G)$, $\mathscr{P}^2(G)$	$\mathscr{P}(G) \cap \mathscr{L}^1_{\mathbf{C}}(G)$, $\mathscr{P}(G) \cap \mathscr{L}^2_{\mathbf{C}}(G)$, respectively: 22.10
$m_{\hat{G}}$	Haar measure on G associated to m_G: 22.10
$\mathscr{P}p(G)$	space of almost periodic functions on G: 22.10, prob. 7
$M(f)$	mean of an almost periodic function: 22.10, prob. 9
${}^t u$	transpose of a homomorphism u: $G \to G'$ of locally compact commutative groups: 22.11
$H^\perp$	annihilator in $\hat{G}$ of a subgroup H of G: 22.11
$m_{G/H}$, m_G/m_H	measure on G/H such that $\int_G f(x)\, dm_G(x) = \int_{G/H} dm_{G/H}(\dot{x}) \int_H f(xt)\, dm_H(t)$: 22.12
$\int_{\mathbf{R}^n} f(x)\, dx$	integral of f with respect to Lebesgue measure: 22.14
$\mathscr{F}\mu$	$\mathscr{F}\mu(y) = \int_{\mathbf{R}^n} \exp(-2\pi i(x \mid y))\, d\mu(x)$: 22.14
$\mathscr{F}f$	$\mathscr{F}f(y) = \int_{\mathbf{R}^n} \exp(-2\pi i(x \mid y)) f(x)\, dx$: 22.14

$\mathbf{Z_a}, \mathbf{T_a}$	monothetic group and solenoidal group: 22.14, prob. 3
W_m	Hermite–Weber function: 22.14, prob. 11
$\mathscr{S}(\mathbf{R}^n)$, $\mathscr{S}$	space of declining functions: 22.16
$q_{s,m}$	seminorms on $\mathscr{S}$: $q_{s,m}(f) = \sup\limits_{x \in \mathbf{R}^n,\ \lvert\nu\rvert \le s} (1 + r^2(x))^m \lvert D^\nu f(x)\rvert$: 22.16
$\mathscr{S}'(\mathbf{R}^n)$, $\mathscr{S}'$	space of tempered distributions: 22.17
$\mathscr{F}\mathrm{T}$, $\bar{\mathscr{F}}\mathrm{T}$	Fourier transform and cotransform of a distribution: 22.17
$\mathop{\ast}\limits_{n=1}^{\infty} \mu_n$	infinite convolution product of probability measures: 22.17, prob. 23
$\Gamma(\mathrm{T})$, Γ	set of $\eta \in \mathbf{R}^n$ such that $\exp(2\pi(\cdot\mid\eta))\cdot\mathrm{T}$ is tempered: 22.18, prob. 4
$\mathscr{F}\mathscr{L}\mathrm{T}$, $\mathscr{L}\mathrm{T}$, $\mathscr{M}\mathrm{T}$	Fourier–Laplace, Laplace and Mellin transforms of a distribution T: 22.18, prob. 4
$\mathscr{C}\mathrm{T}$	Carleman transform of a mean-periodic distribution: 22.18, prob. 22
$\lvert\nu\rvert$	$\sup\limits_k \lvert\nu_k\rvert$, for $\nu = (\nu_1, \ldots, \nu_n)$: 22.19 (*and only in this section*)
$\mathscr{S}(\mathbf{Z}^n)$	space of rapidly decreasing families $(c_\nu)_{\nu \in \mathbf{Z}^n}$: 22.19
q_m	seminorms on $\mathscr{S}(\mathbf{Z}^n)$: $q_m((c_\nu)) = \sup\limits_{\nu \in \mathbf{Z}^n} (1 + \lvert\nu\rvert)^m \lvert c_\nu\rvert$: 22.19
$\mathrm{H}^s(\mathbf{R}^n)$, H^s	Sobolev space: 22.20
$(\mathrm{S}\mid\mathrm{T})_s$, $\lVert\mathrm{T}\rVert_s$	$(\mathrm{S}\mid\mathrm{T})_s = \int_{\mathbf{R}^n} (1 + \lvert\xi\rvert^2)^s \mathscr{F}\mathrm{S}(\xi)\overline{\mathscr{F}\mathrm{T}(\xi)}\,d\xi$,
H^∞, $\mathrm{H}^{-\infty}$	$\mathrm{H}^\infty = \bigcap\limits_s \mathrm{H}^s$, $\mathrm{H}^{-\infty} = \bigcup\limits_s \mathrm{H}^s$: 22.20

CHAPTER XXII

HARMONIC ANALYSIS

We have seen in Chapter XXI how the study of the *finite*-dimensional linear representations of Lie groups leads to a profound analysis of the structure of these groups, and especially of the structure of compact connected Lie groups. By contrast, the present chapter is largely concerned with *infinite*-dimensional linear representations of locally compact groups, which need not be Lie groups, and the focus of interest now is not so much the groups as the representations themselves.

The origin of this subject lies at the heart of classical analysis: it is the theory of *Fourier series and integrals*, whose abundant development and manifold applications since the end of the 18th century are well known. But it was not until about 1925, as a result of the fundamental work of Hermann Weyl, that it came to be realized that the theory of "Fourier expansions" of periodic functions expresses precisely the *decomposition* of the regular representation of the compact group $\mathbf{T} = \mathbf{R}/\mathbf{Z}$, a particular case of the Peter–Weyl theorem (21.2.3). However, the inverse problem, that of the "summation of trigonometric series," together with the whole theory of Fourier integrals, made it clear that this type of "discrete" decomposition, which works so well for compact groups, does not generalize to noncompact groups; it has to be replaced by a "continuous" decomposition, in which series are replaced by integrals. There existed already in the spectral theory of Hilbert (15.11) a model of this type of generalization, and in the end it was spectral theory, in the form given to it by Gelfand and his school, which about 1940 led to the creation of a body of doctrine of an elegant simplicity, namely *commutative harmonic analysis*: one which is valid for all commutative locally compact groups and which, for the groups $\mathbf{R}^n$, $\mathbf{Z}^n$, and $\mathbf{T}^n$, reduces to the classical theories.

The relatively elementary character of this theory is a consequence of the basic fact that the irreducible representations of commutative locally compact groups are all *one-dimensional*. This fact also lies at the root of the remarkable "Pontrjagin duality" for these groups, which dominates their whole theory **(22.10.8)**. Clearly one has no right to expect anything as simple for noncommutative groups, as the example of compact groups already shows. However, here again classical analysis provides the clues: in addition to Fourier series, it offers types of series expansions in terms of "special functions" other than the exponentials $e^{2\pi i n x}$. Élie Cartan, about 1930, was the first to show that one of these, namely expansions in terms of the "spherical harmonics" of Laplace, should be regarded as attached to the representation theory of the orthogonal group $\mathbf{SO}(3)$. This observation became the starting-point of a far-reaching theory of infinite-dimensional representations of Lie groups, which is still now far from complete, and which goes by the name of *noncommutative harmonic analysis*. It is out of the question for us in this treatise to embark on this formidable theory, hedged about as it is with difficulties both conceptual and technical. We shall merely refer the reader to the works [76], [92], and [97] in the bibliography, which will give him an idea of the subject; he will also see in [91] and [96] how it has incorporated the theory of most of the "special functions" of classical analysis (such as the hypergeometric function and its degenerations), which find their natural context there.

However, we have in this chapter gone beyond the confines of commutative harmonic analysis, in order to bring out certain simple ideas which are fundamental in noncommutative harmonic analysis and which do not emerge clearly enough in the commutative case. One of these notions, that of a *function of positive type*, appears fortuitous and somewhat artificial until it is realized that it is the counterpart, for linear representations of groups, of the notion of a *topologically cyclic* representation **(22.1)**. Still more important is the notion of an *induced representation*, due to Frobenius, which provides a powerful tool for constructing representations of a group from representations of its subgroups. Fortunately it is possible to introduce these notions in a way which is not entirely superficial and which nevertheless leads to substantial results, without however requiring more in the way of technical means than commutative harmonic analysis provides. That this is so is due to a discovery of I. Gelfand, namely that a certain "noncommutative" situation involving an induced representation (which occurs rather frequently in practice, and in particular in connection with "spherical harmonics") depends in fact on an involutory algebra of functions which is *commutative* **(22.6.3)** and therefore amenable to the general theorems of Plancherel–Godement and Bochner–Godement, just like classical Fourier analysis. We have therefore preferred to begin by developing all the consequences of these general theorems in the context of "Gelfand pairs" before specializing to the particular case of commutative groups, if only because in this way one sees much more clearly what is special to this particular case and what is not. Moreover, the general notion of "spherical function" for Gelfand pairs (22.6.6) is an important ingredient in the general theory of noncommutative harmonic analysis [97].

Since 1950, the theory of distributions has provided a much wider and more satisfactory framework for classical harmonic analysis (on $\mathbf{R}^n$, $\mathbf{Z}^n$, or $\mathbf{T}^n$) (and moreover it plays an important role in the work of Harish-Chandra on the representations of semisimple Lie groups [97]). Not only does it provide much more freedom in the manipulation of fundamental operations such as differentiation, convolution, and passage to the limit, but also it dissolves a whole range of false "pathologies" such as the distinction between "Fourier series" and "trigonometric series" introduced in the 19th century. We have therefore sought to emphasize the notions attached to the theory of distributions (and hence to the concept of a function as an "operator") rather than the archaic ideas of pointwise convergence or summability (22.16–22.20).

It is also in the context of the theory of distributions that the most spectacular advances have been made during the last twenty years in the theory of linear partial differential equations, the oldest (because it goes back to Laplace and Fourier) and best-known application of harmonic analysis. We shall devote the whole of Chapter XXIII to this topic. However, the usefulness of harmonic analysis does not end there. Already in the 19th century it was applied in the classical theory of modular functions and theta-functions ([75], [84], and [94]), and this domain of application is much more in evidence nowadays, because the general concept of "automorphic form" has been completely absorbed by noncommutative harmonic analysis ([74] and [86]). A much more unexpected application is that which has followed on the introduction of *adèles* and *idèles* in number theory: it has come to be realized that *all* of algebraic number theory, including abelian class-field theory, can be presented as an application of the results of commutative harmonic analysis to the adèle and idèle groups of a number field [98]; and the extension of these ideas to noncommutative adèle groups looks most promising ([36], [74], [78]). There are also other points of contact, still more mysterious, between harmonic analysis and number theory, especially those connected with problems of diophantine approximation ([87], [90]). The limitations of this treatise do not permit us to develop these applications, nor yet many others: to probability theory, holomorphic functions, Tauberian theorems, spectral synthesis, and so on. Only by consulting the works listed in the bibliography will the reader be able to form a just estimate of the extraordinary fecundity, in the most diverse domains, of the concepts and results of harmonic analysis.

1. CONTINUOUS FUNCTIONS OF POSITIVE TYPE

(22.1.1) Let G be a *unimodular* (separable, metrizable) locally compact group, e its identity element, and β a Haar measure on G. We shall denote by A (or A(G)) the involutory Banach subalgebra $L_{\mathbf{C}}^1(G) + \mathbf{C}\varepsilon_e$ of $M_{\mathbf{C}}^1(G)$. The result of **(21.1.7)** may be expressed by saying that there is a canonical one-to-one correspondence between the continuous unitary representations of G on a separable Hilbert space E, and the representations (in the sense of **(15.5)**) of the involutory algebra A, for which the restriction to $L_{\mathbf{C}}^1(G)$ is *nondegenerate*. From **(15.5.6)** we know that if V is a representation of A on E, then E is the Hilbert sum of a sequence (E_n) of closed subspaces stable under V, such that the restriction V_n of V to each E_n is *topologically cyclic*; if the restriction of V to $L_{\mathbf{C}}^1(G)$ is nondegenerate, then by definition **(15.5.5)** the same is true of the restriction of each V_n. We shall say that a continuous unitary representation of G on E is *topologically cyclic* (or *monogenic*) if the corresponding representation of A(G) on E is topologically cyclic. We shall show that the study of topologically cyclic representations of G on E can be canonically reduced to that of certain *continuous functions* on G.

(22.1.2) Consider a continuous unitary representation U of G on a separable Hilbert space E, and for each $x_0 \in E$ and each $s \in G$, put

$$\psi_{U, x_0}(s) = (U(s) \cdot x_0 | x_0). \tag{22.1.2.1}$$

It follows immediately from this definition that ψ_{U, x_0} is a *continuous* complex-valued function on G; also it is *bounded* on G, because $\|U(s) \cdot x_0\| = \|x_0\|$ and therefore

$$|\psi_{U, x_0}(s)| \leqq \|x_0\|^2 = \psi_{U, x_0}(e) \tag{22.1.2.2}$$

for all $s \in G$; also, we have

$$\begin{aligned} \psi_{U, x_0}(s^{-1}) &= ((U(s))^* \cdot x_0 | x_0) \\ &= (x_0 | U(s) \cdot x_0) \\ &= \overline{\psi_{U, x_0}(s)}. \end{aligned} \tag{22.1.2.3}$$

(22.1.3) *Let p be a continuous complex-valued function on* G. *Then the following conditions are equivalent.*

(a) *There exists a continuous unitary representation U of* G *on a separable Hilbert space* E, *and a vector* $x_0 \in E$ *such that* $p = \psi_{U, x_0}$.

(b) *For each function* $f \in \mathscr{K}(G)$, *we have*

(22.1.3.1) $$\langle p, \check{\bar{f}} * f \rangle = \iint p(y^{-1}z)\overline{f(y)}f(z)\, d\beta(y)\, d\beta(z) \geqq 0.$$

(c) *The function p is bounded, and for each bounded measure* μ *on* G *we have*

(22.1.3.2) $$\langle p, \check{\bar{\mu}} * \mu \rangle = \iint p(y^{-1}z)\, d\bar{\mu}(y)\, d\mu(z) \geqq 0.$$

If p satisfies these conditions, there exists a topologically cyclic continuous unitary representation V_1 *of* G *on a separable Hilbert space* E_1, *and a totalizing vector* x_1, *such that* $p = \psi_{V_1, x_1}$. *Moreover, if* V_2 *is a topologically cyclic representation of* G *on a separable Hilbert space* E_2, *with totalizing vector* x_2, *such that* $p = \psi_{V_2, x_2}$, *then there exists an isomorphism T of the Hilbert space* E_1 *onto the Hilbert space* E_2, *such that* $T \cdot x_1 = x_2$ *and* $V_2 = TV_1T^{-1}$.

We shall first show that (a) implies (b). If $p = \psi_{U, x_0}$, then p is bounded (22.1.2.2), hence belongs to $\mathscr{L}_{\mathbf{C}}^{\infty}(G)$; consequently the mapping $\omega : f \mapsto \langle p, f \rangle$ is a continuous linear form on $\mathscr{L}_{\mathbf{C}}^{1}(G)$, and we have

$$\omega(f) = \int f(s)(U(s) \cdot x_0 | x_0)\, d\beta(s) = (U(f) \cdot x_0 | x_0)$$

by virtue of the definition of $U(f)$ (21.1.4.2). The relation (22.1.3.1) then follows from the fact that $\tilde{f} \mapsto \omega(f)$ is a *positive linear form* on the involutory algebra $L_{\mathbf{C}}^{1}(G)$ (15.6.4).

Next we shall prove that (b) implies (c). Suppose first that μ has *compact support*. Then, for every $f \in \mathscr{K}(G)$, $\mu * f$ is a function belonging to $\mathscr{K}(G)$ ((14.9.2) and (14.5.4)); hence we have

(22.1.3.3) $$\begin{aligned} 0 &\leqq \langle p, \check{\bar{f}} * \check{\bar{\mu}} * \mu * f \rangle \\ &= \iiiint p(xyzt)\check{\bar{f}}(x)f(t)\, d\beta(x)\, d\beta(t)\, d\check{\bar{\mu}}(y)\, d\mu(z) \\ &= \iint d\check{\bar{\mu}}(y)\, d\mu(z) \iint p(xyzt)\check{\bar{f}}(x)f(t)\, d\beta(x)\, d\beta(t). \end{aligned}$$

Let (V_n) be a fundamental system of neighborhoods of e in G, and for each n let f_n be a continuous function $\geqq 0$ with support contained in V_n, and such that $\int f_n \, d\beta = 1$. By applying (14.11.1) to the group $G \times G^0$, we see that the sequence of functions

$$(y, z) \mapsto \iint p(xyzt)\check{f}_n(x) f_n(t) \, d\beta(x) \, d\beta(t)$$

converges uniformly to the function $(y, z) \mapsto p(y, z)$ on each compact subset of G. By replacing f by f_n in (22.1.3.3) and then letting $n \to \infty$, we obtain the relation (22.1.3.2) by virtue of (13.19.3).

In particular, take μ to be an atomic measure with *finite* support $\{s_1, \ldots, s_n\}$, defined by a mass ξ_j placed at the point s_j for $1 \leqq j \leqq n$ (13.1.13). The formula (22.1.3.2) then becomes

$$\text{(22.1.3.4)} \qquad \sum_{j, k} p(s_j^{-1} s_k) \bar{\xi}_j \xi_k \geqq 0$$

for all $\xi_j \in \mathbf{C}$ $(1 \leqq j \leqq n)$; in other words, the sesquilinear form Φ on $\mathbf{C}^n$ with matrix $(p(s_j^{-1} s_k))$ must be such that $\Phi(x, x) \geqq 0$ for all vectors x. Since $\Phi(x + y, x + y) = \Phi(x, x) + \Phi(x, y) + \Phi(y, x) + \Phi(y, y)$, this implies first of all that $\Phi(x, y) + \Phi(y, x)$ is real; then, by replacing x by λx where $\lambda \in \mathbf{C}$, that $\Phi(y, x) = \overline{\Phi(x, y)}$; so that Φ is a *positive* (*semidefinite*) *Hermitian form.* In particular, by taking $n = 2$, $s_1 = e$, $s_2 = s$, we see that the matrix

$$\begin{pmatrix} p(e) & p(s) \\ p(s^{-1}) & p(e) \end{pmatrix}$$

must be positive Hermitian, which (6.2.1) implies that

$$\text{(22.1.3.5)} \qquad p(s^{-1}) = \overline{p(s)}, \qquad |p(s)| \leqq p(e)$$

for all $s \in G$.

To complete the proof that (b) implies (c), consider an arbitrary bounded measure μ on G. If (K_n) is an increasing sequence of compact sets whose union is G (3.18.3), then $|\mu|(G - K_n) \to 0$ (13.8.7); hence, if we put $\mu_n = \varphi_{K_n} \cdot \mu$, the sequence of norms $\|\mu - \mu_n\|$ tends to 0 as $n \to \infty$, and so by (14.6.2) the same is true of the sequence of norms $\|\check{\mu} * \mu - \check{\mu}_n * \mu_n\|$. But μ_n has compact support, hence $\langle p, \check{\mu}_n * \mu_n \rangle \geqq 0$ by what has already been proved; and since p is bounded, $\langle p, \check{\mu}_n * \mu_n \rangle$ tends to $\langle p, \check{\mu} * \mu \rangle$, which is therefore $\geqq 0$.

Finally, let us prove that (c) implies (a). The mapping $\omega: \mu \mapsto \langle p, \mu \rangle$ is a positive linear form on the involutory Banach algebra $M_{\mathbf{C}}^1(G)$, which has an identity element; its restriction to the *Banach subalgebra* A is therefore a *continuous* positive linear form on A **(15.6.11)**. Let $\mathfrak{n}$ be the left ideal in A consisting of the measures $f \cdot \beta + \lambda\varepsilon_e = \mu$ such that $\omega(\check{\bar{\mu}} * \mu) = 0$ **(15.6.8)**, and let $\pi: A \to A/\mathfrak{n} = E_0$ be the canonical mapping. Relative to the pre-Hilbert space structure on E_0 defined by ω **(15.6.8)**, we have

$$\|\pi(\mu)\|^2 = \omega(\check{\bar{\mu}} * \mu) \leqq \omega(\varepsilon_e)\|\check{\bar{\mu}} * \mu\| \leqq \omega(\varepsilon_e)\|\mu\|^2$$

(15.6.11), so that π is *continuous*; and since A is separable, so also is E_0. The space E_0 may therefore be identified with a dense subspace of a separable Hilbert space E, and for each $\mu \in A$ the endomorphism $\pi(v) \mapsto \pi(\mu * v)$ of E_0 extends to a continuous endomorphism $x \mapsto V(\mu) \cdot x$ of E, so that $\mu \mapsto V(\mu)$ is a representation of A on E **(15.6.10)**. This representation admits a *totalizing* vector $x_0 = \pi(\varepsilon_e)$ **(15.6.10)**, and we have $\omega(\mu) = (V(\mu) \cdot x_0 \,|\, x_0)$. Finally, the restriction of V to $L_{\mathbf{C}}^1(G)$ is *nondegenerate*: indeed, since π is continuous, it is enough to show that the functions $f * g$ form a total set in $\mathscr{L}_{\mathbf{C}}^1(G)$ as f and g run through $\mathscr{L}_{\mathbf{C}}^1(G)$ **(15.6.10)**, and this is a consequence of regularization **(14.11.1)**. We have therefore established that (c) implies (a) **(22.1.1)**, and moreover that we may assume the representation U to be topologically cyclic and given by

(22.1.3.6) $$U(s) \cdot \pi(v) = \pi(\varepsilon_s * v)$$

for all $s \in G$ and all measures $v \in A$. In particular, $U(s) \cdot x_0 = \pi(\varepsilon_s)$, and by virtue of **(21.1.7)** the vectors $\pi(\varepsilon_s)$ for $s \in G$ form a *total* set in E.

Finally, the uniqueness assertion in **(22.1.3)** is a consequence of **(15.6.7)**.

(22.1.3.7) A continuous function p on G which satisfies the equivalent conditions of **(22.1.3)** is said to be a *continuous function of positive type* on G.

(22.1.4) *Examples.* (i) The constant functions $\geqq 0$ are clearly continuous functions of positive type.

(ii) Consider the left regular representation $s \mapsto R(s)$ of G on the Hilbert space $L_{\mathbf{C}}^2(G)$ **(21.1.9)**. For each function $f \in \mathscr{L}_{\mathbf{C}}^2(G)$, we have

$$\begin{aligned}(R(s) \cdot \check{\bar{f}} \,|\, \check{\bar{f}}) &= \int \overline{f(s^{-1}t)} f(t)\, d\beta(t) \\ &= \int \check{\bar{f}}(t^{-1}s) f(t)\, d\beta(t) = (f * \check{\bar{f}})(s)\end{aligned}$$

(14.10.1); the function $f * \check{\bar{f}}$ is therefore *continuous of positive type.*

(22.1.5) *Let p be a continuous function of positive type on G. Then for all s, t in G, we have*

(22.1.5.1)
$$|p(s) - p(t)|^2 \leqq 2p(e)(p(e) - \mathscr{R}(p(s^{-1}t))).$$

In particular, p is uniformly continuous on G relative to a left- or right-invariant distance **(12.9)**.

Since $p(s) = (U(s) \cdot x_0 | x_0)$ **(22.1.3)**, the Cauchy–Schwarz inequality gives

$$\begin{aligned}
|p(s) - p(t)|^2 &= |((U(s) - U(t)) \cdot x_0 | x_0)|^2 \\
&\leqq \|x_0\|^2 \cdot \|U(s) \cdot x_0 - U(t) \cdot x_0\|^2 \\
&= p(e)(\|U(s) \cdot x_0\|^2 + \|U(t) \cdot x_0\|^2 - 2\mathscr{R}(U(s) \cdot x_0 | U(t) \cdot x_0)) \\
&= p(e)(2\|x_0\|^2 - 2\mathscr{R}(U(s^{-1}t) \cdot x_0 | x_0)) \\
&= 2p(e)(p(e) - \mathscr{R}(p(s^{-1}t))).
\end{aligned}$$

The last assertion follows from the definitions and the relation $p(s^{-1}) = \overline{p(s)}$.

(22.1.6) *Let p be a continuous function of positive type on G. For all a, b in G, the function $s \mapsto p(asb)$ is a linear combination of four continuous functions of positive type.*

With the same notation as in **(22.1.5)**, put $U(b) \cdot x_0 = y_0$ and $U(a)^{-1} \cdot x_0 = z_0$. Then

$$\begin{aligned}
4p(asb) &= 4(U(asb) \cdot x_0 | x_0) \\
&= 4(U(s) \cdot y_0 | z_0) \\
&= (U(s) \cdot (y_0 + z_0) | y_0 + z_0) - (U(s) \cdot (y_0 - z_0) | y_0 - z_0) \\
&\quad + i(U(s) \cdot (y_0 + iz_0) | y_0 + iz_0) - i(U(s) \cdot (y_0 - iz_0) | y_0 - iz_0).
\end{aligned}$$

(22.1.7) Let $p = \psi_{U, x_0}$ be a continuous function of positive type on G, with the notation of **(22.1.2)**, and let (a_n) be a Hilbert basis of E. Then we may write

$$U(s) \cdot x_0 = \sum_n p_n(s) a_n,$$

where each function p_n is continuous on G and we have

$$\sum_n |p_n(s)|^2 = \|x_0\|^2$$

for all $s \in$ G. It follows that

(22.1.7.1) $$p(s^{-1}t) = (U(t) \cdot x_0 \mid U(s) \cdot x_0) = \sum_n \overline{p_n(s)}\, p_n(t)$$

with $\sum_n |p_n(s)p_n(t)| \leqq \|x_0\|^2$.

(22.1.8) It is clear that every linear combination, with coefficients $\geqq 0$, of continuous functions of positive type is again continuous of positive type. Moreover:

(22.1.9) *The product of two continuous functions of positive type on* G *is a continuous function of positive type.*

Let p, q be continuous functions of positive type. Then it follows from (22.1.7.1) that, for each function $f \in \mathscr{K}(\mathrm{G})$, we may write

$$\iint p(s^{-1}t)q(s^{-1}t)\overline{f(s)}\, f(t)\, d\beta(s)\, d\beta(t)$$
$$= \sum_n \iint q(s^{-1}t)\overline{p_n(s)f(s)}\, p_n(t)f(t)\, d\beta(s)\, d\beta(t) \geqq 0,$$

in view of (13.8.5).

(22.1.10) Let $\mathscr{P}_0$ denote the set of continuous functions of positive type on G such that $p(e) = 1$. Since the support of β is the whole of G, $\mathscr{P}_0$ may be identified with a subset of $\mathrm{L}^\infty_{\mathbf{C}}(\mathrm{G})$ consisting of functions such that $\mathrm{N}_\infty(p) = 1$ (22.1.2.2); also $\mathscr{P}_0$ is a closed subset of the Banach space $\mathscr{C}^\infty_{\mathbf{C}}(\mathrm{G})$ (7.2).

(22.1.11) *The topologies induced on* $\mathscr{P}_0$ *by the topology of the Fréchet space* $\mathscr{C}_{\mathbf{C}}(\mathrm{G})$ (12.14.6) *and the weak topology of* $\mathrm{L}^\infty_{\mathbf{C}}(\mathrm{G})$ *considered as the dual of* $\mathrm{L}^1_{\mathbf{C}}(\mathrm{G})$ ((13.17.1) *and* (12.15)) *are the same.*

The proof is in several steps.

(22.1.11.1) *On every subset* B *of* $\mathscr{C}_{\mathbf{C}}(\mathrm{G})$ *consisting of uniformly bounded functions, the topology induced by the weak topology of* $\mathrm{L}^\infty_{\mathbf{C}}(\mathrm{G})$ *is coarser than that induced by the topology of* $\mathscr{C}_{\mathbf{C}}(\mathrm{G})$.

Suppose we have $|p(s)| \leqq M$ for all functions $p \in B$ and all $s \in G$. Let $f_1, \ldots, f_n$ be functions belonging to $\mathscr{L}_C^1(G)$; then for each $\varepsilon > 0$ there exists a compact subset K of G such that $\int_{G-K} |f_j| \, d\beta \leqq \varepsilon/4M$ for $1 \leqq j \leqq n$ (13.9.14). Put $N = 1 + \sup_j \int_K |f_j| \, d\beta$. For each function $p_0 \in B$, the set V of functions $p \in B$ such that $|p(s) - p_0(s)| \leqq \varepsilon/2N$ for all $s \in K$ is a neighborhood of p_0 for the topology induced by that of $\mathscr{C}_C(G)$. But for these functions $p \in V$ we have, for $1 \leqq j \leqq n$,

$$|\langle p - p_0, f_j \rangle| \leqq \int_K |p - p_0| \cdot |f_j| \, d\beta + 2M \int_{G-K} |f_j| \, d\beta \leqq \varepsilon$$

because $|p(s) - p_0(s)| \leqq 2M$ for all $s \in G$; hence V is contained in the neighborhood of p_0 for the weak topology defined by the conditions $|\langle p - p_0, f_j \rangle| \leqq \varepsilon$ for $1 \leqq j \leqq n$.

Next, we have the following proposition:

(22.1.11.2) *Let* B *be a subset of* $\mathscr{C}_C(G)$ *consisting of uniformly bounded functions and having the following property: for each* $p_0 \in B$, *each compact subset* K *of* G *and each* $\varepsilon > 0$, *there exists a neighborhood* U *of* p_0 *in* B *for the weak topology of* $L_C^\infty(G)$ *and a compact neighborhood* W *of e in* G *such that for all functions* $p \in U$ *we have*

$$|(a^{-1}\varphi_W * p)(s) - p(s)| \leqq \varepsilon \tag{22.1.11.3}$$

for all $s \in K$, *where* $a = \beta(W)$.

Then the topology induced on B *by the weak topology of* $L_C^\infty(G)$ *is the same as that induced by the topology of* $\mathscr{C}_C(G)$.

By virtue of (22.1.11.1), it is enough to show that for each $p_0 \in B$ there exists a neighborhood V of p_0 in B for the weak topology such that for all $p \in V$ and $s \in K$ we have

$$|(a^{-1}\varphi_W * p)(s) - (a^{-1}\varphi_W * p_0)(s)| \leqq \varepsilon. \tag{22.1.11.4}$$

This will result from the following lemma:

(22.1.11.5) *For each function* $f \in \mathscr{L}_C^1(G)$ *and each bounded subset* B *of the Banach space* $L_C^\infty(G)$, *the mapping* $\tilde{g} \mapsto f * g$ *is a continuous mapping of* B, *endowed with the weak topology, into the Fréchet space* $\mathscr{C}_C(G)$.

Recall that for $f \in \mathscr{L}^1_{\mathbf{C}}(\mathrm{G})$ and $g \in \mathscr{L}^\infty_{\mathbf{C}}(\mathrm{G})$, the function $f * g$ is *continuous* on G **(14.10.6)**. For each $s \in \mathrm{G}$ we may write

$$(f * g)(s) = \int f(t)g(t^{-1}s)\, d\beta(t) = \int f(st)g(t^{-1})\, d\beta(t) = \langle \check{g}, \gamma(s^{-1})f \rangle.$$

Let K be a compact subset of G. Since the mapping $s \mapsto (\gamma(s^{-1})f)^{\sim}$ of G into the Banach space $\mathrm{L}^1_{\mathbf{C}}(\mathrm{G})$ is *continuous* **(14.10.6.4)**, the image of K under this mapping is a *compact* set $\mathrm{K}_f \subset \mathrm{L}^1_{\mathbf{C}}(\mathrm{G})$.

Also, as $\tilde{g}$ runs through B, the set of functions $\tilde{h} \mapsto \langle \check{g}, h \rangle$ defined on $\mathrm{L}^1_{\mathbf{C}}(\mathrm{G})$ is *equicontinuous* because B is bounded **(12.15.7.1)**. It follows therefore from **(7.5.6)** that, for each $\tilde{g}_0 \in \mathrm{B}$ and each $\varepsilon > 0$, there exists a neighborhood V of $\tilde{g}_0$ in B for the weak topology such that the relation $\tilde{g} \in \mathrm{V}$ implies $|\langle \check{g} - \check{g}_0, \gamma(s^{-1})f \rangle| \leqq \varepsilon$ for all $s \in \mathrm{K}$.

To complete the proof of **(22.1.11)**, it remains to show that the set $\mathscr{P}_0$ satisfies the conditions of **(22.1.11.2)**. Choose the compact neighborhood W of e in G such that

$$|p_0(s) - 1| = |p_0(s) - p_0(e)| \leqq \tfrac{1}{4}\varepsilon^2$$

for all $s \in \mathrm{W}$. Then take U to be the set of $p \in \mathscr{P}_0$ which satisfy

$$|\langle p - p_0, \varphi_{\mathrm{W}} \rangle| \leqq \tfrac{1}{4}a\varepsilon^2,$$

in other words such that

$$\left| \int_{\mathrm{W}} (p(s) - p_0(s))\, d\beta(s) \right| \leqq \tfrac{1}{4}a\varepsilon^2.$$

It then follows that, for $p \in \mathrm{U}$, we have

(22.1.11.6)
$$\left| \int_{\mathrm{W}} (1 - p(s))\, d\beta(s) \right| \leqq \left| \int_{\mathrm{W}} (1 - p_0(s))\, d\beta(s) \right| + \left| \int_{\mathrm{W}} (p(s) - p_0(s))\, d\beta(s) \right| \leqq \tfrac{1}{2}a\varepsilon^2.$$

Hence, for $p \in \mathrm{U}$ and $s \in \mathrm{G}$, we have

$$\begin{aligned} |(a^{-1}\varphi_{\mathrm{W}} * p)(s) - p(s)| &= \left| a^{-1} \int \varphi_{\mathrm{W}}(t)p(t^{-1}s)\, d\beta(t) - p(s) \right| \\ &= \left| a^{-1} \int_{\mathrm{W}} (p(t^{-1}s) - p(s))\, d\beta(t) \right| \\ &\leqq a^{-1} \int_{\mathrm{W}} |p(t^{-1}s) - p(s)|\, d\beta(t). \end{aligned}$$

But now, by virtue of (22.1.5.1) and (22.1.11.6) and the Cauchy–Schwarz inequality, we have

$$\begin{aligned} |(a^{-1}\varphi_{\mathrm{W}} * p)(s) - p(s)| &\leqq \frac{\sqrt{2}}{a}\int_{\mathrm{W}}(1 - \mathscr{R}p(t))^{1/2}\, d\beta(t) \\ &\leqq \frac{\sqrt{2}}{a}\left(\int_{\mathrm{W}}(1 - \mathscr{R}p(t))\, d\beta(t)\right)^{1/2}\left(\int_{\mathrm{W}} 1 \cdot d\beta(t)\right)^{1/2} \\ &\leqq \frac{\sqrt{2}}{a}\cdot \varepsilon \cdot \sqrt{\frac{a}{2}}\cdot \sqrt{a} = \varepsilon. \end{aligned}$$

Q.E.D.

PROBLEMS

1. Let G be a unimodular, separable, metrizable, locally compact group. Show that if a continuous complex-valued function p on G satisfies the condition (22.1.3.4) (in other words, if the mapping $(s, t) \mapsto p(s^{-1}t)$ of $\mathrm{G} \times \mathrm{G}$ into $\mathbf{C}$ is of positive type (Section 6.3, Problem 4)), then p is of positive type. (Use Section 13.4, Problem 12.)

   Deduce that, for each closed subgroup H of G, the restriction $p \mid \mathrm{H}$ of a continuous function of positive type p on G is a function of positive type on H.

2. Let G be a unimodular, separable, metrizable, locally compact group. A continuous complex-valued function q on G is said to be of *restricted negative type* if, for each finite sequence $(s_j)_{1 \leq j \leq n}$ of n points of G, and each sequence $(\xi_j)_{1 \leq j \leq n}$ of n complex numbers *such that* $\sum_j \xi_j = 0$, we have

$$\sum_{j,k} q(s_j^{-1}s_k)\bar{\xi}_j\xi_k \leqq 0$$

   together with $q(s^{-1}) = \overline{q(s)}$ and $q(e) \geqq 0$.

   (a) Show that for q to be of restricted negative type it is necessary and sufficient that, for each finite sequence $(s_j)_{1 \leq j \leq n}$ of n points of G and *each* sequence $(\xi_j)_{1 \leq j \leq n}$ of n complex numbers, we have

$$\sum_{j,k} (\overline{q(s_j)} + q(s_k) - q(s_j^{-1}s_k))\bar{\xi}_j\xi_k \geqq 0.$$

   If q is of restricted negative type, then $\mathscr{R}q(s) \geqq q(e)$ for all $s \in \mathrm{G}$, and

$$|q(s) + q(t) - q(st)|^2 \leqq 4(\mathscr{R}q(s))(\mathscr{R}q(t))$$

   for all $s, t \in \mathrm{G}$.

   (b) If p is a continuous function of positive type on G, then $q(x) = p(e) - p(x)$ is of restricted negative type. Every $\mathbf{R}$-linear mapping of $\mathbf{R}^n$ into $i\mathbf{R}$ is of restricted negative type, and so is every real quadratic form on $\mathbf{R}^n$ which is $\geqq 0$.

   (c) A continuous complex-valued function q on G is of restricted negative type if and only if, for each $\xi > 0$, the function $e^{-\xi q}$ is of positive type. (Use the fact that if A is a Hermitian matrix, e^A is hermitian and positive.)

(d) Let μ be a positive measure on $]0, +\infty[$, not necessarily bounded, such that $t/(1+t)$ is μ-integrable. Show that the function

$$F(u) = \int_0^{+\infty} (1 - e^{-tu})\, d\mu(t)$$

is defined and continuous on the closed half-plane $\mathscr{R}u \geqq 0$, and is holomorphic on the open half-plane $\mathscr{R}u > 0$. If q is a continuous function of restricted negative type on G, then $F \circ q$ is also of restricted negative type.

2. MEASURES OF POSITIVE TYPE

(22.2.1) With the same general hypotheses on G as in **(22.1)**, we shall generalize the notion of continuous functions of positive type on G, as follows. A *complex measure* μ on G is said to be *of positive type* if, for each function $f \in \mathscr{K}(G)$, we have

$$\text{(22.2.1.1)} \qquad \langle \mu, \check{\tilde{f}} * f \rangle = \iint f(s^{-1}x)\overline{f(s^{-1})}\, d\beta(s)\, d\mu(x) \geqq 0.$$

A complex *locally β-integrable* function p on G is said to be of *positive type* if the measure $\mu = p \cdot \beta$ with density p with respect to β is a measure of positive type on G. By replacing μ by $p \cdot \beta$ in **(22.2.1.1)**, using the fact that G is unimodular, and replacing s by y^{-1} and x by $y^{-1}z$, we obtain the same condition as **(22.1.3.1)**:

$$\iint p(y^{-1}z)\overline{f(y)}f(z)\, d\beta(y)\, d\beta(z) \geqq 0.$$

For *continuous* functions on G, the present definition therefore agrees with that of **(22.1.3)**.

For every measure μ of positive type, we have

$$\text{(22.2.1.2)} \qquad \check{\tilde{\mu}} = \mu.$$

For if $f \in \mathscr{K}(G)$, then by **(22.1.3.5)** and **(22.1.4)** we have

$$\overline{(\check{\tilde{f}} * f)(x^{-1})} = (\check{\tilde{f}} * f)(x),$$

and since $\langle \check{\tilde{\mu}}, g \rangle = \overline{\int \check{g}\, d\mu}$ by definition, we obtain

$$\langle \check{\tilde{\mu}}, \check{\tilde{f}} * f \rangle = \langle \mu, \check{\tilde{f}} * f \rangle.$$

By using the identity

(22.2.1.3) $$4\check{\bar{g}} * f = \overline{(f+g)}^{\vee} * (f+g) - \overline{(f-g)}^{\vee} * (f-g) + i\overline{(f+ig)}^{\vee} * (f+ig) - i\overline{(f-ig)}^{\vee} * (f-ig)$$

we obtain

$$\langle \check{\bar{\mu}}, \check{\bar{f}} * g \rangle = \langle \mu, \check{\bar{f}} * g \rangle$$

for all f, g in $\mathscr{K}(G)$, and then by regularization (14.11.1)

$$\langle \check{\bar{\mu}}, h \rangle = \langle \mu, h \rangle$$

for all $h \in \mathscr{K}(G)$, which proves (22.2.1.2).

It is clear that every *vague limit* (13.4.1) of measures of positive type on G is again of positive type. In particular, if (f_n) is a sequence of functions of positive type belonging to $\mathscr{L}^1_C(G, \beta)$ (resp. $\mathscr{L}^2_C(G, \beta)$) and converging to a limit f in this space, then f is a function of positive type.

(22.2.2) For every complex *bounded* measure ν, the measure $\mu = \check{\bar{\nu}} * \nu$ is of positive type. For if $f \in \mathscr{K}(G)$ we calculate

$$\begin{aligned} \langle \check{\bar{\nu}} * \nu, \check{\bar{f}} * f \rangle &= \iiint f(s^{-1}yz)\overline{f(s^{-1})}\, d\beta(s)\, d\check{\bar{\nu}}(y)\, d\nu(z) \\ &= \iiint f(s^{-1}y^{-1}z)\overline{f(s^{-1})}\, d\bar{\nu}(y)\, d\nu(z)\, d\beta(s) \\ &= \iiint f(s^{-1}z)\overline{f(s^{-1}y)}\, d\bar{\nu}(y)\, d\nu(z)\, d\beta(s) \\ &= \int \left| \int f(s^{-1}t)\, d\nu(t) \right|^2 d\beta(s) \geqq 0. \end{aligned}$$

In particular, the *Dirac measure* ε_e is of positive type, because $\varepsilon_e = \check{\bar{\varepsilon}}_e = \check{\bar{\varepsilon}}_e * \varepsilon_e$. Again, for every function $f \in \mathscr{L}^1_C(G)$, the function $\check{\bar{f}} * f$, which belongs to $\mathscr{L}^1_C(G)$ because G is unimodular, is an (in general *unbounded*) function of positive type (take for example $G = \mathbf{R}$ and f to be the function which is equal to $x^{-1/2}$ for $0 < x < 1$, and zero elsewhere).

(22.2.3) Let H be a *closed normal* subgroup of G, and assume that H is *unimodular*. If we identify the Haar measure α on H with its image under the canonical injection of H in G (13.1.7), the measure α *on* G is *of positive type*.

For if $f \in \mathscr{K}(G)$, we have

$$\langle \alpha, \overset{*}{f} * f \rangle = \iint_{G \times H} f(s^{-1}x)\overline{f(s^{-1})}\, d\beta(s)\, d\alpha(x)$$
$$= \iint_{G \times H} f(sx)\overline{f(s)}\, d\beta(s)\, d\alpha(x)$$

because G is unimodular. Since $(G \times H)/(H \times H)$ may be identified with G/H **(12.12.6)**, there exists by virtue of **(14.4.2)** a measure γ on G/H such that

$$\iint_{G \times H} f(sx)\overline{f(s)}\, d\beta(s)\, d\alpha(x) = \int_{G/H} d\gamma(\dot{s}) \iint_{H \times H} f(s\xi x)\overline{f(s\xi)}\, d\alpha(x)\, d\alpha(\xi)$$

($\dot{s}$ being the coset of s mod. H); but by the invariance of the measure α on H, we have

$$\iint_{H \times H} f(s\xi x)\overline{f(s\xi)}\, d\alpha(x)\, d\alpha(\xi) = \iint_{H \times H} f(sx)\overline{f(s\xi)}\, d\alpha(x)\, d\alpha(\xi)$$
$$= \left| \int_H f(sx)\, d\alpha(x) \right|^2 \geqq 0,$$

which proves our assertion.

(22.2.4) Let μ be a measure of positive type on G. The vector space $\mathscr{K}(G)$ is an *involutory algebra* with respect to convolution and the involution $f \mapsto \overset{*}{f}$ **(14.10.5)**. The definition **(22.2.1.1)** then shows that μ is a *positive linear form in the sense of* **(15.6.4)** on this involutory algebra. Let $\mathfrak{n}$ be the left ideal in the algebra $\mathscr{K}(G)$ which consists of the functions f such that $\langle \mu, \overset{*}{f} * f \rangle = 0$, and let $\pi\colon \mathscr{K}(G) \to \mathscr{K}(G)/\mathfrak{n}$ be the canonical mapping; we then define on $E_0 = \mathscr{K}(G)/\mathfrak{n}$ a pre-Hilbert space structure by taking the scalar product $(\xi \mid \eta)$ (or $(\xi \mid \eta)_\mu$) to be given by

(22.2.4.1) $$(\pi(f) \mid \pi(g))_\mu = \langle \mu, \overset{*}{g} * f \rangle = \iint f(t^{-1}s)\overline{g(t^{-1})}\, d\beta(t)\, d\mu(s).$$

(15.6.8). It is immediately seen that if a sequence of functions (f_n) in $\mathscr{K}(G)$ converges uniformly to $f \in \mathscr{K}(G)$, and if the supports of the f_n all lie in a fixed compact set, then $\overset{*}{f}_n * f_n$ converges uniformly to $\overset{*}{f} * f$, and the supports of the $\overset{*}{f}_n * f_n$ all lie in a fixed compact set, hence $\|\pi(f_n) - \pi(f)\|_\mu \to 0$. This shows that the pre-Hilbert space E_0 is separable, hence may be identified with a dense subspace of a separable Hilbert space E (or E_μ). Instead of verifying the condition (U) of **(15.6.9)** and thence obtaining by **(15.6.10)** a representation of the involutory algebra $\mathscr{K}(G)$ on E, it is simpler in the present situation to define directly a *continuous unitary representation* U_μ of G on E_μ, by putting

$$U_\mu(s) \cdot \pi(f) = \pi(\varepsilon_s * f) \tag{22.2.4.2}$$

for $f \in \mathscr{K}(G)$ and $s \in G$; the function $\varepsilon_s * f = \gamma(s)f: x \mapsto f(s^{-1}x)$ belongs to $\mathscr{K}(G)$. Let us show that the linear mapping $U_\mu(s)$ so defined is continuous on E_0 and extends to a unitary transformation of E_μ. We have, by **(22.2.4.1)**,

$$\begin{aligned}\|U_\mu(s) \cdot \pi(f)\|^2 &= \iint f(s^{-1}t^{-1}u)\overline{f(s^{-1}t^{-1})}\, d\beta(t)\, d\mu(u) \\ &= \iint f(t^{-1}u)\overline{f(t^{-1})}\, d\beta(t)\, d\mu(u) \\ &= \|\pi(f)\|^2\end{aligned}$$

from which our assertion follows. Moreover, if a sequence (s_n) converges to s in G, the sequence of functions $\gamma(s_n)f$ converges to $\gamma(s)f$, and the supports of these functions all lie in a fixed compact set; hence it follows from above that the sequence $(U_\mu(s_n) \cdot \pi(f))$ converges to $U_\mu(s) \cdot \pi(f)$ in E_μ. Since E_0 is dense in E_μ, and since the set of mappings $U_\mu(s)$ of E into E_μ, as s runs through G, is equicontinuous **(12.15.7.1)**, the continuity of each mapping $s \mapsto U_\mu(s) \cdot x$ follows from **(7.5.5)**.

PROBLEMS

1. With the notation of **(22.2.4)**, suppose that there exists a neighborhood V of e in G such that, for each function $f \geq 0$ in $\mathscr{K}(G)$ with support contained in V, we have

$$\langle \mu, \check{f} * f \rangle \leq c \cdot \left(\int f(t)\, d\beta(t) \right)^2 \tag{1}$$

where $c \geq 0$ is a constant independent of f.

(a) Let (V_n) be a fundamental system of neighborhoods of e contained in V, and for each n let g_n be a function ≥ 0 belonging to $\mathscr{K}(G)$, with support contained in V_n, and such that $\int g_n(t)\, d\beta(t) = 1$. Show that in the Hilbert space E_μ the sequence $(\pi(g_n))$ converges weakly to an element u such that, for each function $f \in \mathscr{K}(G)$, we have $(u \mid \pi(f))_\mu = \int \overline{f(s^{-1})}\, d\mu(s)$. (Use (1) together with **(14.11.1)** and **(12.15.10)**.) Deduce that for each $t \in G$ we have $(U_\mu(t) \cdot u \mid \pi(f)) = \int \overline{f(ts^{-1})}\, d\mu(s)$.

(b) Put $p(s) = (U_\mu(s) \cdot u \mid u)$, which is a continuous function of positive type on G. Show that $\mu = p \cdot \beta$ (use (a)).
(c) Let p be a continuous function of positive type on G, and let q be a locally β-integrable function of positive type, such that $p - q$ is of positive type. Show that q is equal almost everywhere to a continuous function (use (b)).

In particular, if a locally integrable function p is of positive type and is *bounded in a neighborhood of e*, it is equal almost everywhere to a continuous function.

2. (a) Let G be a unimodular, separable, metrizable, locally compact group. A function $f \in \mathscr{L}^2_{\mathbf{C}}(G)$ is said to be *tame* if there exists a constant $a > 0$ such that $N_2(f * g) \leqq a \cdot N_2(g)$ for all functions $g \in \mathscr{K}(G)$. Then, for every function $g \in \mathscr{L}^2_{\mathbf{C}}(G)$, we have $f * g \in \mathscr{L}^2_{\mathbf{C}}(G)$, and the inequality above still holds; the mapping $\tilde{g} \mapsto (f * g)^\sim$ is a continuous endomorphism of $L^2_{\mathbf{C}}(G)$, which we denote by $R(f)$; and for $g \in \mathscr{K}(G)$ we have

$$(R(f) \cdot \tilde{g} \mid \tilde{g}) = \langle f, \check{\bar{g}} * g \rangle.$$

The functions belonging to $\mathscr{L}^1_{\mathbf{C}}(G) \cap \mathscr{L}^2_{\mathbf{C}}(G)$ are tame, and hence in particular the functions belonging to $\mathscr{K}(G)$ are tame.

If f and g are tame, we have $f * g \in \mathscr{L}^2_{\mathbf{C}}(G)$, and $f * g$ is tame. The classes of tame functions therefore form (with respect to convolution) an algebra T(G) over **C**, and $\tilde{f} \mapsto R(f)$ is an injective homomorphism of this algebra into $\mathscr{L}(L^2_{\mathbf{C}}(G))$.
(b) For a tame function f to be of positive type, it is necessary and sufficient that $R(f)$ should be a *positive self-adjoint* operator (**11.5**). Deduce that if f and g are tame and of positive type, and if $f * g = g * f$, then $f * g$ is continuous, tame and of positive type. (Observe that if A and B are two *commuting* positive self-adjoint operators on a Hilbert space, then $AB = A^{1/2}BA^{1/2}$ (Section **15.11**, Problem 2) and consequently AB is positive and self-adjoint.) Further, we have $(f \mid g) \geqq 0$, and hence $N_2(f - g)^2 \leqq N_2(f)^2 - N_2(g)^2$ if $f - g$ is a function of positive type.
(c) Let f be a continuous function of positive type, belonging to $\mathscr{L}^2_{\mathbf{C}}(G)$. For every polynomial $P \in \mathbf{C}[X]$ which is $\geqq 0$ on the interval $[0, \|f\|]$ and vanishes at $X = 0$, the function $P(f)$ (which is defined by substituting for each X^m the convolution product of m factors equal to f) is continuous, of positive type, and belongs to $\mathscr{L}^2_{\mathbf{C}}(G)$. Let F be a continuous function on $[0, \|f\|]$, with values $\geqq 0$ and such that $F(0) = 0$; and suppose that F is the limit of an increasing sequence (P_n) of polynomials with zero constant terms and with values $\geqq 0$ on the interval $[0, \|f\|]$; suppose moreover that there exists a polynomial Q with zero constant term such that $P_n^2 \leqq Q$ for all n. Show that the sequence of functions $P_n(f)$ converges in $\mathscr{L}^2_{\mathbf{C}}(G)$ to a tame function g of positive type, such that $R(g) = F(R(f))$. This function g is independent of the sequence (P_n) used to construct it, and is denoted by $F(f)$. (Use (b) to establish the convergence of the sequence $(P_n(f))$, together with (**15.11.8**) and (**14.10.4**); observe that for each function $h \in \mathscr{K}(G)$, $(P_n(f) * h)$ converges in $\mathscr{L}^2_{\mathbf{C}}(G)$ to $F(R(f)) \cdot h$.)
(d) Deduce in particular from (c) that for each continuous function f of positive type which belongs to $\mathscr{L}^2_{\mathbf{C}}(G)$, there exists a tame function g of positive type such that $f = g * g$ almost everywhere.

3. Let G be a unimodular, separable, metrizable, locally compact group. Show that for a continuous function p of positive type on G, the following properties are equivalent:

(α) For each compact subset K of G, there exists a sequence (f_n) of functions in $\mathscr{K}(G)$ such that the sequence $(\check{\bar{f}}_n * f_n)$ converges uniformly to p in K.

(β) For each compact subset K of G, there exists a sequence (p_n) of continuous functions of positive type and compact support which converges uniformly to p in K.

(γ) For each compact subset K of G, there exists a sequence (p_n) of continuous functions of positive type and integrable square which converges uniformly to p in K. (To show that (γ) implies (α), use Problem 2(d).)

4. Let G be a unimodular, separable, metrizable, locally compact group. Show that the following properties are equivalent:

(α) For each compact subset K of G, and each $\varepsilon > 0$, there exists a function $f \geqq 0$ belonging to $\mathscr{K}(G)$ such that $N_1(f) = 1$ and $N_1(\gamma(s)f - f) \leqq \varepsilon$ for all $s \in K$.

(β) For a real number $p \geqq 1$, for each compact subset K of G and each $\varepsilon > 0$, there exists a function $f \geqq 0$ belonging to $\mathscr{L}_C^p(G)$ (Section **13.11**, Problem 12) such that $N_p(f) = 1$ and $N_p(\gamma(s)f - f) \leqq \varepsilon$ for all $s \in K$.

(γ) For each continuous function g of positive type on G, and each compact subset K of G, there exists a sequence (f_n) of functions in $\mathscr{K}(G)$ such that the sequence $(\check{\bar{f}}_n * f_n)$ converges uniformly to g in K.

(δ) For each compact subset K of G, there exists a sequence (f_n) of functions in $\mathscr{K}(G)$ such that the sequence $(\check{\bar{f}}_n * f_n)$ converges uniformly to 1 in K.

(To prove that (δ) implies (γ), use Problem **3**. To prove that (γ) is equivalent to (β) for $p = 2$, observe that $N_2(\gamma(s)f - f)^2 = 2((f * \check{\bar{f}})(e) - \mathscr{R}((f * \check{\bar{f}})(s)))$. Finally, to prove the equivalence of (α) and (β), use Hölder's inequality together with the inequalities

$$|a - b|^p \leqq |a^p - b^p| \leqq p|a - b|(a^{p-1} + b^{p-1})$$

for $a, b \geqq 0$.)

A group G satisfying these conditions is said to be *amenable*. If G is amenable, every unimodular closed subgroup of G, and every unimodular quotient group G/N of G by a closed normal subgroup N of G, is amenable.

5. (a) Let G be an amenable group (Problem 4), let K be a compact subset of G, and M a real number > 0. Show that for each $\varepsilon > 0$ there exists a function $f \geqq 0$ belonging to $\mathscr{L}_C^1(G)$, such that $N_1(f) = 1$ and such that, for *each* function $g \in \mathscr{L}_C^1(G)$ which is zero on $\complement K$ and satisfies $N_1(g) \leqq M$, we have

$$N_1\left(g * f - \left(\int g(x)\,d\alpha(x)\right)f\right) \leqq \varepsilon$$

where α is Haar measure on G.

(b) Let G be a separable, metrizable, locally compact group and G′ a unimodular closed normal subgroup. Let $\alpha, \alpha', \alpha''$ denote left Haar measures on G, G′, G″ = G/G′, respectively, satisfying the identity (**14.4.2**). Suppose in addition that G′ is amenable. Let K be a compact subset of G and let $M > 0$. Show that for each $\varepsilon > 0$ there exists a function $h \geqq 0$ belonging to $\mathscr{L}_C^1(G')$, such that $N_1(h) = 1$ and such that, for *all* $g \in \mathscr{K}(G)$ with support contained in K and satisfying $\|g\| \leqq M$, we have

$$\int_{G'}\left|\int_{G'} g(x\eta\xi)h(\xi^{-1})\,d\alpha'(\xi)\right|d\alpha'(\eta) \leqq \left|\int_{G'} g(x\xi)\,d\alpha'(\xi)\right| + \varepsilon$$

for all $x \in KG'$. (Use (a).)

Deduce that, with the same hypotheses on g, we have

$$\int_G \left| \int_{G'} g(x\xi)h(\xi^{-1})\, d\alpha'(\xi) \right| d\alpha(x) \leq \int_{G''} \left| \int_{G'} g(x\xi)\, d\alpha'(\xi) \right| d\alpha''(x) + \varepsilon\alpha''(\pi(K))$$

where $\pi: G \to G''$ is the canonical homomorphism.

(c) Let G be a unimodular, separable, metrizable, locally compact group and G′ a unimodular closed normal subgroup of G, such that $G'' = G/G'$ is unimodular. Show that if G′ and G″ are amenable, then so also is G. (Let $\pi: G \to G''$ be the canonical homomorphism. For each compact subset K of G and each $\varepsilon > 0$, there exists $h \in \mathscr{K}(G'')$ such that $h \geqq 0$, $N_1(h) = 1$ and $N_1(\gamma(\pi(s))h - h) \leqq \varepsilon$ for all $s \in K$. Consider a function $f \in \mathscr{K}(G)$ such that $f^\flat = h$ (Section **14.4**, Problem 2) and apply (b) to the functions $g_s(x) = f(s^{-1}x) - f(x)$, where $s \in K$.)

6. A compact group is amenable. A commutative (metrizable and separable) locally compact group is amenable. (To establish property (β) of Problem 4 with $p = 1$, take for f a function proportional to the square of the Fourier transform of the characteristic function of a suitable neighborhood of the identity element in the dual group $\hat{G}$ **(22.10)**.)

7. Let G be an amenable group, H a compact subgroup of G. Show that for every compact subset K of G, there exists a function $f \geqq 0$ satisfying property (α) of Problem 4, and in addition such that $f(tx) = f(x)$ for all $t \in H$. (Consider the compact set $KH \cup H$.)

8. (a) Let G be a separable, metrizable, locally compact group in which there exists a compact subgroup H and a closed subgroup S such that the mapping $(t, u) \mapsto tu$ of $H \times S$ into G is a homeomorphism of $H \times S$ onto G; denote the inverse mapping by $x \mapsto (h(x), s(x))$. If m_H, m_S are left Haar measures on H and S, respectively, there exists a left Haar measure m_G on G such that, for all functions $f \in \mathscr{K}(G)$, we have

$$\int f(z)\, dm_G(z) = \iint_{H \times S} f(xy)\, \Delta_G(x)\, \Delta_S(y)^{-1}\, dm_H(x)\, dm_S(y)$$

(Section **14.3**, Problem 5). Assume that G, but not S, is unimodular. Let $u_0 \in S$ be such that $\Delta_S(u_0) \neq 1$, and let $f \in \mathscr{K}(G)$ be such that $f(tx) = f(x)$ for all $t \in H$. Then we have

$$N_1(\gamma(u_0)f - f) \geqq c_0 \cdot \left| \int f(x)\, dm_G(x) \right|$$

where

$$c_0 = \int_H |\Delta_S(s(u_0^{-1}t)) - 1|\, dm_H(t) > 0.$$

(b) Deduce from (a) and Problem 7 that G is not amenable. In particular, a noncompact semisimple Lie group with finite center is not amenable (cf. **(21.21.10)** and Section **14.4**, Problem 2).

9. Let G be an amenable group (Problem 4), μ a positive bounded measure on G. Show that the norm of the continuous endomorphism $\tilde{f} \mapsto (\mu * f)^\sim$ of $L^p_{\mathbf{C}}(G)$, where $1 < p < +\infty$ (Section **14.9**, Problem 2) is equal to $\|\mu\|$. (Reduce to the case where the support of μ is compact, and use property (β) of Problem 4.)

10. Let G be a compact group. Every continuous central function f on G can be written as $f = \sum_{\rho} c_{\rho}\chi_{\rho}$, the series being convergent in the space $\mathscr{L}^2_{\mathbf{C}}(G)$. Show that for f to be of positive type it is necessary and sufficient that $c_{\rho} \geqq 0$ for all $\rho \in R(G)$ and $\sum_{\rho} n_{\rho} c_{\rho} < +\infty$. (Use Problem 2(d).)

11. Extend to nonunimodular (separable, metrizable) locally compact groups the results of Sections **22.1** and **22.2** (cf. Section **21.1**, Problem 13).

12. (a) Let G be a unimodular, separable, metrizable, locally compact group. If f is a function of positive type on G which belongs to $\mathscr{L}^p_{\mathbf{C}}(G)$ for some $p \in [1, +\infty]$, show that for every bounded complex measure μ on G the function $\mu * f * \check{\bar{\mu}} \in \mathscr{L}^p_{\mathbf{C}}(G)$ is of positive type. Deduce that if $1 \leqq p < +\infty$, the function f is the limit, in $\mathscr{L}^p_{\mathbf{C}}(G)$, of a sequence of continuous functions of positive type belonging to $\mathscr{L}^p_{\mathbf{C}}(G)$.
(b) Let $p, q \in [1, +\infty[$ be such that $p^{-1} + q^{-1} = r^{-1} \leqq 1$. Show that if $f \in \mathscr{L}^p_{\mathbf{C}}(G)$ and $g \in \mathscr{L}^q_{\mathbf{C}}(G)$ are of positive type, then so also is $fg \in \mathscr{L}^r_{\mathbf{C}}(G)$. (Reduce to **(22.1.9)** with the help of (a).)
(c) Show that if G is amenable (Problem 4), then $\int d\mu(x) \geqq 0$ for all bounded measures μ of positive type on G.

3. INDUCED REPRESENTATIONS

(22.3.1) Let G be a group (not necessarily topologized), H a subgroup of G, X = G/H the corresponding homogeneous space, and E a complex vector space (not topologized). Generalizing the definitions in **(19.1)** for Lie groups, we shall say that G acts *equilinearly* (on the left) on X × E and X if it acts on these two sets in such a way that

(1) $\mathrm{pr}_1(s \cdot (x, z)) = s \cdot x$ for $s \in G$, $x \in X$ and $z \in E$;
(2) for each $s \in G$ and $x \in X$, the mapping $z \mapsto \mathrm{pr}_2(s \cdot (x, z))$ of E into E is *linear*.

As in **(19.1.2)**, we obtain from these data an action of G on the complex vector space $L = E^X$ of *all* mappings of X into E, by defining for all $f \in L$ and all $s \in G$ the mapping $s \cdot f$ by

(22.3.1.1) $$(s \cdot x, (s \cdot f)(s \cdot x)) = s \cdot (x, f(x))$$

or equivalently

(22.3.1.2) $$(s \cdot f)(s \cdot x) = \mathrm{pr}_2(s \cdot (x, f(x))).$$

It follows immediately from condition (2) above that for all $s \in G$ the mapping $\rho(s): f \mapsto s \cdot f$ of L into itself is *linear*, and hence defines a *homomorphism* $s \mapsto \rho(s)$ of G into the group Aut(L) of automorphisms of the vector space L.

(22.3.2) We shall now use the hypothesis that $X = G/H$ to give an alternative description of the equilinear actions of G on $X \times E$ and X. Let $\alpha(s, x)$ denote the automorphism $z \mapsto \mathrm{pr}_2(s \cdot (x, z))$ of E, so that

$$s \cdot (x, z) = (s \cdot x, \alpha(s, x) \cdot z).$$

From the conditions $e \cdot (x, z) = (x, z)$ and $s \cdot (t \cdot (x, z)) = (st) \cdot (x, z)$ (12.10) we obtain the relations

(22.3.2.1) $$\alpha(e, x) = 1_E \qquad (x \in X),$$

(22.3.2.2) $$\alpha(st, x) = \alpha(s, t \cdot x) \circ \alpha(t, x) \qquad (x \in X;\ s, t \in G),$$

which imply in particular, by taking $t = s^{-1}$ in (22.3.2.2),

$$\alpha(s^{-1}, x) = (\alpha(s, s^{-1} \cdot x))^{-1}.$$

A mapping $(s, x) \mapsto \alpha(s, x)$ satisfying (22.3.2.1) and (22.3.2.2) is called a *left 1-cocycle of* G *with values in* Aut(E).

For each $x \in X = G/H$, let r_x be a representative of the coset x mod H, so that every element of $x = r_x H$ is uniquely expressible in the form $r_x u$, $u \in H$. When $x = x_0 = H$, we take $r_{x_0} = e$. Since $x = r_x \cdot x_0$, we obtain from (22.3.2.2) the relation

(22.3.2.3) $$\alpha(s, x) = \alpha(sr_x, x_0) \circ (\alpha(r_x, x_0))^{-1}$$

which shows that knowledge of the automorphisms $\alpha(s, x_0) = \alpha(s)$ of E determines the $\alpha(s, x)$ for all $x \in X$. Also it follows from (22.3.2.1) and (22.3.2.2) that $\alpha(e) = 1_E$ and $\alpha(su, x_0) = \alpha(s, u \cdot x_0) \circ \alpha(u, x_0)$ for $s \in G$ and $u \in H$, and *because* $u \cdot x_0 = x_0$,

(22.3.2.4) $$\alpha(su) = \alpha(s) \circ \alpha(u) \qquad (s \in G,\ u \in H)$$

and in particular $\alpha(r_x u) = \alpha(r_x) \circ \alpha(u)$. Put $\sigma(u) = \alpha(u) \in \mathrm{Aut}(E)$ for $u \in H$, and $\beta(x) = \alpha(r_x)$. Observe next that for $s \in G$ and $x \in X$ we may write $sr_x = r_{s,x} u(s, x)$, where $u(s, x) \in H$; it then follows from above that we have

(22.3.2.5) $$\alpha(s, x) = \beta(s \cdot x) \circ \sigma(u(s, x)) \circ \beta(x)^{-1}$$

with

(22.3.2.6) $$u(s, x) = r_{s \cdot x}^{-1} s r_x.$$

Furthermore, we have

(22.3.2.7) $$\sigma(uv) = \sigma(u) \circ \sigma(v)$$

for u, v in H: in other words, σ is a *homomorphism* of H into the group Aut(E).

Conversely, it is immediately verified that if we are given *arbitrarily* a mapping β of X into Aut(E) and a homomorphism σ of H into Aut(E), the formulas (22.3.2.5) and (22.3.2.6) define a mapping of $G \times X$ into Aut(E) which satisfies the conditions (22.3.2.1) and (22.3.2.2), and consequently defines an action of G on $X \times E$ such that G acts equilinearly on $X \times E$ and X; from this action we derive a homomorphism ρ: $G \to \text{Aut}(L)$ as in (22.3.1): explicitly,

(22.3.2.8) $$\begin{aligned}(\rho(s) \cdot f)(x) &= \alpha(s, s^{-1} \cdot x) \cdot f(s^{-1} \cdot x) \\ &= \alpha(s^{-1}, x)^{-1} \cdot f(s^{-1} \cdot x).\end{aligned}$$

(22.3.3) Let $x \mapsto c(x)$ be a mapping of X into the group Aut(E) such that $c(x_0) = 1_E$; it gives rise to an element γ of Aut(L) by the formula

(22.3.3.1) $$(\gamma \cdot f)(x) = c(x) \cdot f(x) \qquad (x \in X, f \in L).$$

If we put

(22.3.3.2) $$\rho'(s) = \gamma \circ \rho(s) \circ \gamma^{-1} \qquad (s \in G)$$

it is clear that $s \mapsto \rho'(s)$ is another homomorphism of G into Aut(L), and the formula (22.3.2.7) shows that we have

(22.3.3.3) $$(\rho'(s) \cdot f)(x) = (c(x) \circ \alpha(s, s^{-1} \cdot x) \circ c(s^{-1} \cdot x)^{-1}) \cdot f(s^{-1} \cdot x).$$

In other words, ρ' is obtained by the same procedure as ρ, but with $\alpha(s, x)$ replaced by

(22.3.3.4) $$\alpha'(s, x) = c(s \cdot x) \circ \alpha(s, x) \circ c(x)^{-1}.$$

If we put $\alpha'(s) = \alpha'(s, x_0)$ and $\beta'(x) = \alpha'(r_x)$ as above, the hypothesis $c(x_0) = 1_E$ implies that $\alpha'(u) = \sigma(u)$ for $u \in H$, and one verifies immediately that

(22.3.3.5) $$\beta'(x) = c(x) \circ \beta(x).$$

Hence if we take $c(x) = \beta(x)^{-1}$ (which is possible because $\beta(x_0) = 1_E$ by (22.3.2.1)) we shall have $\beta'(x) = 1_E$ for all $x \in X$, and consequently

(22.3.3.6) $$\alpha'(s, x) = \sigma(u(s, x)),$$

where $u(s, x)$ is determined by (22.3.2.6).

Since a homomorphism of a group G into the group Aut(V) of automorphisms of a vector space V is the same thing as a *linear representation of* G *on* V, and since two such homomorphisms ρ, ρ' are *equivalent linear representations* when $\rho'(s) = \gamma \circ \rho(s) \circ \gamma^{-1}$ for all $s \in G$ (where γ is a fixed element of AutV), we may say that a given *linear representation* σ *of* H *on* E determines a *class of equivalent linear representations of* G *on* $L = E^X$. It is immediately verified that this class is not changed by replacing σ by an equivalent representation of H on E. The representation ρ of G on L defined by (22.3.2.8), for a function α satisfying (22.3.2.1) and (22.3.2.2), is said to be *induced* by the representation $\sigma: u \mapsto \alpha(u, x_0)$ of H on E. If M is a vector subspace of L which is *stable* under all the automorphisms $\rho(s)$, the linear subrepresentation $s \mapsto \rho(s)|M$ of G on M is also said to be *induced* by σ.

(22.3.4) We shall now show how the function α satisfying (22.3.2.1) and (22.3.2.2) enables us to define an *isomorphism* $\tau: f \mapsto f^\alpha$ of $L = X^E$ onto a vector subspace L^α of the vector space of *all mappings of* G *into* E. For each $f \in L$, put

(22.3.4.1) $$\begin{aligned} f^\alpha(s) &= \alpha(s^{-1}, s \cdot x_0) \cdot f(s \cdot x_0) \\ &= \alpha(s, x_0)^{-1} \cdot f(s \cdot x_0) \in E \end{aligned}$$

for all $s \in G$. The function f^α so defined satisfies the relation

(22.3.4.2) $$f^\alpha(su) = \sigma(u^{-1}) \cdot f^\alpha(s) \qquad (u \in H)$$

which follows directly from (22.3.2.2) and the fact that $u \cdot x_0 = x_0$. *Conversely*, if L^α denotes the vector space of all functions $g: G \to E$ such that

(22.3.4.3) $$g(su) = \sigma(u^{-1}) \cdot g(s) \qquad (u \in H),$$

then for each such $g \in L^\alpha$ there exists one and only one function $f \in L$ such that $g = f^\alpha$. For by virtue of (22.3.4.3) and (22.3.2.2), the element $\alpha(s, x_0) \cdot g(s)$ is *unaltered* by replacing s by su with $u \in H$, and therefore depends only on the coset $x = s \cdot x_0$ of s in X; and one checks immediately that if $f(x)$ denotes this element we have $g = f^\alpha$.

If now we identify L and L^α by means of the isomorphism τ (which comes to the same thing as replacing $\rho(s)$ by $\tau \circ \rho(s) \circ \tau^{-1}$) we see that ρ may be considered as a *linear representation of* G *on* L^α, defined by

$$(\rho(s) \cdot g)(t) = g(s^{-1}t) \qquad (s, t \in G) \tag{22.3.4.4}$$

since the function $t \mapsto g(s^{-1}t)$ evidently satisfies the same relation (22.3.4.3) as g does. We remark also that L^α *depends only on the representation* σ of H on E, and therefore we denote L^α also by L^σ.

(22.3.5) Suppose now that G is a (separable, metrizable) *locally compact* group, and H a *closed* subgroup of G; then X = G/H is separable, metrizable and locally compact ((12.10.9), (12.10.10), and (12.11.2)). For each positive measure μ on X and each $s \in G$, let $\gamma(s)\mu$ denote the image of μ under the homeomorphism $x \mapsto s \cdot x$ of X onto itself (13.1.6), so that we have

$$\int h(x)\, d(\gamma(s)\mu)(x) = \int h(s \cdot x)\, d\mu(x) \tag{22.3.5.1}$$

for all functions $h \in \mathscr{K}(X)$. We shall show that, under certain conditions, given a *continuous unitary representation* U of the group H on a separable complex Hilbert space E (21.1.1), we can construct a *continuous unitary representation* of G on another separable Hilbert space, by using the methods described above.

Let $(\mathbf{a}_n)$ be a Hilbert basis of E, so that every mapping $\mathbf{f}$ of X into E may be written uniquely in the form

$$\mathbf{f} = \sum_n f_n \mathbf{a}_n,$$

where the f_n are complex-valued functions on X, such that the series $\sum_n |f_n|^2 = \|\mathbf{f}\|^2$ converges everywhere. For $\mathbf{f}$ to be μ-measurable, it is necessary and sufficient that the f_n should be μ-measurable. Indeed, the condition is clearly necessary, because $f_n = (\mathbf{f} | \mathbf{a}_n)$ (13.9.6). Conversely, if the f_n are measurable, it follows from Egoroff's theorem (13.9.10) that $\mathbf{f}$, being the simple limit of the measurable functions $\sum_{n=1}^{N} f_n \mathbf{a}_n$ as $N \to +\infty$, is measurable.

We denote by $\mathscr{L}^2_{\mathrm{E}}(\mathrm{X}, \mu)$ or $\mathscr{L}^2_{\mathrm{E}}(\mu)$ the complex vector space of all μ-measurable mappings $\mathbf{f}\colon \mathrm{X} \to \mathrm{E}$ such that the function $\|\mathbf{f}\|^2 = \sum_n |f_n|^2$ is μ-integrable; this implies that each of the functions f_n belongs to $\mathscr{L}^2_{\mathbf{C}}(\mu)$ and that $\int \|\mathbf{f}\|^2\, d\mu = \sum_n \int |f_n|^2\, d\mu$ ((13.9.13) and (13.8.5)), and it follows that for $\mathbf{f} = \sum_n f_n \mathbf{a}_n$ and $\mathbf{g} = \sum_n g_n \mathbf{a}_n$ in $\mathscr{L}^2_{\mathrm{E}}(\mu)$, the complex-valued function $x \mapsto (\mathbf{f}(x) | \mathbf{g}(x))$ is integrable, and that

$$\int (\mathbf{f}(x) | \mathbf{g}(x))\, d\mu(x) = \sum_n \int f_n(x)\overline{g_n(x)}\, d\mu(x).$$

To say that a function $\mathbf{f} \in \mathscr{L}^2_{\mathrm{E}}(\mu)$ is negligible is equivalent to saying that $\|\mathbf{f}\|^2$ is negligible; the classes $\tilde{\mathbf{f}}$ of the functions $\mathbf{f} \in \mathscr{L}^2_{\mathrm{E}}(\mu)$ therefore form a pre-Hilbert space, denoted by $\mathrm{L}^2_{\mathrm{E}}(\mathrm{X}, \mu)$ or $\mathrm{L}^2_{\mathrm{E}}(\mu)$, for the scalar product

(22.3.5.2) $$(\tilde{\mathbf{f}} | \tilde{\mathbf{g}}) = \int (\mathbf{f}(x) | \mathbf{g}(x))\, d\mu(x)$$

and the norm

(22.3.5.3) $$\mathrm{N}_2(\tilde{\mathbf{f}}) = (\tilde{\mathbf{f}} | \tilde{\mathbf{f}})^{1/2}$$

(which is also denoted by $\mathrm{N}_2(\mathbf{f})$).

In fact, $\mathrm{L}^2_{\mathrm{E}}(\mu)$ is *complete*, and therefore a Hilbert space. For suppose that the sequence of functions $\mathbf{f}^{(m)} = \sum_n f_n^{(m)} \mathbf{a}_n$ is such that $(\tilde{\mathbf{f}}^{(m)})$ is a Cauchy sequence in $\mathrm{L}^2_{\mathrm{E}}(\mu)$; then for each $\varepsilon > 0$ there exists an integer m_0 such that we have $\sum_n \int |f_n^{(p)} - f_n^{(q)}|^2\, d\mu \leqq \varepsilon$ for all $p, q \geqq m_0$, from which it follows that for each n the sequence $(\tilde{f}_n^{(m)})$ is a Cauchy sequence in $\mathrm{L}^2_{\mathbf{C}}(\mu)$, and therefore tends to a limit $\tilde{g}_n$. Also, for each integer $\mathrm{N} > 0$, by letting $q \to \infty$ we obtain $\sum_{n=1}^{\mathrm{N}} (\mathrm{N}_2(g_n - f_n^{(p)}))^2 \leqq \varepsilon$. This shows that the series $\sum_{n=1}^{\infty} (\mathrm{N}_2(g_n))^2$ converges, hence that $\mathbf{g} = \sum_n g_n \mathbf{a}_n$ belongs to $\mathrm{L}^2_{\mathrm{E}}(\mu)$, and then that $(\mathrm{N}_2(\mathbf{g} - \mathbf{f}^{(p)}))^2 \leqq \varepsilon$ for all $p \geqq m_0$, so that $\tilde{\mathbf{g}}$ is the limit of the sequence $(\tilde{\mathbf{f}}^{(m)})$ in $\mathrm{L}^2_{\mathrm{E}}(\mu)$.

Finally, the space $\mathrm{L}^2_{\mathrm{E}}(\mu)$ is *separable*. For if D is a denumerable dense set in $\mathrm{L}^2_{\mathbf{C}}(\mu)$, one sees immediately that the set of classes of mappings $\mathbf{f} = \sum_n f_n \mathbf{a}_n$ such that $f_n \in \mathrm{D}$ for all n, and $f_n = 0$ for all but finitely many n, is denumerable and dense in $\mathrm{L}^2_{\mathrm{E}}(\mu)$.

(22.3.6) This being so, suppose first of all that the measure μ is G-*invariant*, i.e., that $\gamma(s)\mu = \mu$ for all $s \in G$. Let E be a separable Hilbert space, and let U be a continuous unitary representation of H on E. Suppose now that for all $(s, x) \in G \times X$, $\alpha(s, x)$ satisfies (22.3.2.1) and (22.3.2.2), and that in addition it has the following properties:

(1) $\mathbf{z} \mapsto \alpha(s, x)\mathbf{z}$ is a *unitary* operator on E, and $\alpha(u, x_0) = U(u)$ for all $u \in H$;

(2) for each $s \in G$, the mapping $x \mapsto \alpha(s, x) \cdot \mathbf{f}(x)$ of X into E is μ-measurable for each function $\mathbf{f} \in \mathscr{L}_E^2(\mu)$.

These two conditions already imply that, for each function $\mathbf{f} \in \mathscr{L}_E^2(\mu)$, the function $\rho(s) \cdot \mathbf{f}$ defined by (22.3.2.8) belongs to $\mathscr{L}_E^2(\mu)$ and that

(22.3.6.1) $$N_2(\rho(s) \cdot \mathbf{f}) = N_2(\mathbf{f})$$

for all $s \in G$, because $\|\alpha(s, s^{-1} \cdot x) \cdot \mathbf{f}(s^{-1} \cdot x)\| = \|\mathbf{f}(s^{-1} \cdot x)\|$, and the invariance of μ implies that $\int \|\mathbf{f}(s^{-1} \cdot x)\|^2 \, d\mu(x) = \int \|\mathbf{f}(x)\|^2 \, d\mu(x)$.

Suppose also that:

(3) for each function $\mathbf{f} \in L_E^2(\mu)$, the mapping

(22.3.6.2) $$s \mapsto (\rho(s) \cdot \mathbf{f})^\sim$$

of G into $L_E^2(\mu)$ is *continuous*.

Then it is clear that, if we put $\rho(s) \cdot \tilde{\mathbf{f}} = (\rho(s) \cdot \mathbf{f})^\sim$, ρ is a *continuous unitary representation* of G on the separable Hilbert space $L_E^2(\mu)$, called the representation *induced* by U, and denoted by U^{ind}, or $U^{\text{ind}}_{G/H}$.

For example, all the conditions above are satisfied when U is the *trivial* representation (21.1.1) of H on E, by taking $\alpha(s, x) = 1_E$ for all $(s, x) \in G \times X$. Only condition (3) requires verification: and since the mappings $\tilde{\mathbf{f}} \mapsto (\rho(s) \cdot \mathbf{f})^\sim$ form an *equicontinuous* set (12.15.7.1), it is enough to verify (3) for functions $\mathbf{f}$ of the form $f_n \mathbf{a}_n$, with $f_n \in \mathscr{L}_C^2(\mu)$ (7.5.5), and in this case the conclusion follows from (14.10.6.3).

The subspace of L^σ corresponding to $\mathscr{L}_E^2(\mu)$ (22.3.4) is in this case the space of functions $f \circ \pi$, where $f \in \mathscr{L}_E^2(\mu)$ and $\pi: G \to G/H = X$ is the canonical mapping.

(22.3.7) An important particular case of (22.3.6) is that in which H is a *compact* subgroup of G; then there exists on X = G/H a G-invariant positive measure $\mu \neq 0$, which is unique up to a constant factor. If $\pi: G \to G/H$ is the canonical mapping, in order to define such a measure μ it is enough to observe that for each function $f \in \mathscr{K}(\mathrm{G/H})$ the function $f \circ \pi$ belongs to $\mathscr{K}(\mathrm{G})$, and to define

(22.3.7.1) $$\int f\,d\mu = \int (f \circ \pi)\,dm_{\mathrm{G}}$$

where m_{G} is a left Haar measure on G. Conversely, if the positive measure $\nu \neq 0$ on $\mathrm{X} = \mathrm{G/H}$ is G-invariant, and if m_{H} is a Haar measure on H, it is immediately verified that the positive measure on G

(22.3.7.2) $$h \mapsto \int_{\mathrm{X}} d\nu(z) \int_{\mathrm{H}} h(st)\,dm_{\mathrm{H}}(t),$$

where $z = \pi(s)$, is left-invariant and therefore proportional to m_{G}; and by taking in particular $h = f \circ \pi$ in the formula above, where $f \in \mathscr{K}(\mathrm{G/H})$, we conclude that ν is proportional to μ.

With H *compact*, suppose that α satisfies the conditions of (22.3.6) and also that the mapping $s \mapsto \mathbf{f}^{\alpha}(s) = \alpha(s^{-1}, s \cdot x_0) \cdot \mathbf{f}(s \cdot x_0)$ of G into E is m_{G}-measurable for each function $\mathbf{f} \in \mathscr{L}^2_{\mathrm{E}}(\mu)$. Then we have $\|\mathbf{f}^{\alpha}(su)\| = \|\mathbf{f}^{\alpha}(s)\| = \|\mathbf{f}(s \cdot x_0)\|$ for all $s \in \mathrm{G}$ and all $u \in \mathrm{H}$, so that $\mathbf{f}^{\alpha}$ belongs to the space $\mathscr{L}^2_{\mathrm{E}}(m_{\mathrm{G}})$ by virtue of (22.3.7.1). Conversely, if $\mathbf{g} \in \mathscr{L}^2_{\mathrm{E}}(m_{\mathrm{G}})$ is such that

(22.3.7.3) $$\mathbf{g}(su) = U(u^{-1}) \cdot \mathbf{g}(s) \qquad (s \in \mathrm{G},\ u \in \mathrm{H})$$

and if the mapping $s \mapsto \alpha(s, x_0) \cdot \mathbf{g}(s)$ is m_{G}-measurable, then we may write this mapping in the form $\mathbf{f} \circ \pi$ with $\mathbf{f} \in \mathscr{L}^2_{\mathrm{E}}(\mu)$, and we have $\mathbf{g} = \mathbf{f}^{\alpha}$.

In this case, therefore, we may consider U^{ind} (up to equivalence) as the continuous unitary representation of G on the closed subspace F of the Hilbert space $\mathrm{L}^2_{\mathrm{E}}(m_{\mathrm{G}})$ generated by the classes of functions $\mathbf{g}$ satisfying (22.3.7.3); we have then, by definition,

(22.3.7.4) $$U^{\mathrm{ind}}(s) \cdot \tilde{\mathbf{g}} = (\gamma(s)\mathbf{g})^{\sim}$$

for $\tilde{\mathbf{g}} \in \mathrm{F}$. When $\mathrm{E} = \mathbf{C}$, U^{ind} is a *subrepresentation* of the *regular* representation of G.

If we take for U the trivial representation of H on E, the functions $\mathbf{g}$ satisfying (22.3.7.3) are the functions that are constant on each coset $s\mathrm{H}$. The linear representation U^{ind} of G on $\mathrm{L}^2_{\mathrm{E}}(\mu)$ is then called the *canonical representation* of G corresponding to the compact subgroup H and its trivial representation on E. When $\mathrm{H} = \{e\}$ and $\mathrm{E} = \mathbf{C}$, we are back to the regular representation of G (21.1.9).

(22.3.7.5) The definition **(22.3.7.4)** of U^{ind} involves neither μ nor α. It may be shown that this definition can be extended to any closed subgroup H of G, with an appropriate definition of the space F (cf. Problem 6).

(22.3.7.6) Suppose in addition that G is a *Lie group* and H a *compact connected* Lie subgroup of G. The Haar measures m_G and m_H then correspond canonically **(16.24.1)** to volume forms υ_G and υ_H, respectively, invariant under left translations. Then there exists a G-invariant volume form $\upsilon_{G/H}$ on G/H such that for each $x \in H$ we have

(22.3.7.7)
$$\upsilon_H(x) = \upsilon_G(x)/\upsilon_{G/H}(x_0)$$

(where $x_0 = \pi(e)$) in the notation of **(16.21.7)**. We have only to repeat the argument of **(21.15.3)**, replacing T by H and $\mathfrak{m}$ by a supplement of $\mathfrak{h} = \text{Lie}(H)$ in $\mathfrak{g} = \text{Lie}(G)$, orthogonal to $\mathfrak{h}$ with respect to some Ad(H)-invariant scalar product on $\mathfrak{g}$ **(20.11.3.1)**. This argument establishes the existence of the G-invariant volume form $\upsilon_{G/H}$, and the construction of this form establishes the formula **(22.3.7.7)**, when we compare this construction with the right-hand side of **(16.21.7)**. We remark that the existence of the volume form $\upsilon_{G/H}$ shows that, under the given conditions, G/H is *orientable* (compare with **(16.21.12)**). Furthermore, the positive measure $m_{G/H}$ on G/H corresponding to $\upsilon_{G/H}$ **(16.24.1)** is such that the positive measure defined by **(22.3.7.2)**, namely

$$h \mapsto \int_{G/H} dm_{G/H}(z) \int_H h(st)\, dm_H(t),$$

where $z = \pi(s)$, is equal to m_G. This follows from **(22.3.7.7)** and the integration formula **(16.24.8.1)** applied to the form $\upsilon = h \cdot \upsilon_G$ on G and the form $\zeta = \upsilon_{G/H}$ on G/H.

(22.3.8) It happens often that there exists no G-invariant measure μ on X, but only a *quasi-invariant* measure μ, i.e., such that for each $s \in G$ the measure $\gamma(s)\mu$ is *equivalent* to μ **(13.15.6)**; this is always the case when G is a Lie group and μ a Lebesgue measure on X, because $x \mapsto s \cdot x$ is then a diffeomorphism of X **(16.22.1)**. For each $s \in G$, we have then $\gamma(s)\mu = J_s \cdot \mu$, where J_s is a function everywhere > 0, and both J_s and J_s^{-1} are locally μ-integrable. For each function $h \in \mathscr{K}(X)$, we have then

(22.3.8.1)
$$\int h(s \cdot x)\, d\mu(x) = \int h(x) J_s(x)\, d\mu(x)$$

from which it follows, for s and t in G, that

$$\begin{aligned}\int h(x)\mathrm{J}_{st}(x)\,d\mu(x) &= \int h(st\cdot x)\,d\mu(x)\\ &= \int h(s\cdot x)\mathrm{J}_t(x)\,d\mu(x)\\ &= \int h(x)\mathrm{J}_s(x)\mathrm{J}_t(s^{-1}\cdot x)\,d\mu(x),\end{aligned}$$

which in turn implies that, almost everywhere with respect to μ **(13.14.4)**

(22.3.8.2) $$\mathrm{J}_{st}(x) = \mathrm{J}_s(x)\mathrm{J}_t(s^{-1}\cdot x).$$

We therefore replace condition (1) of **(22.3.6)** by

(1 *bis*) $\mathbf{z} \mapsto (\mathrm{J}_{s^{-1}}(x))^{1/2}\alpha(s, x)\cdot \mathbf{z}$ is a *unitary* operator on E which reduces to $U(u)$ for $s = u \in \mathrm{H}$ and $x = x_0$.

If this condition and condition (2) of **(22.3.6)** are satisfied, then for each function $\mathbf{f} \in \mathscr{L}^2_{\mathrm{E}}(\mu)$ the function $\rho(s)\cdot\mathbf{f}$ belongs to $\mathscr{L}^2_{\mathrm{E}}(\mu)$ and the relation **(22.3.6.1)** holds for each $s \in \mathrm{G}$: for we have

$$\begin{aligned}\int \|\alpha(s, s^{-1}\cdot x)\cdot\mathbf{f}(s^{-1}\cdot x)\|^2\,d\mu(x) &= \int \mathrm{J}_{s^{-1}}(x)\|\alpha(s, x)\cdot\mathbf{f}(x)\|^2\,d\mu(x)\\ &= \int \|\mathbf{f}(x)\|^2\,d\mu(x).\end{aligned}$$

If condition (3) of **(22.3.6)** is also satisfied, we shall have defined a *continuous unitary representation* **(22.3.6.2)** of G on $\mathrm{L}^2_{\mathrm{E}}(\mu)$.

For example, for each real number r we may take

(22.3.8.3) $$\alpha(s, x) = (\mathrm{J}_{s^{-1}}(x))^{-\frac{1}{2}+ri}\cdot 1_{\mathrm{E}}$$

because by virtue of **(22.3.8.2)** this function satisfies the conditions **(22.3.2.1)** and **(22.3.2.2)**, and $(\mathrm{J}_{s^{-1}}(x))^{1/2}\alpha(s, x)$ is the operator of multiplication by a complex-valued function of absolute value 1, hence is unitary. Condition (2) of **(22.3.6)** is satisfied because each function J_s is μ-measurable. Finally, since $\rho(s) = \rho(ss_0^{-1})\rho(s_0)$, it is enough to establish the continuity of **(22.3.6.2)** at the point e; by using the equicontinuity of the set of unitary operators **(12.15.7.1)**, we reduce by virtue of **(7.5.5)** to the case where $\mathbf{f}$ is a function of the form $h\mathbf{a}_n$ with $h \in \mathscr{K}(\mathrm{X})$. Now by **(22.3.8.2)** we have $\mathrm{J}_{s^{-1}}(s^{-1}\cdot x)^{-1} = \mathrm{J}_s(x)$ almost everywhere; writing $j_s(x) = \mathrm{J}_s(x)^{\frac{1}{2}-ri}$, in order that condition (3) of **(22.3.6)** be satisfied it is enough that for each compact $K \subset X$ the mapping $s \mapsto j_s \varphi_K$ of G into $\mathscr{L}^2_C(X)$ should be *continuous* at e. Consider a compact neighborhood K of Supp(h) in X, and let V_0 be a neighborhood of e in G such that $\text{Supp}(\gamma(s)h) \subset K$ for all $s \in V_0$; also put $m = N_\infty(h)$. Then we may write

$$\int |j_s(x)h(s^{-1} \cdot x) - h(x)|^2 \, d\mu(x)$$
$$\leqq m \cdot \int |j_s - 1|^2 \varphi_K \, d\mu + \int |\gamma(s)h - h|^2 \, d\mu$$

and the condition, together with **(14.10.6.3)**, shows that the right-hand side is arbitrarily small provided that s stays inside a sufficiently small neighborhood V of e contained in V_0.

Examples

(22.3.9) Let $G = \mathbf{SL}(2, \mathbf{R})$. The Iwasawa decomposition **(21.21.10)** enables us to write $G = KS_0$, where K is the group $\mathbf{SO}(2, \mathbf{R})$ (isomorphic to **U**) consisting of the matrices

$$\begin{pmatrix} \cos\theta & \sin\theta \\ -\sin\theta & \cos\theta \end{pmatrix}$$

for $0 \leqq \theta \leqq 2\pi$, and S_0 is the solvable group of triangular matrices

$$\begin{pmatrix} a & b \\ 0 & a^{-1} \end{pmatrix}$$

with b real and $a > 0$. If S is the group of matrices

$$\begin{pmatrix} a & b \\ 0 & a^{-1} \end{pmatrix}$$

with b real and a real $\neq 0$, *of either sign*, then clearly we have also $G = KS$.

The group G acts differentiably and transitively on the real projective line $\mathbf{P}_1(\mathbf{R})$, the point x with homogeneous coordinates (x_1, x_2) being mapped by

$$s = \begin{pmatrix} a & b \\ c & d \end{pmatrix} \in \mathbf{SL}(2, \mathbf{R})$$

to the point with homogeneous coordinates $(ax_1 + bx_2, cx_1 + dx_2)$; we may also say that $x \in \mathbf{P}_1(\mathbf{R})$ is sent to $(ax + b)/(cx + d)$, with the understanding that this point is a/c (with $a/0 = \infty$) when $x = \infty$, and that it is ∞ when

$x = -d/c$. The stabilizer of the point $x = \infty$ is the subgroup S, and G/S may therefore be identified with $\mathbf{P}_1(\mathbf{R})$ (16.10.8). Consider the diffuse measure μ on $\mathbf{P}_1(\mathbf{R})$ which induces Lebesgue measure λ on $\mathbf{R}$; we have then $J_s(x) = (cx - a)^{-2}$. The procedure of (22.3.8) (with $E = \mathbf{C}$) then provides, for each real number r, a continuous unitary representation V_{ri} of G on $L^2_{\mathbf{C}}(\lambda)$: the unitary operator $V_{ri}(s)$ maps the class of a function $f \in \mathscr{L}^2_{\mathbf{C}}(\lambda)$ to the class of the function

$$x \mapsto |cx - a|^{-1+ri} f\left(\frac{b - dx}{cx - a}\right).$$

It can be shown that as r runs through the set of real numbers > 0, the representations V_{ri} are irreducible and pairwise inequivalent.

(22.3.10) We can also write $G = S_0 K$; the group G acts differentiably and transitively on the upper half-plane P: $\mathscr{I}z > 0$ in $\mathbf{C}$, by the rule $s \cdot z = (az + b)/(cz + d)$; the stabilizer of the point i is the subgroup K, and therefore P may be identified with G/K (16.10.8). One verifies immediately that the differential 2-form $y^{-2}\, dx \wedge dy = \frac{1}{2} i y^{-2}\, dz \wedge d\bar{z}$ on P is G-invariant, and the measure $\mu: h \mapsto \iint_{\mathrm{P}} h(x + iy) y^{-2}\, dx\, dy$ on P is therefore G-*invariant* (cf. (22.3.7)). For each point $z = x + iy$ in P, we may take as representative of the coset $s\mathrm{K} \subset \mathrm{G}$ which sends i to z the matrix

$$r_z = \begin{pmatrix} y^{1/2} & x/y^{1/2} \\ 0 & 1/y^{1/2} \end{pmatrix} \in S_0.$$

For each element

$$s = \begin{pmatrix} a & b \\ c & d \end{pmatrix}$$

in G, the imaginary part of $s \cdot z = (az + b)/(cz + d)$ is $y|cz + d|^{-2}$ (since a, b, c, d are real); hence we have

$$r_{s \cdot z} = \begin{pmatrix} y^{1/2}/|cz + d| & * \\ 0 & |cz + d|/y^{1/2} \end{pmatrix},$$

and the relation $sr_z = r_{s \cdot z} u(s, z)$, where

$$u(s, z) = \begin{pmatrix} \cos\theta & \sin\theta \\ -\sin\theta & \cos\theta \end{pmatrix},$$

leads in particular to the relations

$$cy^{1/2} = -y^{-1/2}|cz + d| \sin\theta, \qquad y^{-1/2}(cx + d) = y^{-1/2}|cz + d| \cos\theta,$$

whence $e^{-i\theta} = (cz+d)/|cz+d|$. Consider now a one-dimensional unitary representation σ of K, which is necessarily of the form

$$u \mapsto \sigma_n(u) = e^{-ni\theta}$$

where n is a *rational integer* (21.3.9). The method of (22.3.3) shows that by taking $\alpha(s, z)$ to be multiplication by the function $(cz+d)^n/|cz+d|^n$ we obtain the unitary representation ρ_n of G on $L^2_{\mathbf{C}}(\mu)$ *induced* by σ_n: the operator $\rho_n(s)$, where

$$s = \begin{pmatrix} a & b \\ c & d \end{pmatrix} \in \mathrm{G},$$

maps the class of a function $f \in \mathscr{L}^2_{\mathbf{C}}(\mu)$ to the class of the function

$$z \mapsto f\left(\frac{b-dz}{cz-a}\right) \frac{(cz-a)^{-n}}{|cz-a|^{-n}}.$$

We can define an equivalent representation by taking $c(x+iy) = y^{-n/2}$, in the notation of (22.3.3); the image of $\mathscr{L}^2_{\mathbf{C}}(\mu)$ under the mapping $f \mapsto c \cdot f$ is the space of complex-valued functions $z \mapsto g(z)$ defined on P for which $z \mapsto y^n g(z)^2$ is μ-integrable. The equivalent representation on the Hilbert space E_n consisting of the classes of these functions is then described as follows: the operator corresponding to

$$s = \begin{pmatrix} a & b \\ c & d \end{pmatrix} \in \mathrm{G}$$

takes the class of a function g to that of the function

$$z \mapsto g\left(\frac{b-dz}{cz-a}\right)(cz-a)^{-n}.$$

This representation is not always irreducible, because the subspace H^2_n of E_n consisting of the classes of the functions g which are *holomorphic* on P is clearly stable under the representation ρ_n and is not zero for $n > 1$ (Problem 4). It can be shown that H^2_n is closed and that the representation $s \mapsto \rho_n(s) | \mathrm{H}^2_n$ is irreducible.

Another way of defining ρ_n, up to equivalence, is to apply the method of (22.3.7). The mapping $g \mapsto g^{\alpha}$ sends a function g with class $\tilde{g} \in \mathrm{E}_n$ to the function on G defined by

$$\begin{pmatrix} a & b \\ c & d \end{pmatrix} \mapsto (-1)^n (ci+d)^{-n} g\left(\frac{ai+b}{ci+d}\right).$$

The representation ρ_n then makes G act by left translations on the space of classes of functions $f \in L^2_{\mathbf{C}}(m_G)$ such that $f(tu) = \sigma_n(u^{-1})f(t)$ for all $t \in G$ and $u \in K$.

(22.3.11) Everything we have done in this section for *left* actions of G has of course a counterpart for *right* actions. If we denote by $\alpha(s, x)$ the automorphism $z \mapsto \mathrm{pr}_2((x, z) \cdot s)$, the relation **(22.3.2.2)** is replaced by

(22.3.11.1) $$\alpha(st, x) = \alpha(t, x \cdot s) \circ \alpha(s, x).$$

Such a mapping α is called a *right* 1-*cocycle of* G *with values in* Aut(E). The linear representation ρ of G on L corresponding to α is defined by

(22.3.11.2) $$\begin{aligned}(\rho(s) \cdot f)(x) &= \alpha(s^{-1}, x \cdot s) \cdot f(x \cdot s)\\ &= \alpha(s, x)^{-1} \cdot f(x \cdot s).\end{aligned}$$

We leave it to the reader to transcribe the other definitions and results.

PROBLEMS

1. Let G be a Lie group, H a Lie subgroup of G. With the notation of **(22.3.2)**, suppose that the mapping $x \mapsto r_x$ of $X = G/H$ into G is of class C^∞. If also σ is a continuous (but not necessarily unitary) representation of H on a finite-dimensional complex vector space E, then by taking $\alpha(s, x) = \sigma(u(s, x))$ we obtain a continuous linear representation ρ of G on the Fréchet space $\mathscr{E}(X; E)$ of C^∞-mappings of X in E, by the formula **(22.3.2.8)**.

2. With the notation of **(22.3.9)**, the homogeneous space $X = G/S_0$ may be identified with the circle $\mathbf{S}_1 = \mathbf{U}$, and the mapping $x \mapsto r_x$ may be taken to be

$$e^{i\theta} \mapsto \begin{pmatrix} \cos\theta & \sin\theta \\ -\sin\theta & \cos\theta \end{pmatrix}.$$

For

$$s = \begin{pmatrix} a & b \\ c & d \end{pmatrix}$$

and $x = e^{i\theta}$, we have then

$$u(s, x) = \begin{pmatrix} \lambda & \mu \\ 0 & \lambda^{-1} \end{pmatrix}$$

with $\lambda > 0$ and

(*) $$\lambda^2 = (a\cos\theta - b\sin\theta)^2 + (c\cos\theta - d\sin\theta)^2,$$

and $s \cdot x = e^{i\theta'}$, where θ' is given by $\lambda \cos \theta' = a \cos \theta - b \sin \theta$ and $-\lambda \sin \theta' = c \cos \theta - d \sin \theta$. The result of Problem 1 is then applicable: taking $\mathrm{E} = \mathbf{C}$, every continuous representation σ of S_0 on $\mathbf{C}$ is such that

$$\sigma\begin{pmatrix} \lambda & \mu \\ 0 & \lambda^{-1} \end{pmatrix} = \lambda^{1-\zeta}$$

where ζ is an arbitrary complex number; we therefore have $\alpha(s, x) = \lambda(s, \theta)^{1-\zeta}$, where $\lambda(s, \theta)$ is given by the formula $(*)$.

Show that the induced representation ρ corresponding to σ and α is equivalent to the representation V_ζ of G defined as follows: consider the set $\mathscr{E}_\zeta$ of complex-valued C^∞-functions on $\mathbf{R}^2 - \{0\}$ which satisfy the relation

$$f(cx_1, cx_2) = c^\zeta f(x_1, x_2)$$

for all $c > 0$, and give $\mathscr{E}_\zeta$ the topology induced by that of $\mathscr{E}(\mathbf{R}^2 - \{0\})$; then for

$$s = \begin{pmatrix} a & b \\ c & d \end{pmatrix}$$

the function $V_\zeta(s) \cdot f$ is defined by

$$(x_1, x_2) \mapsto f(dx_1 - bx_2, -cx_1 + ax_2).$$

The space $\mathscr{E}_\zeta$ is the direct sum of two subspaces $\mathscr{E}_{(\zeta, \varepsilon)}$ (where $\varepsilon = 0$ or 1) consisting of the functions $f \in \mathscr{E}_\zeta$ which satisfy

$$f(-x_1, -x_2) = (-1)^\varepsilon f(x_1, x_2).$$

These subspaces are stable under V_ζ, and we denote by $V_{(\zeta, \varepsilon)}$ ($\varepsilon = 0, 1$) the corresponding representations of G on the spaces $\mathscr{E}_{(\zeta, \varepsilon)}$.

Show that the space $\mathscr{E}_{(\zeta, \varepsilon)}$ may be identified with the space of functions $g \in \mathscr{E}(\mathbf{R})$ such that the function

$$x \mapsto |x|^{\zeta-1}(\operatorname{sgn} x)^\varepsilon g(-x^{-1}),$$

defined for $x \neq 0$, extends to a function belonging to $\mathscr{E}(\mathbf{R})$. Then $V_{(\zeta, \varepsilon)}(s) \cdot g$, for

$$s = \begin{pmatrix} a & b \\ c & d \end{pmatrix},$$

is the function

$$x \mapsto |cx - a|^{\zeta-1}(\operatorname{sgn}(cx - a))^\varepsilon g\left(\frac{b - dx}{cx - a}\right).$$

If $\zeta = ri$, where r is real, the representation $V_{(\zeta, \varepsilon)}$ extends to a unitary representation of G on $\mathrm{L}^2_{\mathbf{C}}(\lambda)$, which for $\varepsilon = 0$ is the representation denoted by V_{ri} in (22.3.9).

If $-1 < \zeta < 0$, show that $V_{(\zeta, 0)}$ extends to a unitary representation of G on the Hilbert space of classes of λ-measurable functions g on $\mathbf{R}$ satisfying

$$\iint_{\mathbf{R}^2} |y - z|^{-\zeta-1} |g(y)g(z)| \, dy \, dz < +\infty,$$

the scalar product being defined by

$$B(g, h) = \iint_{\mathbf{R}^2} |y - z|^{-\zeta - 1} g(y)\overline{h(z)}\, dy\, dz.$$

(To show that $B(g, g) \geqq 0$, put $y = \tan\theta$, $z = \tan\theta'$ and expand g as a Fourier series; use the formula

$$\int_0^\pi \sin^{-\zeta-1}\theta \cos 2k\theta\, d\theta = \frac{2^{\zeta+1}\pi \cdot \Gamma(-\zeta) \cdot (-1)^k}{\Gamma(\frac{1}{2} - \frac{1}{2}\zeta + k)\Gamma(\frac{1}{2} - \frac{1}{2}\zeta - k)},$$

where k is any integer $\geqq 0$, which can be deduced from the properties of the Eulerian integrals and the formula

$$\int_0^\pi f(\cos\theta)\cos 2n\theta\, d\theta = \frac{1}{1 \cdot 3 \cdot 5 \cdots (2n-1)} \int_0^\pi f^{(n)}(\cos\theta)\sin^{2n}\theta\, d\theta.$$

3. Let α be a real number and let μ be the measure on the upper half-plane $P: \mathscr{I}z > 0$ defined in **(22.3.10)**. Let H_α^2 be the set of holomorphic functions on P for which the integral $\iint_P |f(z)|^2 y^\alpha\, d\mu(z)$ is finite. Show that H_α^2 is a Hilbert space with respect to the scalar product

$$(f \mid g) = \iint_P f(z)\overline{g(z)}\, y^\alpha\, d\mu(z).$$

(Same method as in the problem in Section **9.13**). Moreover, there exists a reproducing kernel $K(z, \zeta)$ in H_α^2 (Section **6.3**, Problem 4), such that for each function $f \in H_\alpha^2$ and each $z \in P$ we have

$$(*) \qquad f(z) = \iint_P f(\zeta)\overline{K(\zeta, z)}\,\eta^\alpha\, d\mu(\zeta)$$

(where $\eta = \mathscr{I}\zeta$) with $K(\zeta, z) = \overline{K(z, \zeta)}$, and the function $z \mapsto K(z, \zeta)$ belongs to H_α^2 for each $\zeta \in P$.

Show that for each

$$s = \begin{pmatrix} a & b \\ c & d \end{pmatrix} \in G$$

we have

$$K(s \cdot z, s \cdot \zeta) = (cz + d)^\alpha (c\bar{\zeta} + d)^\alpha K(z, \zeta)$$

for those values of α such that $H_\alpha^2 \neq \{0\}$ (use $(*)$). For any complex number z such that $\mathscr{I}(z) \neq 0$, put $z^\alpha = e^{\alpha \log z}$, taking the principal branch of the logarithm. Deduce that

$$K(z, \zeta) = C_\alpha (z - \bar{\zeta})^{-\alpha}$$

where C_α is a constant. (Observe that if $F(z, \zeta) = (z - \bar{\zeta})^\alpha K(z, \zeta)$, we have $F(s \cdot z, s \cdot \zeta) = F(z, \zeta)$ for all $s \in G$.)

Show that $H_\alpha^2 \neq \{0\}$ if and only if $\alpha > 1$. (Express that the function $(z + i)^{-\alpha}$ belongs to H_α^2.)

4. Deduce from Problem 3 that the representation ρ_n of G on the space H_n^2 defined in (**22.3.10**) is irreducible for each integer $n > 1$. (Let T be a continuous operator on H_n^2 such that $\rho_n(s)T = T\rho_n(s)$ for all $s \in G$. Let $K_z(\zeta) = K(\zeta, z)$, so that $K(\zeta, z) = (K_z | K_\zeta)$, and consider the function $K_T(z, \zeta) = (T \cdot K_\zeta | K_z)$; by considering the function $K_T(s \cdot z, s \cdot \zeta)$, show that $K_T = \lambda \cdot K$ for some constant $\lambda \in \mathbf{C}$. Complete the proof by observing that the functions K_ζ, $\zeta \in P$, form a total set in the Hilbert space H_n^2.)

5. Let $\tilde{G}$ be the Lie group which is the universal covering of the group $\mathbf{SL}(2, \mathbf{R})$, and let $\pi\colon \tilde{G} \to G$ be the canonical homomorphism. For each $s \in \tilde{G}$, put

$$\pi(s) = \begin{pmatrix} a_s & b_s \\ c_s & d_s \end{pmatrix}.$$

Show that there exists a C^∞-mapping $(s, z) \mapsto \theta(s, z)$ of $\tilde{G} \times P$ into $\mathbf{R}$, defined up to an integer multiple of 2π, such that

$$\frac{c_s z + d_s}{|c_s z + d_s|} = e^{i\theta(s, z)}$$

on $\tilde{G} \times P$ (cf. (**16.28.8**)). With the notation of Problem 3, deduce that a continuous unitary representation ρ of $\tilde{G}$ is defined on the Hilbert space H_α^2, for each $\alpha > 1$, by the rule that $\rho(s) \cdot g$ is the function

$$z \mapsto |c_s z - a_s|^\alpha \cdot e^{\alpha i\theta(s, z)} g(s^{-1} \cdot z).$$

These representations are irreducible but not square-integrable (Section **21.4**, Problem 5(c)).

6. Let G be a (separable, metrizable) locally compact group and H a closed subgroup of G; suppose that G and H are unimodular, and denote by m_G and m_H Haar measures on G and H, respectively.

Then there exists on G/H a unique G-invariant measure $m_{G/H}$ (**22.3.6**) such that for each function $f \in \mathcal{K}(G)$ we have

$$\int_G f(s)\, dm_G(s) = \int_{G/H} dm_{G/H}(\dot{s}) \int_H f(st)\, dm_H(t),$$

where $\dot{s} = sH$. Moreover, there exists a continuous function $h \geq 0$ on G whose support has compact intersection with every set KH, where K is a compact subset of G, and which is such that $\int_H h(st)\, dm_H(t) = 1$ for all $s \in G$. (Use Section **14.4** Problem 2, and a partition of unity on G/H.)

(a) Let U be a continuous unitary representation of H on a separable Hilbert space E. If a mapping $\mathbf{g}$ of G into E is such that $\mathbf{g}(st) = U(t^{-1}) \cdot \mathbf{g}(s)$ for all $s \in G$ and $t \in H$, then $\|\mathbf{g}(s)\|$ depends only on the coset $\dot{s} = sH$. Let F denote the vector space of classes of m_G-measurable functions $\mathbf{g}$ satisfying this condition and such that the mapping $\dot{s} \mapsto \|\mathbf{g}(s)\|^2$ is $m_{G/H}$-integrable; it comes to the same thing to say that the function $s \mapsto h(s)\|\mathbf{g}(s)\|^2$ is m_G-integrable and that $\int_{G/H} \|\mathbf{g}(s)\|^2\, dm_{G/H}(\dot{s}) = \int_G h(s)\|\mathbf{g}(s)\|^2\, dm_G(s)$. Show that F is a Hilbert space for the scalar product

$$(\tilde{\mathbf{g}}_1 | \tilde{\mathbf{g}}_2) = \int_G h(s)(\mathbf{g}_1(s) | \mathbf{g}_2(s))\, dm_G(s).$$

(b) Show that for each function $f \in \mathscr{K}(G)$ and each vector $\mathbf{v} \in E$, the integral **(13.10)**

$$(\mathbf{B}(f, \mathbf{v}))(s) = \int_H f(st)(U(t) \cdot \mathbf{v})\, dm_H(t)$$

is defined for all $s \in G$; the mapping $\mathbf{B}(f, \mathbf{v})$ of G into E so defined is continuous, and if $\text{Supp}(f) = K$ the support of $\mathbf{B}(f, \mathbf{v})$ is contained in KH. Show that the functions $\mathbf{B}(f, \mathbf{v})$ belong to F and that, for each function $\mathbf{g}$ such that $\tilde{\mathbf{g}} \in F$, we have

$$(\mathbf{B}(f, \mathbf{v}) \mid \tilde{\mathbf{g}}) = \int_G f(s)(\mathbf{v} \mid \mathbf{g}(s))\, dm_G(s).$$

Deduce that the functions $\mathbf{B}(f, \mathbf{v})$ with $f \in \mathscr{K}(G)$ and $\mathbf{v} \in E$ form a *total* set in the Hilbert space F.

(c) If $U^{\text{ind}}(s)$ is defined by the formula **(22.3.7.4)** for $s \in G$ and $\tilde{\mathbf{g}} \in F$, show that U^{ind} is a continuous unitary representation of G on F; it is called the representation *induced* by U. (Let F_0 be the subspace of F consisting of the functions $\mathbf{g}$ which are continuous on G and whose support is contained in a set of the form KH, where $K \subset G$ is compact. For $\mathbf{f}, \mathbf{g} \in F_0$ and s in a sufficiently small neighborhood of e, show that

$$(U^{\text{ind}}(s) \cdot \mathbf{f} \mid \mathbf{g}) = \int_G k(x)(\mathbf{f}(s^{-1}x) \mid \mathbf{g}(x))\, dm_G(x),$$

where $k \in \mathscr{K}(G)$ depends only on the supports of $\mathbf{f}$ and $\mathbf{g}$, and then use (b).)

(d) Let N be a closed normal subgroup of G and let V be a continuous unitary representation of G on a Hilbert space F, such that the restriction of V to N is trivial, so that on passing to the quotient V determines a continuous unitary representation V' of G/N on F, such that $V'(\pi(s)) = V(s)$, where $\pi: G \to G/N$ is the canonical homomorphism. If $V' = U'^{\text{ind}}$, where U' is a continuous unitary representation of a subgroup H′ of G/N on a Hilbert space E, show that V is equivalent to U^{ind}, where U is the continuous unitary representation of $H = \pi^{-1}(H')$ on E defined by $U(t) = U'(\pi(t))$.

7. Let G be a (separable, metrizable) locally compact group and H a closed subgroup of G. Suppose that G and H are unimodular, and let m_G and m_H denote Haar measures on G and H, respectively.

(a) Let μ be a measure of positive type *on* H. If μ is identified with its image under the canonical injection of H in G, show that μ is of positive type as a measure on G. (For each function $f \in \mathscr{K}(G)$, show that

$$\iint_{G \times H} f(sx)\overline{f(s)}\, dm_G(s)\, d\mu(x) = \iiint_{G \times H \times H} h(s) f(sy^{-1}x)\overline{f(sy^{-1})}\, dm_G(s)\, dm_H(y)\, d\mu(x)$$

where $h \in \mathscr{K}(G)$ is such that

$$\int_H h(su)\, dm_H(u) = 1$$

for $s \in \text{Supp}(f)$.

(b) Let U_μ (resp U'_μ) be the continuous unitary representation of H (resp G) on E_μ (resp. E'_μ) defined by the method of **(22.2.4)** from μ, considered as a measure of positive type on H (resp. G). Show that the representation U'_μ is equivalent to U_μ^{ind} as defined in **Problem 6**, (Observe that each function $f \in \mathscr{K}(G)$ defines canonically a mapping $f': G \to E_\mu$, by considering the function $u \mapsto f(su)$ in $\mathscr{K}(H)$, for each $s \in G$, and then use Problem 6.)

8. Let G be a (separable, metrizable) locally compact group, and let H_1, H_2 be two closed subgroups of G such that $H_1 \subset H_2$. Suppose that G, H_1, H_2 are all unimodular, and let m_G, m_{H_1}, m_{H_2} denote Haar measures on these three groups.

Let U be a continuous unitary representation of H_1 on a separable Hilbert space E. We propose to show that the representations U^{ind}_{G/H_1} and $(U^{\text{ind}}_{H_2/H_1})^{\text{ind}}_{G/H_2}$ are equivalent ("transitivity of induction"). Put $V = U^{\text{ind}}_{H_2/H_1}$. Let F_0 denote the pre-Hilbert space of functions $\mathbf{g}: G \to E$ which are continuous on G, with support contained in a set of the form KH_1, where K is compact, and which satisfy the relation $\mathbf{g}(su_1) = U(u_1^{-1}) \cdot \mathbf{g}(s)$ for all $s \in G$ and $u_1 \in H_1$, and are such that the mapping $sH_1 \mapsto \|\mathbf{g}(s)\|^2$ is m_{G/H_1}-integrable. Denote by $F^{(1)}$ the space of the representation V, and by $F_0^{(2)}$ the pre-Hilbert space of functions $\mathbf{g}: G \to F^{(1)}$ which are continuous on G, with support contained in a set of the form KH_2, where K is compact, and which satisfy the relation $\mathbf{g}(su_2) = V(u_2^{-1}) \cdot \mathbf{g}(s)$ for all $s \in G$ and $u_2 \in H_2$, and are such that the mapping $sH_2 \mapsto \|\mathbf{g}(s)\|^2$ is m_{G/H_2}-integrable.

For each function $\mathbf{f} \in F_0$, put $\mathbf{f}_1(s, u_2) = \mathbf{f}(su_2)$ for $s \in G$ and $u_2 \in H_2$, and $\mathbf{f}_2(s) = \mathbf{f}_1(s, \cdot)$ Show that $\mathbf{f}_2(s) \in F^{(1)}$ for each $s \in G$, and that $\mathbf{f}_2 \in F_0^{(2)}$. Then prove that the mapping $\mathbf{f} \mapsto \mathbf{f}_2$ is an isometry of F_0 onto a dense subspace of $F_0^{(2)}$. To prove that the image is dense, define as in Problem 6 a function $\mathbf{B}_1(g, \mathbf{v}) \in F^{(1)}$ for each $g \in \mathscr{K}(H_2)$ and $\mathbf{v} \in E$, and a function $\mathbf{B}_2(h, \mathbf{w}) \in F_0^{(2)}$ for each $h \in \mathscr{K}(G)$ and $\mathbf{w} \in F^{(1)}$; and consider the functions $\mathbf{B}_2(h, \mathbf{B}_1(g, \mathbf{v}))$.

9. Let G be a (separable, metrizable) locally compact group and Γ a closed subgroup of G. Suppose that both G and Γ are unimodular, and retain the notation of Problem 6.
(a) Let V be the unitary representation of G on $L^2_{\mathbf{C}}(m_{G/\Gamma})$ induced by the *trivial* representation of Γ on $\mathbf{C}$. Show that for each function $f \in \mathscr{L}^1_{\mathbf{C}}(m_G)$ and each function $g \in \mathscr{L}^2_{\mathbf{C}}(m_{G/\Gamma})$, $V(f) \cdot \tilde{g}$ (cf. **(21.1.4)**) is the class of the function

$$\dot{x} \mapsto \int_G f(s) g(s^{-1} \cdot \dot{x})\, dm_G(s),$$

defined for almost all $\dot{x} \in G/\Gamma$. (Show that if $h \in \mathscr{L}^2_{\mathbf{C}}(m_{G/\Gamma})$, the function

$$(s, \dot{x}) \mapsto f(s) g(s^{-1} \cdot \dot{x}) h(\dot{x})$$

is integrable with respect to $m_G \otimes m_{G/\Gamma}$, and that

$$\iint |f(s) g(s^{-1} \cdot \dot{x}) h(\dot{x})|\, dm_G(s)\, dm_{G/\Gamma}(\dot{x}) \leqq N_1(f) N_2(g) N_2(h).$$

This function is denoted by $f * g$. If $\pi: G \to G/\Gamma$ is the canonical mapping, then for almost all $x \in G$ we have

$$(f * g)(\pi(x)) = \int_G f(xs^{-1}) g(\pi(s))\, dm_G(s).$$

(b) Suppose that Γ is *discrete* and m_Γ normalized **(14.3.5)**. For any $f \in \mathscr{K}(G)$ and $x, y \in G$, the sum $\sum_{\xi \in \Gamma} f(x\xi y^{-1})$ is defined, and depends only on the cosets $\dot{x} = \pi(x)$ and $\dot{y} = \pi(y)$; the function $K_f: (\dot{x}, \dot{y}) \mapsto \sum_{\xi \in \Gamma} f(x\xi y^{-1})$ is continuous on $(G/\Gamma) \times (G/\Gamma)$; moreover, for $g \in \mathscr{K}(G/\Gamma)$ we have

$$(*) \qquad (f * g)(\dot{x}) = \int_{G/\Gamma} K_f(\dot{x}, \dot{y}) g(\dot{y})\, dm_{G/\Gamma}(\dot{y}).$$

Deduce that the continuous function K_f is bounded, and that $\|K_f\| \leqq N_1(f)$.

(c) Suppose that Γ is discrete and that G/Γ is *compact*. Show that for each function $f \in \mathscr{L}^1_{\mathbf{C}}(m_G)$ the operator $V(f)$ on $L^2_{\mathbf{C}}(m_{G/\Gamma})$ is *compact*. (Cf. (**11.2.8**) and (**11.2.10**), and use (b).)

10. Let G be a (separable, metrizable) unimodular locally compact group and m_G a Haar measure on G. Consider $\mathscr{K}(G)$ as an involutory normed algebra, the multiplication being the *usual* product of complex-valued functions, the involution $f \mapsto \bar{f}$ and the norm $\|f\|$ the norm of $\mathscr{B}(G)$ (**7.1**).

(a) Let E be a separable complex Hilbert space. For each function $f \in \mathscr{K}(G)$, let $M(f)$ denote the linear mapping $\tilde{\mathbf{g}} \mapsto (f\mathbf{g})^{\sim}$ in the Hilbert space $L^2_E(m_G)$ (**22.3.5**). The mapping $f \mapsto M(f)$ is a *representation* (**15.5**) of $\mathscr{K}(G)$ on $L^2_E(m_G)$, such that $\|M(f)\| \leqq \|f\|$. Also let R be the continuous unitary representation of G on $L^2_E(m_G)$ for which $R(s)$ is the linear mapping $\tilde{\mathbf{g}} \mapsto (\gamma(s)\mathbf{g})^{\sim}$. Then we have

$$R(s)M(f)R(s^{-1}) = M(\gamma(s)f)$$

for all $s \in G$ and all $f \in \mathscr{K}(G)$.

(b) Conversely, let F be a separable complex Hilbert space; $f \mapsto L(f)$ a nondegenerate representation (in the sense of (**15.5**)) of $\mathscr{K}(G)$ on F such that $\|L(f)\| \leqq \|f\|$; U a continuous unitary representation of G on F; and suppose that we have

$$(*) \qquad U(s)L(f)U(s^{-1}) = L(\gamma(s)f)$$

for all $s \in G$ and all $f \in \mathscr{K}(G)$. We propose to show that there exists a separable Hilbert space E and an *isometry* T of $L^2_E(m_G)$ onto F such that $U(s) = TR(s)T^{-1}$ and $L(f) = TM(f)T^{-1}$ for all $s \in G$ and all $f \in \mathscr{K}(G)$. We may proceed as follows:

(c) Let x, y be two points of F, and u a function belonging to $\mathscr{K}(G \times G)$. Put $u_s(t) = u(st, t)$ for $s, t \in G$. Show that the function $s \mapsto (x \mid U(s) \cdot (L(u_s) \cdot y))$ belongs to $\mathscr{K}(G)$ and that the linear form

$$\nu_{x,y} \colon u \mapsto \int_G (x \mid U(s) \cdot (L(u_s) \cdot y)) \, dm_G(s)$$

is a measure on $G \times G$. For $f, g \in \mathscr{K}(G)$ we have

$$(**) \qquad \int_{G \times G} f(st^{-1})g(t) \, d\nu_{x,y}(s, t) = \int_G (x \mid U(s) \cdot (L(g) \cdot y)) f(s) \, dm_G(s)$$

and (by virtue of $(*)$)

$$(***) \qquad \int_{G \times G} f(s)g(t) \, d\nu_{x,y}(s, t) = \int_G (x \mid (L(f)U(s)L(g)) \cdot y) \, dm_G(s).$$

Deduce that for all $f, g \in \mathscr{K}(G)$ there exists an operator $A(f, g) \in \mathscr{L}(F)$ such that

$$(x \mid A(f, g) \cdot y) = \int_{G \times G} f(s)g(t) \, d\nu_{x,y}(s, t);$$

deduce from $(**)$ that, for a given $x \in F$, it cannot happen that $\nu_{x,y} = 0$ for all $y \in F$, and consequently that the vectors $A(f, g) \cdot y$ (where $y \in F$ and $f, g \in \mathscr{K}(G)$) generate a dense subspace of F.

(d) Fix y and g so that the linear mapping $f \mapsto A(f, g) \cdot y$ of $\mathscr{K}(G)$ into F is not identically zero, and denote it by $T_{g,y}$. Show that

$$T_{g,y}(\gamma(s)f) = U(s) \cdot T_{g,y}(f), \qquad T_{g,y}(f_1 f_2) = L(f_1) \cdot T_{g,y}(f_2).$$

(e) Put $B(f_1, f_2) = (T_{g,y}(f_1) | T_{g,y}(f_2))$ for $f_1, f_2 \in \mathscr{K}(G)$. Show that $B(\gamma(s)f_1, \gamma(s)f_2) = B(f_1, f_2)$ for all $s \in G$, and that $B(f_1 f_2, f_3) = B(f_1, \bar{f}_2 f_3)$. Deduce that if $u_j, v_j \in \mathscr{K}(G)$ $(1 \leqq j \leqq m)$ are such that $\sum_j \bar{u}_j v_j = 0$, then $\sum_j B(u_j, v_j) = 0$. Show also that $B(f, f) \geqq 0$, and deduce that there exists a positive measure μ on G such that $B(f_1, f_2) = \mu(f_1 \bar{f}_2)$; finally show that μ is proportional to m_G, and may be taken to be equal to m_G by multiplying y by a suitable scalar.
(f) Complete the proof by showing that there exists a countable family of pairs (g_n, y_n) such that the sum of the spaces $T_{g_n, y_n}(\mathscr{K}(G))$ is dense in F.

11. (a) Let E be a separable complex Hilbert space, and let $t \mapsto P(t)$ be a mapping of $\mathbf{R}$ into $\mathscr{L}(E)$ such that $P(t)$ is an orthogonal projection in E, for each $t \in \mathbf{R}$ (in other words (**11.5.3**), $(P(t))^2 = P(t) = (P(t))^*$). Suppose also that (1) for each $x \in E$, the mapping $t \mapsto (x | P(t) \cdot x)$ is increasing and continuous on the right, tends to 0 as $t \to -\infty$, and to $(x | x)$ as $t \to +\infty$; (2) $P(s)P(t) = P(\inf(s, t))$ for all $s, t \in \mathbf{R}$. Let $\mu_{x,x}$ denote the unique positive measure on $\mathbf{R}$ such that $\mu_{x,x}(]a, b]) = (x | P(b) \cdot x) - (x | P(a) \cdot x)$ for every half-open interval $]a, b]$ (Section **13.18**, Problem 6). For x, y in E, put

$$4\mu_{x,y} = \mu_{x+y, x+y} - \mu_{x-y, x-y} + i\mu_{x+iy, x+iy} - i\mu_{x-iy, x-iy}.$$

Show that for each function $u \in \mathscr{U}_C(\mathbf{R})$ there exists one and only one continuous operator $L(u)$ on E such that

$$(L(u) \cdot x | y) = \int_{\mathbf{R}} u(t) \, d\mu_{x,y}(t),$$

and that $u \mapsto L(u)$ is a representation of the involutory Banach algebra $\mathscr{U}_C(\mathbf{R})$ on E (cf. Section **15.10**, Problem 1). We have $L(\varphi_{]-\infty, t]}) = P(t)$ for all $t \in \mathbf{R}$.
(b) Let U be a continuous unitary representation of $\mathbf{R}$ on the Hilbert space E, and suppose that we have

$$U(s)P(t)U(s)^{-1} = P(s + t)$$

for all $s, t \in \mathbf{R}$. Show that

$$U(s)L(f)U(s)^{-1} = L(\gamma(s)f)$$

for all $f \in \mathscr{K}(\mathbf{R})$ and all $s \in \mathbf{R}$.

12. Let G, Γ be two unimodular locally compact groups, and let m_G, m_Γ be Haar measures on G, Γ, respectively. Let U be a continuous unitary representation of Γ on $L^2_E(m_G)$, where E is a separable Hilbert space; also, for each function $f \in \mathscr{K}(G)$, let $M(f)$ denote the linear mapping $\tilde{\mathbf{g}} \mapsto (f\mathbf{g})^\sim$ in $L^2_E(m_G)$. Suppose that $U(\sigma)M(f) = M(f)U(\sigma)$ for all $\sigma \in \Gamma$ and all $f \in \mathscr{K}(G)$. Show that there exists an $(m_\Gamma \otimes m_G)$-measurable mapping $(\sigma, t) \mapsto K(\sigma, t)$ of $\Gamma \times G$ into the Banach space $\mathscr{L}(E)$ such that, for each $t \in G$, $\sigma \mapsto K(\sigma, t)$ is a continuous unitary representation of Γ on E and such that, for each function $\mathbf{g} \in \mathscr{L}^2_E(m_G)$, the class $U(\sigma) \cdot \tilde{\mathbf{g}}$ is that of the function $t \mapsto K(\sigma, t) \cdot \mathbf{g}(t)$. (First observe that for each compact subset S of G and each function $\mathbf{g}$ which vanishes almost everywhere in $\complement S$, $U(\sigma) \cdot \tilde{\mathbf{g}}$ is the class of a function vanishing almost everywhere in $\complement S$: in other words, $L^2_E(S)$ is stable under $U(\sigma)$ for all $\sigma \in \Gamma$. If $(\mathbf{e}_n)$ is a Hilbert basis of E and $(\tilde{u}_m)$ is a Hilbert basis of $L^2_C(S)$, the classes $\tilde{u}_m \mathbf{e}_n$ form a Hilbert basis of $L^2_E(S)$, and we may write $U(\sigma) \cdot (\tilde{\varphi}_S \mathbf{e}_n) = \sum_{m,p} c_{mnp}(\sigma)\tilde{u}_m \mathbf{e}_p$. Show that we may define $K(\sigma, t)$ by

$$K(\sigma, t) \cdot \mathbf{e}_n = \sum_{m,p} c_{mnp}(\sigma)u_m(t)\mathbf{e}_p$$

for $\sigma \in \Gamma$ and $t \in S$, and cover G by an increasing sequence of compact sets.)

13. Let G be a (separable, metrizable) locally compact group, K a compact subgroup of G, $X = G/K$ the corresponding homogeneous space, and Γ a *discrete* subgroup of G, which therefore acts *properly* on X on the left (Section **12.10**, Problem 1). With the notation of **(22.3.2)**, a mapping $\mathbf{f}\colon X \to E$ is said to be an *automorphic form* relative to Γ and α if it is *invariant* under the linear representation of Γ on $L = X^E$ which is the restriction to Γ of the representation ρ of G on L defined by **(22.3.2.8)**: this means that for each $\gamma \in \Gamma$ and each $x \in E$ we have $\mathbf{f}(\gamma \cdot x) = \alpha(\gamma, x) \cdot \mathbf{f}(x)$.

The restriction of α to $\Gamma \times X$ is called a *factor of automorphy* with values in Aut(E).

If we put $\mathbf{f}^\alpha(s) = \alpha(s^{-1}, s \cdot x_0) \cdot \mathbf{f}(s \cdot x_0) = \alpha(s, x_0)^{-1} \cdot \mathbf{f}(s \cdot x_0)$ **(22.3.4.1)**, to say that **f** is an automorphic form relative to Γ and α signifies that $\mathbf{f}^\alpha$ satisfies the relations

$$\mathbf{f}^\alpha(su) = \sigma(u^{-1}) \cdot \mathbf{f}^\alpha(s) \qquad (s \in G,\ u \in K),$$

$$\mathbf{f}^\alpha(\gamma s) = \mathbf{f}^\alpha(s) \qquad (s \in G,\ \gamma \in \Gamma).$$

Suppose that G is unimodular, E is finite-dimensional and that $\sigma\colon K \to \mathbf{GL}(E)$ is a continuous linear representation. Let $L^1(\alpha)$ be the subspace of $L^1_E(G)$ consisting of the classes of functions **g** such that, for almost all $s \in G$, we have $\mathbf{g}(su) = \sigma(u^{-1}) \cdot \mathbf{g}(s)$ for all $u \in K$. Also let U be a symmetric neighborhood of e in G such that $U^2 \cap \Gamma = \{e\}$, and let $h \in \mathscr{K}(G)$ be a function $\geqq 0$ with support contained in U. Show that if $\tilde{\mathbf{g}} \in L^1(\alpha)$, the family $((h * \mathbf{g})(\gamma s))_{\gamma \in \Gamma}$ is absolutely summable in E for each $s \in G$, and that the function

$$s \mapsto \mathbf{F}(s) = \sum_{\gamma \in \Gamma} (h * \mathbf{g})(\gamma s)$$

is continuous and bounded on G, and satisfies $\mathbf{F}(\gamma s) = \mathbf{F}(s)$ for all $\gamma \in \Gamma$ and $\mathbf{F}(su) = \sigma(u^{-1}) \cdot \mathbf{F}(s)$ for all $u \in K$, hence is of the form $\mathbf{f}^\alpha$ for some automorphic form **f**. If $s \mapsto \alpha(s, x_0)$ is continuous and bounded on G, then **f** is a bounded continuous automorphic form. (Observe that if $\gamma' \neq \gamma''$ are two distinct elements of Γ, then $\gamma'U$ and $\gamma''U$ are disjoint subsets of G.)

14. (a) With the notation of **(22.3.10)**, let H^1_n be the space of functions g which are *holomorphic* on the half-plane P and such that the function $z \mapsto y^{n/2}g(z)$ is integrable with respect to the invariant measure μ on P. If $\alpha_n(s, z) = (cz + d)^n$, where

$$s = \begin{pmatrix} a & b \\ c & d \end{pmatrix},$$

the mapping $g \mapsto g^{\alpha_n}$ is a bijection of H^1_n onto a subspace $H^1(\alpha_n)$ of $L^1(\alpha_n) \subset L^1(G)$ (Problem 13). For brevity, the functions belonging to $H^1(\alpha_n)$ will be called the *holomorphic functions in* $L^1(\alpha_n)$.

(b) If we put

$$g(z) = (z + i)^{-n} g_0\left(\frac{z - i}{z + i}\right)$$

for $g \in \mathrm{H}_n^1$, the mapping $g \mapsto g_0$ is a bijection of H_n^1 onto the space $\mathrm{H}_n'^1$ of functions $\zeta \mapsto g_0(\zeta)$ holomorphic in the disc D: $|\zeta| < 1$, such that the function $\zeta \mapsto (1 - |\zeta|^2)^{(n-4)/2} g_0(\zeta)$ is Lebesgue-integrable on D. Show that the space $\mathrm{H}_n'^1$ (and therefore also H_n^1) is nonzero if and only if $n \geqq 3$. (Use Section **22.19**, Problem 16(g).)

(c) By transport of structure, the representation ρ_n of G on H_n^1 gives rise to an equivalent representation ρ_n' of G on $\mathrm{H}_n'^1$, defined as follows: if

$$s = \begin{pmatrix} a & b \\ c & d \end{pmatrix}$$

and $f \in \mathrm{H}_n'^1$, the function $\rho_n'(s) \cdot f$ is the function

$$\zeta \mapsto f\left(\frac{\mathrm{A}\zeta + \mathrm{B}}{\bar{\mathrm{B}}\zeta + \bar{\mathrm{A}}}\right)(\bar{\mathrm{B}}\zeta + \bar{\mathrm{A}})^{-n},$$

where $\mathrm{A} = \frac{1}{2}(a + d - i(b - c))$ and $\mathrm{B} = \frac{1}{2}(d - a + i(b + c))$. In particular, for

$$u = \begin{pmatrix} \cos\theta & \sin\theta \\ -\sin\theta & \cos\theta \end{pmatrix} \in \mathrm{K},$$

$\rho_n'(u) \cdot f$ is the function $\zeta \mapsto e^{-2ni\theta} f(e^{-2i\theta}\zeta)$.

Deduce that the *finite*-dimensional subrepresentations of the restriction of ρ_n' to K are the one-dimensional representations in the subspaces $\mathrm{H}_{n,p}'^1$ of $\mathrm{H}_n'^1$ generated by the monomials ζ^p, where p is an integer $\geqq 0$. The character of the subrepresentation on $\mathrm{H}_{n,p}'^1$ is $\sigma_{2(n+p)}$.

(d) For each character χ of K, let $\mathrm{H}_\chi^1(\alpha_n)$ be the subspace of $\mathrm{H}^1(\alpha_n)$ consisting of the holomorphic functions F such that $\mathrm{F}(us) = \chi(u)\mathrm{F}(s)$ for all $u \in \mathrm{K}$. Each of these spaces which is nonzero has dimension 1 and corresponds to a character of the form $\sigma_{2(n+p)}$ with $p \geqq 0$ (use (c) by transport of structure). If h is any compactly supported C^∞-function on G, and χ is any of the characters above, define $h_\chi(s) = \int_{\mathrm{K}} h(us)\overline{\chi(u)}\, dm_{\mathrm{K}}(u)$; this is a function belonging to $\mathscr{D}(\mathrm{G})$, such that $h_\chi(us) = \chi(u)h_\chi(s)$ for $s \in \mathrm{G}$ and $u \in \mathrm{K}$. For each function $\mathrm{F} \in \mathrm{H}^1(\alpha_n)$, show that $h_\chi * \mathrm{F} \in \mathrm{H}^1(\alpha_n)$, and deduce that if also $\mathrm{F} \in \mathrm{H}_\chi^1(\alpha_n)$ and $h_\chi * \mathrm{F} \neq 0$, there exists a complex number $c \neq 0$ such that $\mathrm{F} = c h_\chi * \mathrm{F}$. Finally, show that if $h_\chi * \mathrm{F} = 0$ for all $h \in \mathscr{D}(\mathrm{G})$, then $\mathrm{F} = 0$. (Observe that the hypothesis implies that

$$\int_{\mathrm{K}} \mathrm{F}(y^{-1}ux)\overline{\chi(u)}\, dm_{\mathrm{K}}(u) = 0$$

for all $x, y \in \mathrm{G}$, which is absurd when $y = e$ unless $\mathrm{F} = 0$.)

15. Hypotheses and notation are the same as in Problem 14. Let $\mathrm{H}_{n,\Gamma}$ denote the space of automorphic forms f on P which are *holomorphic* on P and such that for each $\gamma \in \Gamma$ we have

$$f(z) = \mathrm{J}_\gamma(z)^n f(\gamma \cdot z)$$

where $\mathrm{J}_\gamma(z) = (cz + d)^{-1}$ if

$$\gamma = \begin{pmatrix} a & b \\ c & d \end{pmatrix}.$$

Since $\gamma \cdot z = (-\gamma) \cdot z$, we cannot have $f \neq 0$ unless $n = 2k$ is *even*; in which case f is said to be an *automorphic form of weight* k relative to Γ.

(a) Let f be any holomorphic function on P, and $F = f^{\alpha_n}$. Show that for each compact subset A of G there exists a compact neighborhood B of A in G such that

$$\sup_{s \in A} |F(s)| \leqq M \int_B |F(s)| \, dm_G(s)$$

where M is a constant independent of f. (Apply Cauchy's formula to f and the projection of A on $G/K = P$.) Deduce that if $F \in H^1(\alpha_n)$, the family $\left(\sup_{s \in A} |F(\gamma s)|\right)_{\gamma \in \Gamma}$ is summable (use the fact that $BB^{-1} \cap \Gamma$ is finite). Consequently the family $(F(\gamma s))_{\gamma \in \Gamma}$ is absolutely summable for each $s \in G$, and the function

$$F_\Gamma(s) = \sum_{\gamma \in \Gamma} F(\gamma s)$$

is of the form $f_\Gamma^{\alpha_n}$, where f_Γ is an automorphic form of weight k on P which is *holomorphic* in P.

(b) The half-plane $P = G/K$ is naturally endowed (**21.18**) with a structure of a noncompact symmetric Riemannian space, for which the geodesics are the semicircles with centres on the real axis $\mathscr{I}z = 0$. For each of these geodesics C, there exists an involutory isometry $\neq 1$ of P which fixes each point of C (reduce to the case where C is the imaginary axis $\mathscr{R}z = 0$). If d is the distance in P for the Riemannian structure (and hence G-invariant), show that for any two points $z_1, z_2 \in P$ the set of points $z \in P$ such that $d(z, z_1) = d(z, z_2)$ is a geodesic. Now let $z_0 \in P$, and for each $\gamma \in \Gamma$ let C_γ be the geodesic consisting of the points z such that $d(z, z_0) = d(z, \gamma \cdot z_0)$. Show that the family $(C_\gamma)_{\gamma \in \Gamma}$ is locally finite (use the fact that Γ acts properly on P). Deduce that the set Δ of points $z \in P$ such that $d(z, z_0) \leqq d(z, \gamma \cdot z_0)$ for *all* $\gamma \in \Gamma$ is a closed set *in* P whose frontier is μ-negligible. The interior $\mathring{\Delta}$ of Δ is a *fundamental domain* for the action of Γ on P. If Δ_0 is the inverse image of Δ in G, then the frontier of Δ_0 is m_G-negligible, and the sets $\gamma\Delta_0$ ($\gamma \in \Gamma$) have disjoint interiors. For each function $h \in \mathscr{L}^1(m_G)$, we have

$$\int_G h(s) \, dm_G(s) = \sum_{\gamma \in \Gamma} \int_{\Delta_0} h(\gamma s) \, dm_G(s)$$

and consequently, if $m_{\Gamma\backslash G}$ is the image of m_G under the canonical mapping $\pi: G \to \Gamma\backslash G$, for each function $g \in \mathscr{L}^1(m_{\Gamma\backslash G})$ we have

$$\int_{\Gamma\backslash G} g(\dot{s}) \, dm_{\Gamma\backslash G}(s) = \int_{\Delta_0} g(\pi(s)) \, dm_G(s).$$

Deduce from these observations that if $\dot{F}_\Gamma$ is the function on $\Gamma\backslash G$ such that $\dot{F}_\Gamma(\pi(s)) = F_\Gamma(s)$, where F_Γ is the function defined in (a) above, $\dot{F}_\Gamma$ is $m_{\Gamma\backslash G}$-integrable and

$$N_1(\dot{F}_\Gamma) \leqq N_1(F);$$

in other words, $F \mapsto \dot{F}_\Gamma$ is a continuous mapping of the Banach subspace $H^1(\alpha_n)$ of $L^1(m_G)$ into $L^1(m_{\Gamma\backslash G})$.

(c) Deduce from (b) that, for each function $f \in H_n^1$, the family $(J_\gamma(z)^n f(\gamma \cdot z))_{\gamma \in \Gamma}$ is absolutely summable for each $z \in P$. The function

$$f_\Gamma(z) = \sum_{\gamma \in \Gamma} J_\gamma(z)^n f(\gamma \cdot z)$$

is called the *Poincaré series* of weight $\frac{1}{2}n$ determined by f; it is an automorphic form of weight $\frac{1}{2}n$ relative to Γ which is holomorphic in P and such that the function $z \mapsto y^{n/2} f_\Gamma(z)$ is *μ-integrable on a fundamental domain* Δ for the action of Γ on P.

Let B be a subset of P defined by the inequalities $|\mathscr{R}z| \leqq a$, $\mathscr{I}z > b$, where a and b are constants > 0, and suppose that B meets only finitely many translates $\gamma\Delta$ of Δ, where $\gamma \in \Gamma$. Show that as z tends to infinity in B, $f_\Gamma(z) \to 0$ if $n \geqq 4$. (More precisely, for each $\varepsilon > 0$, there exists $N > 0$ such that $|f_\Gamma(z)| \leqq \varepsilon$ for all $z \in B$ such that $y \geqq N$. To prove this, use Cauchy's formula.)

(d) Show that if we take $f \in H^1_{n,p}$ (corresponding by transport of structure to the subspace $H'^1_{n,p}$ of Problem 14(c)), then the corresponding function F_Γ is *bounded* on G. (Use the fact that there exists a function $h \in \mathscr{D}(G)$ such that $F = h * F$ (Problem 14(d)) to majorize $F(\gamma s)$.)

(e) Suppose that Δ is *compact* (or, equivalently, that $\Gamma\backslash G$ is compact). Show that the space $H_{n,\Gamma}$ then has *finite* dimension. (By considering this space as a subspace of $L^1(m_{\Gamma\backslash G})$, show that every bounded subset of $H_{n,\Gamma}$ is relatively compact, by arguing as in (a) and using (9.13.1).) Furthermore, $H_{n,\Gamma}$ is identical with the set of Poincaré series f_Γ with $f \in H^1_n$. (It is enough to show that if $v \in H_{n,\Gamma}$ is such that the function $z \mapsto y^{n/2}v(z)$ is bounded on P and if, putting $V = v^{\alpha_n}$, we have $\int_{\Delta_0} V(s)\overline{F_\Gamma(s)}\, dm_G(s) = 0$ for all $F \in H^1(\alpha_n)$, then $v = 0$. First show that this relation is equivalent to $\int_G V(s)\overline{F(s)}\, dm_G(s) = 0$, then to $\int_P y^n v(z)\overline{f(z)}\, d\mu(z) = 0$, and finally use Problem 14(c) and transport of structure.) Show finally that all the Poincaré series are linear combinations of those corresponding to functions in the spaces $H^1_{n,p}$ (observe that the polynomials are dense in H'^1_n).

16. Let G be a Lie group, H a Lie subgroup of G, $X = G/H$ the corresponding homogeneous space, and E a complex *vector bundle* with base X and projection π (16.15.1). Let x_0 denote the coset $eH = H$ in X, and E_0 the fibre $\pi^{-1}(x_0)$ in E. Suppose that G acts *equilinearly* (on the left) on E and X (19.1), the action of G on X being the canonical one. For each $\mathbf{y} \in E_0$ and $t \in H$, we have $t \cdot \mathbf{y} \in E_0$, and the mapping $\mathbf{y} \mapsto t \cdot \mathbf{y}$ is an automorphism $\sigma(t)$ of the vector space E_0. The mapping $t \mapsto \sigma(t)$ is a Lie group homomorphism of H into $\mathbf{GL}(E_0)$, that is to say a continuous *linear representation* of H on E_0.

Let L be the vector space of *all* sections (continuous or not) of E over X; the group G acts on L by the rule

$$(s \cdot \mathbf{u})(x) = s \cdot (\mathbf{u}(s^{-1} \cdot x)) \qquad (s \in G,\ x \in X)$$

for all sections $\mathbf{u} \in L$ (19.1.2). For each section $\mathbf{u} \in L$, let $\mathbf{u}^0$ be the mapping of G into E_0 defined by

$$\mathbf{u}^0(s) = s^{-1} \cdot \mathbf{u}(s \cdot x_0).$$

Show that $\mathbf{u} \mapsto \mathbf{u}^0$ is a linear bijection of L onto the vector subspace L^0 of E_0^G consisting of all mappings $\mathbf{f}\colon G \to E_0$ such that

$$\mathbf{f}(st) = \sigma(t^{-1}) \cdot \mathbf{f}(s) \qquad (s \in G,\ t \in H).$$

If we put $\rho(s) \cdot \mathbf{u}^0 = (s \cdot \mathbf{u})^0$ for each $\mathbf{u} \in L$, show that ρ is the (in general not continuous) linear representation of G on L^0 *induced* by σ.

4. INDUCED REPRESENTATIONS AND RESTRICTIONS OF REPRESENTATIONS TO SUBGROUPS

(22.4.1) With the notation and hypotheses of **(22.3.1)**, suppose that $\rho: \mathrm{G} \to \mathrm{Aut}(\mathrm{E})$ is a linear representation of G on E, and that there exists a vector subspace F_0 of E such that, for each $u \in \mathrm{H}$, the automorphism $\rho(u)$ of E leaves F_0 *stable*. Put $\mathrm{F} = \mathrm{E}/\mathrm{F}_0$, and let $\pi: \mathrm{E} \to \mathrm{F}$ be the canonical mapping. We can then define a *linear representation* σ *of* H *on* F by the rule

$$\sigma(u) \cdot \pi(z) = \pi(\rho(u) \cdot z) \tag{22.4.1.1}$$

for all $u \in \mathrm{H}$ and $z \in \mathrm{E}$, since it follows immediately from the assumptions that the right-hand side depends only on $\pi(z) \in \mathrm{F}$.

Let us now construct the representation ρ' of G *induced* by σ by the method of **(22.3.4)**: we consider the vector space L^σ of mappings $g: \mathrm{G} \to \mathrm{F}$ satisfying $g(su) = \sigma(u^{-1}) \cdot g(s)$ for all $s \in \mathrm{G}$ and $u \in \mathrm{H}$, and the representation ρ' is defined by the formula $(\rho'(s) \cdot g)(t) = g(s^{-1}t)$. Now define a *linear mapping* $z \mapsto f_z$ *of* E *into* L^σ by the rule

$$f_z(s) = \pi(\rho(s^{-1}) \cdot z) \qquad (s \in \mathrm{G}). \tag{22.4.1.2}$$

The fact that $f_z \in \mathrm{L}^\sigma$ follows from the calculation

$$\begin{aligned} f_z(su) &= \pi(\rho(u^{-1}) \cdot (\rho(s^{-1}) \cdot z)) = \sigma(u^{-1}) \cdot \pi(\rho(s^{-1}) \cdot z) \\ &= \sigma(u^{-1}) \cdot f_z(s). \end{aligned}$$

The linear mapping so defined is *compatible* with the actions of G on E and L^σ defined by ρ and ρ', that is to say

$$\rho'(s) \cdot f_z = f_{\rho(s) \cdot z} \tag{22.4.1.3}$$

because $(\rho'(s) \cdot f_z)(t) = f_z(s^{-1}t) = \pi(\rho(t^{-1}) \cdot (\rho(s) \cdot z)) = f_{\rho(s) \cdot z}(t)$.

In particular, the *image* E′ of E under the mapping $z \mapsto f_z$ is *stable* under the representation ρ'. If in addition ρ is (algebraically) *irreducible* (i.e., if the only vector subspaces of E stable under ρ are $\{0\}$ and E), then each of the mappings f_z for $z \neq 0$ is such that $f_z(\mathrm{G})$ *generates* the vector space F, because the $\rho(s) \cdot z$ generate E; *a fortiori* we have $f_z \neq 0$. It follows that the mapping $z \mapsto f_z$ is *injective*, because its kernel is a subspace of E stable under ρ, by virtue of **(22.4.1.3)**, and we have just seen that it is distinct from E. The linear representation $s \mapsto \rho'(s) | \mathrm{E}'$ is therefore *equivalent* to the irreducible representation ρ, and we have thus obtained ρ, up to equivalence, as a subrepresentation of the representation *induced* by a linear representation of H.

(22.4.2) Suppose now that G is a locally compact group, H a closed subgroup of G; let E be a *finite*-dimensional vector space and ρ a *continuous* linear representation (21.1.1) of G on E; then it follows from (22.4.1.1) that σ is a continuous linear representation of H on F. The subspace L_{cont}^{σ} of L^{σ} consisting of the mappings $g \in L^{\sigma}$ which are *continuous* on G is therefore stable under ρ'. Also, it follows from (22.4.1.2) that for each $z \in E$ the function f_z is continuous, and hence belongs to L_{cont}^{σ}; since E is *finite*-dimensional, the linear mapping $z \mapsto f_z$ is continuous for any structure of Hausdorff locally convex topological vector space on L_{cont}^{σ}. The proof of (22.4.1.3) and the continuity of ρ then imply that $s \mapsto \rho'(s) | E'$ is a *continuous* linear representation of G on E'.

We see therefore that *every* irreducible continuous linear representation of *finite* dimension may be obtained by starting with a *finite*-dimensional continuous linear representation σ of H, and then considering the finite-dimensional subspaces of L_{cont}^{σ} which are stable under the representation *induced* by σ, and the subrepresentations of the latter on these stable subspaces. These subrepresentations are said to be obtained by "decomposing" the representation induced by σ.

We shall next examine this decomposition in more detail, in the situation where G is compact and σ is the trivial representation.

5. PARTIAL TRACES AND INDUCED REPRESENTATIONS OF COMPACT GROUPS

(22.5.1) Let G be a (metrizable) compact group and let H be a closed (and therefore compact) subgroup of G. Consider the irreducible linear representation $s \mapsto M_{\mathfrak{p}}(s)$ of G, of class $\mathfrak{p}$ (21.4.2), the space $E_{\mathfrak{p}}$ of the representation being $\mathbf{C}^{n_{\mathfrak{p}}}$. We may assume that $M_{\mathfrak{p}}$ is *unitary* with respect to the canonical Hermitian scalar product $(x | y) = \sum_j \xi_j \bar{\eta}_j$.

We shall keep to the notation of (21.4) for the irreducible representations of the subgroup H. For a class $\boldsymbol{\sigma} \in R(H)$, we shall therefore denote by $t \mapsto M_{\boldsymbol{\sigma}}(t)$ the representation of class $\boldsymbol{\sigma}$ defined by the two-sided ideal $\mathfrak{a}_{\boldsymbol{\sigma}}$ of $L_{\mathbf{C}}^2(H)$, by $n_{\boldsymbol{\sigma}}$ the order of the matrix $M_{\boldsymbol{\sigma}}(t)$, and by $u_{\boldsymbol{\sigma}}$ the unit element of $\mathfrak{a}_{\boldsymbol{\sigma}}$, which is equal to $n_{\boldsymbol{\sigma}}^{-1}\chi_{\boldsymbol{\sigma}}$, where $\chi_{\boldsymbol{\sigma}}$ is the character of $\boldsymbol{\sigma}$. Consider now the *restriction* of $M_{\mathfrak{p}}$ to the subgroup H, and for each class $\boldsymbol{\sigma} \in R(H)$ denote by $(\mathfrak{p} : \boldsymbol{\sigma})$ the *multiplicity d* of the class $\boldsymbol{\sigma}$ in the class of this restriction (21.4.7). From (21.4.1), the linear mapping

(22.5.1.1) $$M_{\mathfrak{p}}(u_{\bar{\boldsymbol{\sigma}}}) = \frac{1}{n_{\mathfrak{p}}} \int_H M_{\mathfrak{p}}(t)\overline{\chi_{\boldsymbol{\sigma}}(t)}\, dm_H(t),$$

where m_{H} is the Haar measure on H with total mass equal to 1, is an *orthogonal projection* of $\mathbf{C}^{n_\rho}$ onto a subspace E_σ of dimension $p = dn_\sigma$, the Hilbert sum of d subspaces $\mathrm{F}_1, \dots, \mathrm{F}_d$ of dimension n_σ, each stable under $M_\rho(t)$ for all $t \in \mathrm{H}$, and such that the restriction of $M_\rho(t)$ to each F_j is of class $\boldsymbol{\sigma}$. We may therefore choose a new orthonormal basis of $\mathbf{C}^{n_\rho}$, consisting of the union of orthonormal bases of each F_j and a basis of the orthogonal supplement F' of E_σ, in such a way that relative to the basis of E_σ formed by the first p vectors of this basis, the matrix of the restriction of $M_\rho(t)$ to E_σ is the diagonal tableau

(22.5.1.2)
$$\begin{pmatrix} M_\sigma(t) & 0 & \cdots & 0 \\ 0 & M_\sigma(t) & \cdots & 0 \\ \cdots & \cdots & \cdots & \\ 0 & 0 & \cdots & M_\sigma(t) \end{pmatrix}$$

for all $t \in \mathrm{H}$.

(22.5.2) Consider the endomorphism $M_\rho(u_{\bar\sigma})M_\rho(s)M_\rho(u_{\bar\sigma})$ of $\mathbf{C}^{n_\rho}$, which we may envisage as an endomorphism $P_{\rho,\sigma}(s)$ of E_σ, whose matrix with respect to the basis chosen above is

$$\frac{1}{n_\rho}\begin{pmatrix} m_{11}^{(\rho)}(s) & \cdots & m_{1p}^{(\rho)}(s) \\ \cdots & \cdots & \cdots \\ m_{p1}^{(\rho)}(s) & \cdots & m_{pp}^{(\rho)}(s) \end{pmatrix}.$$

For $t \in \mathrm{H}$, the matrix of $P_{\rho,\sigma}(t)$ reduces to (22.5.1.2). It follows directly that

(22.5.2.1)
$$P_{\rho,\sigma}(tst') = P_{\rho,\sigma}(t)P_{\rho,\sigma}(s)P_{\rho,\sigma}(t')$$

for all $s \in \mathrm{G}$ and $t, t' \in \mathrm{H}$, in view of the relations $M_\rho(tst') = M_\rho(t)M_\rho(s)M_\rho(t')$ and

$$M_\rho(t)M_\rho(u_{\bar\sigma}) = M_\rho(u_{\bar\sigma})M_\rho(t) = \begin{pmatrix} P_{\rho,\sigma}(t) & 0 \\ 0 & 0 \end{pmatrix}.$$

The *partial trace* of $\boldsymbol{\rho}$ with respect to $\boldsymbol{\sigma}$ is defined to be the function

(22.5.2.2)
$$\theta_{\rho,\sigma}: s \mapsto \mathrm{Tr}(P_{\rho,\sigma}(s)) = \mathrm{Tr}(M_\rho(u_{\bar\sigma})M_\rho(s)M_\rho(u_{\bar\sigma})) = \mathrm{Tr}(M_\rho(u_{\bar\sigma})M_\rho(s))$$

which in the preceding notation takes the form

(22.5.2.3) $$\theta_{\rho,\sigma}(s) = n_\rho^{-1}(m_{11}^{(\rho)}(s) + \cdots + m_{pp}^{(\rho)}(s)).$$

It is therefore a continuous function on G which is not central, and is nonzero only when σ is contained in the class of the restriction of M_ρ to H; clearly it depends only on the classes ρ and σ, and we have

(22.5.2.4) $$\theta_{\rho,\sigma}(t) = (\rho : \sigma)\chi_\sigma(t) \qquad (t \in \mathrm{H}),$$

(22.5.2.5) $$\chi_\rho(s) = \sum_\sigma \theta_{\rho,\sigma}(s) \qquad (s \in \mathrm{G}).$$

Furthermore

(22.5.3) (i) *We have*

(22.5.3.1) $$\theta_{\rho,\sigma}(tst^{-1}) = \theta_{\rho,\sigma}(s)$$

for $s \in \mathrm{G}$ *and* $t \in \mathrm{H}$.

(ii) *As* ρ *runs through* R(G) *and* σ *through* R(H), *the partial traces* $\theta_{\rho,\sigma}$ *are pairwise orthogonal; in particular,* $\theta_{\rho,\sigma}$ *and* $\theta_{\rho',\sigma'}$ *cannot be proportional unless* $\rho' = \rho$ *and* $\sigma' = \sigma$.

(iii) *The partial traces* $\theta_{\rho,\sigma}$ *are continuous functions of positive type on* G.

The formula (22.5.3.1) follows immediately from (22.5.2.1) and the commutativity of the trace. Assertion (ii) is a direct consequence of the formula (22.5.2.3) and the orthogonality properties of the $m_{ij}^{(\rho)}$ (21.2.5). Finally, assertion (iii) comes from the relations $m_{ii}^{(\rho)} = \check{m}_{ii}^{(\rho)} = m_{ii}^{(\rho)} * m_{ii}^{(\rho)}$ (21.2.5) and from (22.1.4(ii)), which shows that each of the $m_{ii}^{(\rho)}$ is of positive type; hence so also is $\theta_{\rho,\sigma}$ by virtue of (22.5.2.3).

(22.5.4) Since there exists a G-invariant measure μ on G/H (22.3.7) we may construct, as in (22.3.7), the continuous unitary representation of G on $\mathrm{L}_{\mathrm{F}_1}^2(\mu)$ *induced* by a representation of H of class σ; it is easy in the present situation to describe all the representations so obtained (Problem 1). We shall restrict ourselves to the case of the *canonical representation* of G on $\mathrm{L}_{\mathbf{C}}^2(\mathrm{G/H}, \mu)$, induced by the trivial representation of H on **C** (22.3.7) (corresponding therefore to the trivial class σ_0). With the preceding notation, we have $n_{\sigma_0} = 1$ and $p = d$. The space $\mathscr{L}_{\mathbf{C}}^2(\mathrm{G/H}, \mu)$ may be identified with the subspace of $\mathscr{L}_{\mathbf{C}}^2(\mathrm{G})$ consisting of the functions g such that

(22.5.4.1) $$g(st) = g(s) \qquad (s \in \mathrm{G},\ t \in \mathrm{H}),$$

which implies **(14.8.4)**

(22.5.4.2)
$$(g * \varepsilon_t)^\sim = \tilde{g}$$

for all $t \in \mathrm{H}$. It follows immediately that if $\tilde{g}$ satisfies this relation, then so also does the class of the function $\nu * g \in \mathscr{L}^2_{\mathbf{C}}(\mathrm{G})$, for any measure ν on the compact group G, by virtue of the associativity of convolution; the space $\mathrm{L}^2_{\mathbf{C}}(\mathrm{G/H}, \mu)$ (which for simplicity we shall denote by $\mathrm{L}^2_{\mathbf{C}}(\mathrm{G/H})$) is therefore a *left ideal* in the ring $\mathrm{M}_{\mathbf{C}}(\mathrm{G})$ of measures on G, and *a fortiori* a *closed left ideal* in the Banach algebra $\mathrm{L}^2_{\mathbf{C}}(\mathrm{G})$. In particular, for each class $\boldsymbol{\rho} \in \mathrm{R}(\mathrm{G})$, the continuous function $u_{\boldsymbol{\rho}} * g$ belongs to $\mathscr{L}^2_{\mathbf{C}}(\mathrm{G/H}, \mu)$ for each $g \in \mathscr{L}^2_{\mathbf{C}}(\mathrm{G/H}, \mu)$, and consequently $\mathrm{L}^2_{\mathbf{C}}(\mathrm{G/H})$ is the *Hilbert sum* of its orthogonal projections $\mathrm{L}_{\boldsymbol{\rho}}$ on the two-sided ideals $\mathfrak{a}_{\boldsymbol{\rho}}$, $\boldsymbol{\rho} \in \mathrm{R}(\mathrm{G})$. Hence it is enough to determine each of the spaces $\mathrm{L}_{\boldsymbol{\rho}}$.

Now, if we express that a function $g = \sum_{i,j} c_{ij} m_{ij}^{(\boldsymbol{\rho})}$ in $\mathfrak{a}_{\boldsymbol{\rho}}$ satisfies **(22.5.4.1)** we obtain, by taking note of the expression **(22.5.1.2)** of the restriction of $M_{\boldsymbol{\rho}}(t)$ to the space $\mathrm{E}_{\boldsymbol{\sigma}_0}$, the relation

(22.5.4.3)
$$\sum_{i=1}^{n_{\boldsymbol{\rho}}} \sum_{j>d} \sum_{k>d} c_{ij} m_{ik}^{(\boldsymbol{\rho})}(s) m_{kj}^{(\boldsymbol{\rho})}(t) = \sum_{i=1}^{n_{\boldsymbol{\rho}}} \sum_{j>d} c_{ij} m_{ij}^{(\boldsymbol{\rho})}(s),$$

because $m_{kj}^{(\boldsymbol{\rho})}(t) = \delta_{kj}$ for $k \leqq d$ or $j \leqq d$. Since the functions $m_{ij}^{(\boldsymbol{\rho})}$ are linearly independent, this relation implies, for $1 \leqq i \leqq n_{\boldsymbol{\rho}}$ and $k > d$,

(22.5.4.4)
$$\sum_{j>d} c_{ij} m_{kj}^{(\boldsymbol{\rho})}(t) = c_{ik}$$

for all $t \in \mathrm{H}$. Now we can decompose the orthogonal supplement F' of $\mathrm{E}_{\boldsymbol{\sigma}_0}$ in $\mathbf{C}^{n_{\boldsymbol{\rho}}}$ as a direct sum of subspaces $\mathrm{F}'_{\boldsymbol{\sigma}_h}$, stable under $M_{\boldsymbol{\rho}}(t)$ for all $t \in \mathrm{H}$, and such that the subrepresentation $t \mapsto M_{\boldsymbol{\rho}}(t) | \mathrm{F}'_{\boldsymbol{\sigma}_h}$ of H is irreducible of class $\boldsymbol{\sigma}_h \neq \boldsymbol{\sigma}_0$. We may therefore assume that $\mathrm{F}'_{\boldsymbol{\sigma}_h}$ has a basis consisting of vectors of the chosen basis of $\mathbf{C}^n$ with indices k satisfying $k'_h \leqq k \leqq k''_h$, say, where $d < k'_h < k''_h$. Then for $k \in [k'_h, k''_h]$ we have $m_{kj}^{(\boldsymbol{\rho})}(t) = 0$ unless also $j \in [k'_h, k''_h]$, and moreover $\int_{\mathrm{H}} m_{kj}^{(\boldsymbol{\rho})}(t)\, dm_{\mathrm{H}}(t) = 0$, because $\boldsymbol{\sigma}_h \neq \boldsymbol{\sigma}_0$ **(21.2.7)**. Integrating both sides of **(22.5.4.4)** over H, we therefore obtain $c_{ik} = 0$ for $k'_h \leqq k \leqq k''_h$, and by applying this result for each $\mathrm{F}'_{\boldsymbol{\sigma}_h}$ we see finally that $c_{ij} = 0$ *for all* $j > d$. The space $\mathrm{L}_{\boldsymbol{\rho}}$ is therefore the subspace of $\mathfrak{a}_{\boldsymbol{\rho}}$, of dimension $dn_{\boldsymbol{\rho}}$, which is the direct sum of the *first d columns of the matrix* $M_{\boldsymbol{\rho}}$. For each $j \leqq d$, the jth column $\mathrm{L}_{\boldsymbol{\rho},j}$ of $M_{\boldsymbol{\rho}}$ is a *subspace of* $\mathrm{L}_{\boldsymbol{\rho}}$ *of dimension* $n_{\boldsymbol{\rho}}$, *stable under the regular representation of* G; and the subrepresentation of the regular representation, of G on $L_{\rho,j}$, is irreducible of class $\bar{\rho}$ (21.4.1); but by definition, this subrepresentation is also the subrepresentation (on $L_{\rho,j}$) of the representation *induced* by the trivial representation of H. We conclude therefore that the subrepresentation of this induced representation, on the space L_ρ, is the *Hilbert sum of* $(\rho : \sigma_0)$ *irreducible representations of* G *of class* $\bar{\rho}$.

(22.5.5) The Haar measure m_G on G gives rise also to a G-invariant measure μ' on H\G (with G now acting on the right), and the space $\mathscr{L}^2_{\mathbf{C}}(\mathrm{H}\backslash\mathrm{G}, \mu')$ may be identified with the subspace of $\mathscr{L}^2_{\mathbf{C}}(\mathrm{G})$ consisting of the functions g such that $g(ts) = g(s)$ for $t \in \mathrm{H}$ and $s \in \mathrm{G}$; the subspace $\mathrm{L}^2_{\mathbf{C}}(\mathrm{H}\backslash\mathrm{G})$ of $\mathrm{L}^2_{\mathbf{C}}(\mathrm{G})$ consisting of the classes of these functions has as its elements the classes $\tilde{g}$ such that $(\varepsilon_t * g)^\sim = \tilde{g}$ for all $t \in \mathrm{H}$, and is the image of $\mathrm{L}^2_{\mathbf{C}}(\mathrm{G}/\mathrm{H})$ under the mapping $\tilde{g} \mapsto (\check{g})^\sim$. Hence $\mathrm{L}^2_{\mathbf{C}}(\mathrm{H}\backslash\mathrm{G})$ is a *closed right ideal* of the Banach algebra $\mathrm{L}^2_{\mathbf{C}}(\mathrm{G})$, and is the direct sum of the images $\check{\mathrm{L}}_\rho$ of the spaces L_ρ under this mapping; $\check{\mathrm{L}}_\rho$ is the subspace of $\mathfrak{a}_\rho$ which is the direct sum of the *first d rows* of M_ρ.

The intersection $\mathrm{L}^2_{\mathbf{C}}(\mathrm{G}/\mathrm{H}) \cap \mathrm{L}^2_{\mathbf{C}}(\mathrm{H}\backslash\mathrm{G})$ is a closed self-adjoint subalgebra of $\mathrm{L}^2_{\mathbf{C}}(\mathrm{G})$, hence is a *complete Hilbert algebra* (15.8); its elements are the classes of functions $g \in \mathscr{L}^2_{\mathbf{C}}(\mathrm{G})$ such that

(22.5.5.1) $$g(tst') = g(s) \qquad (s \in \mathrm{G};\ t, t' \in \mathrm{H})$$

or equivalently the classes $\tilde{g}$ satisfying

(22.5.5.2) $$(\varepsilon_t * g * \varepsilon_{t'})^\sim = \tilde{g} \qquad (t, t' \in \mathrm{H}).$$

The functions satisfying (22.5.5.1) may be identified with functions defined on the set H\G/H of *double cosets* of G with respect to H (12.10.2); consequently we denote by $\mathscr{L}^2_{\mathbf{C}}(\mathrm{H}\backslash\mathrm{G}/\mathrm{H})$ the space of functions $g \in \mathscr{L}^2_{\mathbf{C}}(\mathrm{G})$ satisfying (22.5.5.1), and by $\mathrm{L}^2_{\mathbf{C}}(\mathrm{H}\backslash\mathrm{G}/\mathrm{H})$ the algebra $\mathrm{L}^2_{\mathbf{C}}(\mathrm{G}/\mathrm{H}) \cap \mathrm{L}^2_{\mathbf{C}}(\mathrm{H}\backslash\mathrm{G})$ consisting of their classes. This algebra is the *Hilbert sum* of minimal two-sided ideals $\mathfrak{a}_{\rho,\sigma_0} = \mathrm{L}_\rho \cap \check{\mathrm{L}}_\rho$. The ideal $\mathfrak{a}_{\rho,\sigma_0}$ has as a basis the $m^{(\rho)}_{ij}$ such that $1 \leqq i, j \leqq d$. The *centre* of $\mathrm{L}^2_{\mathbf{C}}(\mathrm{H}\backslash\mathrm{G}/\mathrm{H})$ is the Hilbert sum of the subspaces $\mathbf{C}u_{\rho,\sigma_0}$, where $u_{\rho,\sigma_0} = n_\rho \theta_{\rho,\sigma_0}$ is the unit element of the algebra $\mathfrak{a}_{\rho,\sigma_0}$; the mapping $\tilde{g} \mapsto (u_{\rho,\sigma_0} * g)^\sim = (g * u_{\rho,\sigma_0})^\sim$ is the *orthogonal projection* of $\mathrm{L}^2_{\mathbf{C}}(\mathrm{H}\backslash\mathrm{G}/\mathrm{H})$ onto $\mathfrak{a}_{\rho,\sigma_0}$.

The subspace $\mathfrak{l}_{\rho,1} = \mathrm{L}_{\rho,1} \cap \mathfrak{a}_{\rho,\sigma_0}$ generated by the $m^{(\rho)}_{i1}$ for $1 \leqq i \leqq d$ is a *minimal left ideal* of $\mathrm{L}^2(\mathrm{H}\backslash\mathrm{G}/\mathrm{H})$, to which there corresponds an *irreducible representation* W_ρ of the algebra $\mathrm{L}^2_{\mathbf{C}}(\mathrm{H}\backslash\mathrm{G}/\mathrm{H})$ (15.8.1), of dimension d; in this way we obtain all the irreducible representations of this algebra, up to equivalence. The representation W_ρ may be described explicitly as follows: for each function $g \in L_C^2(H\backslash G/H)$, we may write

$$g * u_{\rho,\sigma_0} = \sum_{1 \le i, j \le d} c_{ij}(g) m_{ij}^{(\rho)} \in \mathfrak{a}_{\rho,\sigma_0}$$

and $W_\rho(\tilde{g})$ is the $d \times d$ matrix $(c_{ij}(g))$.

From these remarks, the following proposition is clear:

(22.5.6) *For the algebra* $L_C^2(H\backslash G/H)$ *to be commutative, it is necessary and sufficient that* $(\rho : \sigma_0) \leqq 1$ *for each class of representations* $\rho \in R(G)$.

When this condition is satisfied, for each $\rho \in R(G)$ such that $(\rho : \sigma_0) = 1$, the ideal $\mathfrak{a}_{\rho,\sigma_0}$ is 1-dimensional, generated by the single function

(22.5.6.1) $$\omega_\rho(s) = \theta_{\rho,\sigma_0}(s) = n_\rho^{-1} m_{11}^{(\rho)}(s)$$

which is *continuous, of positive type* **(22.5.3)** and such that

(22.5.6.2) $$\omega_\rho(tst') = \omega_\rho(s) \qquad (s \in G;\ t, t' \in H),$$

(22.5.6.3) $$\omega_\rho(e) = 1.$$

The irreducible representation W_ρ of $L_C^2(H\backslash G/H)$ is here one-dimensional: in other words, for each function $g \in L_C^2(H\backslash G/H)$ we have (because $g * \omega_\rho$ is continuous by **(14.10.7)**)

(22.5.6.4) $$g * \omega_\rho = \zeta(\tilde{g})\omega_\rho$$

where ζ is a *Hermitian character* **(15.9.1)** of $L_C^2(H\backslash G/H)$.

The left ideal L_ρ is then *minimal*; it follows therefore from **(21.2.3.2)** that L_ρ is also the vector subspace generated by the *left-translates* $\gamma(x)\omega_\rho = \varepsilon_x * \omega_\rho$ of the function ω_ρ, for all $x \in G$.

In the next section we shall see how all these properties remain valid in a much more general context.

PROBLEMS

1. Let G be a compact group, H a closed subgroup of G, and $\boldsymbol{\sigma}$ a class of irreducible representations of H. We propose to describe the induced representation $M_{\boldsymbol{\sigma}}^{\text{ind}}$. The space $J(\boldsymbol{\sigma})$ of this representation is the subspace of $(L_C^2(G))^{n_\sigma}$ consisting of the classes of vector-valued functions $\mathbf{g} = (g_\alpha)_{1 \le \alpha \le n_\sigma}$ such that $\mathbf{g}(st) = \overline{M_\sigma(t^{-1})} \cdot \mathbf{g}(s)$ for all $s \in G$ and $t \in H$. If we denote by

$J_\rho(\bar{\sigma})$ the space of projections $u_\rho * \mathbf{g} = (u_\rho * g_\alpha)_{1 \le \alpha \le n_\sigma}$ of the functions in $J(\bar{\sigma})$ onto $(\mathfrak{a}_\rho)^{n_\sigma}$, then $J(\bar{\sigma})$ is the Hilbert sum of the $J_\rho(\bar{\sigma})$ for $\rho \in R(G)$. With the notation of (22.5.4), show that $J_\rho(\bar{\sigma})$ is the space of vector-valued functions $\mathbf{h} = (h_\alpha)_{1 \le \alpha \le n_\sigma}$, where

$$h_\alpha = \sum_{i=1}^{n_\rho} \sum_{k=0}^{d-1} c_{ik} m^{(\rho)}_{i,\, kn_\sigma + \alpha},$$

the dn_ρ complex coefficients c_{ik} being arbitrary. (Proceed as in (22.5.4), using Schur's lemma.) Deduce that the subrepresentation of $M^{\text{ind}}_{\bar{\sigma}}$ on the space $J_\rho(\bar{\sigma})$ is the Hilbert sum of $(\rho : \sigma) = (\bar{\rho} : \bar{\sigma})$ irreducible representations of G of class $\bar{\rho}$ (*Frobenius' theorem*).

2. For a finite group G, a linear representation of G on a complex vector space E of finite dimension determines a structure of left **C**[G]-*module* on E, and conversely: two representations are equivalent if and only if the corresponding **C**[G]-modules are isomorphic, and a representation is irreducible if and only if the corresponding **C**[G]-module is *simple* (A.24.2).
(a) Let H be a subgroup of G, and let W be a left **C**[H]-module, corresponding to a linear representation of H on W. Show that the representation of G induced by this representation of H corresponds to the left **C**[G]-module $V = \mathbf{C}[G] \otimes_{\mathbf{C}[H]} W$, where **C**[G] is considered as a free right **C**[H]-module.
(b) Consider a linear representation of G on a vector space V, so that G acts linearly on V. Suppose that V is the direct sum of a finite sequence $(W_j)_{1 \le j \le m}$ of subspaces such that, for each $s \in G$ and each $j \in [1, m]$, the subspace $s \cdot W_j$ is one of the W_k. The group G therefore acts on the set of subspaces W_j; suppose that this action is *transitive* (which will certainly be the case if the representation of G on V is irreducible). Then, if H is the subgroup of G which leaves W_1 stable, the given representation of G on V is induced by the linear representation defined by the action of H on W.
(c) With the same notation as in (a), let E be a **C**[G]-module. For each **C**[G]-homomorphism f: V → E, the restriction $f \mid W$ of f to W is a **C**[H]-homomorphism if E is considered as a **C**[H]-module by restriction of scalars. Show that the mapping $f \mapsto f \mid W$ is a bijection of $\text{Hom}_{\mathbf{C}[G]}(V, E)$ onto $\text{Hom}_{\mathbf{C}[H]}(W, E)$.

3. (a) With the same hypotheses and notation as in Problem 2(a), denote by $\text{Tr}_V(s)$ for $s \in G$ (resp $\text{Tr}_W(t)$ for $t \in H$) the trace of the endomorphism $x \mapsto s \cdot x$ (resp $y \mapsto t \cdot y$) of V (resp W). Show that

$$\text{Tr}_V(s) = (\text{Card H})^{-1} \sum_{\substack{u \in G \\ u^{-1}su \in H}} \text{Tr}_W(u^{-1}su). \tag{1}$$

(Consider the elements $u \in G$ such that $s \cdot (u \cdot W) = u \cdot W$.)
(b) For each central function f on H, show that the following formula defines a central function f^{ind} on G:

$$f^{\text{ind}}(s) = (\text{Card H})^{-1} \sum_{\substack{u \in G \\ u^{-1}su \in H}} f(u^{-1}su).$$

(c) Let $(s_1 | s_2)_G$ and $(t_1 | t_2)_H$ denote the scalar products on $\mathbf{C}[G] = L^2_{\mathbf{C}}(G)$ and $\mathbf{C}[H] = L^2_{\mathbf{C}}(H)$. Show that if f (resp. g) is a central function on G (resp. H), we have

$$(g \mid (f \mid H))_H = (g^{\text{ind}} \mid f)_G. \tag{2}$$

(Reduce to the case where f and g are traces of linear representations. By use of Schur's lemma, show that if V_1 and V_2 are two $\mathbf{C}[G]$-modules, we have $(\mathrm{Tr}_{V_1} | \mathrm{Tr}_{V_2})_G = \dim(\mathrm{Hom}_{\mathbf{C}[G]}(V_1, V_2))$, by decomposing V_1 and V_2 into direct sums of simple $\mathbf{C}[G]$-modules, and then use Problem 2(c).)

(d) With the same notation, show that

$$(g(f \mid H))^{\mathrm{ind}} = (g^{\mathrm{ind}})f.$$

4. Let G be a finite group, H and K two subgroups of G, and U a linear representation of H on a vector space W. Let U^{ind} be the induced representation of G on the vector space $V = \mathbf{C}[G] \otimes_{\mathbf{C}[H]} W$, the direct sum of the subspaces $x \cdot W$ as x runs through a system of representatives of the cosets mod. H.

(a) Let S be a system of representatives of the double cosets KxH, so that G is the disjoint union of the KsH with $s \in S$. Let V_s be the subspace of V generated by the $x \cdot W$ for $x \in KsH$ and fixed $s \in S$, and let $H_s = sHs^{-1} \cap K$. Show that V_s is the direct sum of the subspaces $x \cdot (s \cdot W)$, where x runs through a system of representatives of the cosets of H_s in K. Deduce that the restriction of U^{ind} to K is equivalent to the Hilbert sum of the representations U_s^{ind}, where s runs through S, and U_s is the linear representation of H_s on W defined by $U_s(x) = U(s^{-1}xs)$ for $x \in H_s$.

(b) In particular, take H = K. Show that

$$(\mathrm{Tr}_V | \mathrm{Tr}_V)_G = \sum_{s \in S} (\mathrm{Tr}(U | H_s) | \mathrm{Tr}(U_s))_{H_s}.$$

Deduce that in order that U^{ind} should be irreducible it is necessary and sufficient that U should be irreducible and that, for each $s \in G \cap \complement H$, the two representations $U \mid H_s$ and U_s of H_s have no irreducible component in common, or equivalently that their traces are orthogonal. (Observe that the condition $(\mathrm{Tr}_V | \mathrm{Tr}_V)_G = 1$ is necessary and sufficient for the irreducibility of U^{ind}.) (*Mackey's criterion.*)

5. Let G be a finite group, N a normal subgroup of G, and U an irreducible representation of G on a vector space V. Then either $U \mid N$ is isotypic (Section **21.4**, Problem 3) or else there exists a subgroup H of G, containing N properly, such that U is induced by an irreducible linear representation of H. (Consider the isotypic components of $U \mid N$: observe that each $U(s)$ permutes these components, and apply Problem **2**(b).)

6. (a) Let G_1, G_2 be two finite groups and let U_1 (resp. U_2) be a linear representation of G_1 (resp. G_2) on a vector space V_1 (resp. V_2). Put $U(s_1, s_2) = U_1(s_1) \otimes U_2(s_2)$ for $(s_1, s_2) \in G_1 \times G_2$. Then U is a linear representation of $G_1 \times G_2$ on $V_1 \otimes V_2$: it is denoted by $U_1 \otimes U_2$. If U_1 and U_2 are irreducible, then so is $U_1 \otimes U_2$ (cf. Section **21.4**, Problem 8).

(b) Let G be a finite group, C its center, and let U be an irreducible representation of G on a vector space V. For each $t \in C$, the linear transformation $U(t)$ is a homothety. For each integer $n > 0$, consider the representation $U^{\otimes n} = U \otimes \cdots \otimes U$ (n factors) on $V^{\otimes n} = V \otimes \cdots \otimes V$ (n factors). Let H be the subgroup of C^n consisting of all $x = (x_1, \ldots, x_n)$ such that $x_1 x_2 \cdots x_n = e$, so that $U^{\otimes n}(x) = I$; then $U^{\otimes n}$ determines a representation of G/H on $V^{\otimes n}$. Deduce from (a) and Section **21.3**, Problem 3, that if $g = \mathrm{Card}\ G$, $c = \mathrm{Card}\ C$, $d = \dim V$, then d^n divides g^n/c^{n-1}. Hence show that d divides $g/c = (G : C)$.

(c) Let G be a finite group. N a *commutative normal* subgroup of G. Show that the degree of every irreducible representation of G divides (G : N). (Proceed by induction on the order of G, using Problem 5. If U is an irreducible representation of G on a vector space V, such that $U \mid N$ is isotypic, observe that $U(t)$ is a homothety for each $t \in N$; then apply (b) above.)

7. Let G be a finite group, A a commutative subgroup of G, not necessarily normal in G. For each irreducible representation U of G on a vector space V, show that $\dim(V) \leqq (G : A)$. (Use Frobenius' theorem (Problem 1) by observing that for each irreducible representation T of A (of degree 1), the degree of T^{ind} is $(G : A)$.)

6. GELFAND PAIRS AND SPHERICAL FUNCTIONS

(22.6.1) Let G be a *unimodular* locally compact group (separable and metrizable) and K a *compact* subgroup of G. The continuous complex-valued functions on G/K are in one-to-one correspondence, via the mapping $f \mapsto f \circ \pi$ (where $\pi : G \to G/K$ is the canonical mapping) with the continuous complex-valued functions g on G such that $g(st) = g(s)$ for all $s \in G$ and all $t \in K$ **(12.10.6)**, or equivalently such that

(22.6.1.1) $$g * \varepsilon_t = g \qquad (t \in K).$$

We shall *identify* the vector space $\mathscr{C}(G/K)$ of continuous functions on G/K with the subspace of functions $g \in \mathscr{C}(G)$ which satisfy **(22.6.1.1)**. Likewise, we shall identify $\mathscr{C}(K \backslash G)$ with the subspace of $\mathscr{C}(G)$ consisting of functions which satisfy

(22.6.1.2) $$\varepsilon_t * g = g \qquad (t \in K)$$

(or equivalently $g(ts) = g(s)$ for all $s \in G$ and $t \in K$). We denote by $\mathscr{C}(K \backslash G/K)$ the intersection $\mathscr{C}(G/K) \cap \mathscr{C}(K \backslash G)$, which consists of the continuous functions which are constant on each *double coset* KsK.

Since K is compact, the inverse image $\pi^{-1}(A)$ of every compact subset A of G/K is compact; the mapping $f \mapsto f \circ \pi$ is therefore a bijection of the subspace $\mathscr{K}(G/K)$ of $\mathscr{C}(G/K)$, consisting of the continuous functions on G/K with compact support, onto a vector subspace of $\mathscr{K}(G)$. We shall therefore identify $\mathscr{K}(G/K)$ with this subspace, which can be written as $\mathscr{K}(G) \cap \mathscr{C}(G/K)$. Likewise, we denote by $\mathscr{K}(K \backslash G)$ the intersection $\mathscr{K}(G) \cap \mathscr{C}(K \backslash G)$, and by $\mathscr{K}(K \backslash G/K)$ the intersection

$$\mathscr{K}(G/K) \cap \mathscr{K}(K \backslash G) = \mathscr{K}(G) \cap \mathscr{C}(K \backslash G/K).$$

From **(14.10.5)**, $\mathscr{K}(G)$ is a *self-adjoint subalgebra* (with respect to convolution) of the algebra $L^1_{\mathbf{C}}(G)$ (because the support of Haar measure on G is the whole of G). It follows from **(22.6.1.1)** and **(22.6.1.2)** that $\mathscr{K}(G/K)$ is a *left ideal* and $\mathscr{K}(K \backslash G)$ a *right ideal* in $\mathscr{K}(G)$, and the involution $f \mapsto \overset{*}{f}$ transforms $\mathscr{K}(G/K)$ into $\mathscr{K}(K \backslash G)$. The intersection $\mathscr{K}(K \backslash G/K)$ is a *self-adjoint subalgebra* of $\mathscr{K}(G)$ (and therefore also of $L^1_{\mathbf{C}}(G)$).

Let m_{K} be the Haar measure on K with total mass equal to 1. If we put

(22.6.1.3) $$f^{\natural}(s) = \iint_{\mathrm{K}\times\mathrm{K}} f(tst')\, dm_{\mathrm{K}}(t)\, dm_{\mathrm{K}}(t')$$

for all functions $f \in \mathscr{K}(\mathrm{G})$, the mapping $f \mapsto f^{\natural}$ is a *projection* of the vector space $\mathscr{C}(\mathrm{G})$ onto the vector space $\mathscr{C}(\mathrm{K}\backslash\mathrm{G}/\mathrm{K})$: this follows directly from **(14.1.5.6)** and the left- and right-invariance of m_{K}. It is clear that, for $f \in \mathscr{C}(\mathrm{K}\backslash\mathrm{G}/\mathrm{K})$ and $g \in \mathscr{C}(\mathrm{G})$, we have

(22.6.1.4) $$(fg)^{\natural} = fg^{\natural}.$$

Furthermore, the projection $f \mapsto f^{\natural}$ maps $\mathscr{K}(\mathrm{G})$ *onto* $\mathscr{K}(\mathrm{K}\backslash\mathrm{G}/\mathrm{K})$, and for all functions $f \in \mathscr{K}(\mathrm{K}\backslash\mathrm{G}/\mathrm{K})$ and $g \in \mathscr{C}(\mathrm{G})$ we have

(22.6.1.5) $$(f * g)^{\natural} = f * g^{\natural}, \qquad (g * f)^{\natural} = g^{\natural} * f.$$

For if m_{G} is a Haar measure on G, we have

$$\begin{aligned}(f * g)^{\natural}(x) &= \iiint g(s^{-1}txt')f(s)\, dm_{\mathrm{G}}(s)\, dm_{\mathrm{K}}(t)\, dm_{\mathrm{K}}(t') \\ &= \iint g(s^{-1}xt')f(s)\, dm_{\mathrm{G}}(s)\, dm_{\mathrm{K}}(t')\end{aligned}$$

by reason of the left-invariance of m_{G} and the relation $f(ts) = f(s)$; likewise

$$\begin{aligned}(f * g^{\natural})(x) &= \iiint g(ts^{-1}xt')f(s)\, dm_{\mathrm{G}}(s)\, dm_{\mathrm{K}}(t)\, dm_{\mathrm{K}}(t') \\ &= \iint g(s^{-1}xt')f(s)\, dm_{\mathrm{G}}(s)\, dm_{\mathrm{K}}(t'),\end{aligned}$$

this time by reason of the right-invariance of m_{G} and the relation $f(st) = f(s)$. This establishes the first of the formulas **(22.6.1.5)**; the second one is proved in the same way.

(22.6.2) Under the general hypotheses of **(22.6.1)**, we say that (G, K) is a *Gelfand pair* if the algebra $\mathscr{K}(\mathrm{K}\backslash\mathrm{G}/\mathrm{K})$ (with respect to convolution) is *commutative*. Clearly this will always be the case if G itself is commutative.

The usefulness of this definition is to be found in the following theorem:

(22.6.3) (Gelfand) *Let* G *be a unimodular locally compact group and let* σ *be an involutory automorphism of* G *such that* (1) *the subgroup* K *of elements of*

G fixed by σ is compact; (2) *each $x \in G$ can be written (in at least one way) in the form $x = yz$ where $y \in K$ and $\sigma(z) = z^{-1}$. Then (G, K) is a Gelfand pair.*

Since $\sigma^2 = 1_G$, we have $(\operatorname{mod}(\sigma))^2 = 1$ **(14.3.8)** and hence $\operatorname{mod}(\sigma) = 1$, so that σ leaves invariant a Haar measure m_G on G. For each function $f \in \mathscr{K}(G)$, let f^σ denote the function $x \mapsto f(\sigma(x)) = f(\sigma^{-1}(x))$. Clearly $f \mapsto f^\sigma$ is an involutory automorphism of the vector space $\mathscr{K}(G)$, and the preceding remark shows that, by transport of structure, $f \mapsto f^\sigma$ is an automorphism of the *algebra* $\mathscr{K}(G)$ (with respect to convolution). Since $\sigma(x) = x$ for all $x \in K$, it is clear that $f \mapsto f^\sigma$ leaves invariant the subalgebra $\mathscr{K}(K\backslash G/K)$, and its restriction to $\mathscr{K}(K\backslash G/K)$ is an automorphism of this subalgebra. We shall show that for $f, g \in \mathscr{K}(K\backslash G/K)$ we have $f^\sigma * g^\sigma = g^\sigma * f^\sigma$, which will establish the commutativity of $\mathscr{K}(K\backslash G/K)$.

For each $x \in G$, we can write $x = yz$ with $y \in K$ and $\sigma(z) = z^{-1}$, whence $\sigma(x) = yz^{-1} = y(z^{-1}y^{-1})y$; for any $f \in \mathscr{K}(K\backslash G/K)$ we therefore have

$$f(\sigma(x)) = f(z^{-1}y^{-1}) = f(x^{-1}) = \check{f}(x),$$

in other words $f^\sigma = \check{f}$. But because $L^1_{\mathbf{C}}(G)$ is an algebra with involution, we have

$$(\check{f} * \check{g})^\vee = g * f$$

for $f, g \in \mathscr{K}(G)$. Hence, for f and g in $\mathscr{K}(K\backslash G/K)$, we have

$$f^\sigma * g^\sigma = \check{f} * \check{g} = (g * f)^\vee = (g * f)^\sigma = g^\sigma * f^\sigma. \qquad \text{Q.E.D.}$$

The most important applications of **(22.6.3)** correspond to the following three cases:

(1) G is a compact connected semisimple Lie group and σ is an involutory automorphism of G **(21.18.13)**;

(2) G is a noncompact real form, with finite center, of the complexification of a compact connected semisimple Lie group G_u, corresponding to an involutory automorphism σ_0 of G_u, and K is the subgroup of elements of G_u fixed by σ_0 **(21.18.8)**;

(3) G is unimodular and contains a *commutative normal* closed subgroup A such that the relation $s^2 = e$ for $s \in A$ implies $s = e$, and a compact subgroup K such that the mapping $(t, s) \mapsto ts$ of $K \times A$ into G is a *homeomorphism* (which implies that G is the *semidirect product* of K and A). If we put $\sigma(ts) = ts^{-1}$ for $t \in K$ and $s \in A$, the conditions of **(22.6.3)** are satisfied. For it is clear that σ is continuous and $\sigma^2 = 1_G$; also we have

$\sigma(tst's') = \sigma((tt')(t'^{-1}st's')) = tt'(s'^{-1}t'^{-1}s^{-1}t')$, and on the other hand $\sigma(ts)\sigma(t's') = ts^{-1}t's'^{-1} = tt'(t'^{-1}s^{-1}t's'^{-1})$; since A is commutative, it follows that σ is an automorphism.

(22.6.4) *From now on until the end of Section* **(22.9)**, *we shall assume that* (G, K) *is a Gelfand pair.*

We denote by $L^1_{\mathbf{C}}(K\backslash G/K)$ the closure in $L^1_{\mathbf{C}}(G)$ of the subalgebra $\mathscr{K}(K\backslash G/K)$. It is clearly a *commutative self-adjoint Banach subalgebra* of $L^1_{\mathbf{C}}(G)$; moreover, the projection $f \mapsto f^{\natural}$ of $\mathscr{K}(G)$ onto $\mathscr{K}(K\backslash G/K)$ extends to a *continuous projection* of $L^1_{\mathbf{C}}(G)$ onto the subspace $L^1_{\mathbf{C}}(K\backslash G/K)$, which we denote by $\tilde{f} \mapsto \tilde{f}^{\natural}$. For it is enough to remark that for $f \in \mathscr{K}(G)$ we have

$$\text{(22.6.4.1)} \qquad N_1(f^{\natural}) = \int_G \left| \iint_{K \times K} f(sts')\, dm_K(t)\, dm_K(t') \right| dm_G(s)$$

$$\leqq \iiint |f(sts')|\, dm_G(s)\, dm_K(t)\, dm_K(t')$$

$$= N_1(f)$$

by virtue of the left- and right-invariance of m_G. For $f \in \mathscr{L}^1_{\mathbf{C}}(G)$, the class $\tilde{f}^{\natural}$ is also the class of the function $f^{\natural} \in \mathscr{L}^1_{\mathbf{C}}(G)$ given almost everywhere (with respect to m_G) by the same formula **(22.6.1.3)**: for $\tilde{f}$ is the limit in $L^1_{\mathbf{C}}(G)$ of a sequence $(\tilde{f}_n)$, where the f_n belong to $\mathscr{K}(G)$ and $f_n(x) \to f(x)$ almost everywhere **(13.11.4)**. The function $(s, t, t') \mapsto f(tst')$ is therefore measurable on $G \times K \times K$ **(13.9.10)**, and the invariance of m_G shows that

$$\iiint^* |f(tst')|\, dm_G(s)\, dm_K(t)\, dm_K(t') = \int |f(s)|\, dm_G(s) < +\infty$$

by use of **(13.21.8)**. Hence the function $(s, t, t') \mapsto f(tst')$ is integrable with respect to $m_G \otimes m_K \otimes m_K$ **(13.9.13)**, and the Lebesgue–Fubini theorem **(13.21.7)** then shows that, for almost all $s \in G$, the integral

$$f^{\natural}(s) = \iint f(tst')\, dm_K(t)\, dm_K(t')$$

exists and that the function $f^{\natural}$ belongs to $\mathscr{L}^1_{\mathbf{C}}(G)$ and satisfies **(22.6.4.1)**; moreover, replacing f by $f - f_n$ in **(22.6.4.1)**, we see that

$$N_1(f^{\natural} - f_n^{\natural}) \leqq N_1(f - f_n),$$

which establishes our assertion since $\tilde{f}^{\natural}$ is the limit of the sequence $(\tilde{f}_n^{\natural})$.

Again, by virtue of the Cauchy–Schwarz inequality, for $f \in \mathscr{K}(G)$ we have

(22.6.4.2)
$$N_2(f^{\natural})^2 = \int_G \left| \iint_{K \times K} f(tst')\, dm_K(t)\, dm_K(t') \right|^2 dm_G(s)$$
$$\leq \iiint |f(tst')|^2\, dm_G(s)\, dm_K(t)\, dm_K(t')$$
$$= N_2(f)^2,$$

so that the projection $f \mapsto f^{\natural}$ extends to a continuous projection $\tilde{f} \mapsto \tilde{f}^{\natural}$ of $L^2_{\mathbf{C}}(G)$ onto the closure $L^2_{\mathbf{C}}(K\backslash G/K)$ of $\mathscr{K}(K\backslash G/K)$ in $L^2_{\mathbf{C}}(G)$. Moreover, if $f \in L^2_{\mathbf{C}}(G)$, the integral

$$f^{\natural}(s) = \iint f(tst')\, dm_K(t)\, dm_K(t')$$

again exists for almost all $s \in G$, and $\tilde{f}^{\natural}$ is the class of $f^{\natural}$. For $\tilde{f}$ is the limit in $L^2_{\mathbf{C}}(G)$ of a sequence $(\tilde{f}_n)$, where the f_n belong to $\mathscr{K}(G)$ and $f_n(x) \to f(x)$ almost everywhere **(13.11.4)**, which as before implies that the function $(s, t, t') \mapsto f(tst')$ is measurable. Also it follows from **(13.11.2.2)** and **(13.21.8)** that

$$\int_G^* \left(\iint_{K \times K}^* |f(tst')|\, dm_K(t)\, dm_K(t') \right)^2 dm_G(s)$$
$$\leq \iiint^* |f(tst')|^2\, dm_G(s)\, dm_K(t)\, dm_K(t') = N_2(f)^2.$$

This shows that for almost all $s \in G$ we have

$$\iint^* |f(tst')|\, dm_K(t)\, dm_K(t') < +\infty$$

(13.6.4), and since for almost all $s \in G$ the function $(t, t') \mapsto f(tst')$ is measurable with respect to $m_K \otimes m_K$ **(13.21.6)**, it follows that the integral $f^{\natural}(s)$ exists for almost all $s \in G$. By applying **(22.6.4.2)** to $f - f_n$, we see that $N_2(f^{\natural} - f_n^{\natural}) \leq N_2(f - f_n)$. If $\tilde{g}$ is the limit of the Cauchy sequence $(\tilde{f}_n^{\natural})$ in $L^2_{\mathbf{C}}(G)$, it follows from **(13.11.2)** that we have $N_2(|f^{\natural} - g|) = 0$; hence $f^{\natural} \in \mathscr{L}^2_{\mathbf{C}}(G)$ and $\tilde{g}$ is the class of $f^{\natural}$, which proves our assertion.

For $p = 1$ or 2, we denote by $\mathscr{L}^p_{\mathbf{C}}(G/K)$ (resp $\mathscr{L}^p_{\mathbf{C}}(K\backslash G)$) the subspace of $\mathscr{L}^p_{\mathbf{C}}(G)$ consisting of the functions f such that, for almost all $s \in G$, we have $f(st) = f(s)$ (resp. $f(ts) = f(s)$) for *all* $t \in K$; and we put $\mathscr{L}^p_{\mathbf{C}}(K\backslash G/K) = \mathscr{L}^p_{\mathbf{C}}(G/K) \cap \mathscr{L}^p_{\mathbf{C}}(K\backslash G)$. If $f \in \mathscr{L}^p_{\mathbf{C}}(G)$, then $f^\natural \in \mathscr{L}^p_{\mathbf{C}}(K\backslash G/K)$, and $L^p_{\mathbf{C}}(K\backslash G/K)$ is the canonical image of $\mathscr{L}^p_{\mathbf{C}}(K\backslash G/K)$ in $L^p_{\mathbf{C}}(G)$. If we denote by $L^p_{\mathbf{C}}(G/K)$ and $L^p_{\mathbf{C}}(K\backslash G)$ the canonical images of $\mathscr{L}^p_{\mathbf{C}}(G/K)$ and $\mathscr{L}^p_{\mathbf{C}}(K\backslash G)$ in $L^p_{\mathbf{C}}(G)$, then we have $L^p_{\mathbf{C}}(K\backslash G/K) = L^p_{\mathbf{C}}(G/K) \cap L^p_{\mathbf{C}}(K\backslash G)$.

(22.6.5) *Every character of the commutative Banach algebra* $L^1_{\mathbf{C}}(K\backslash G/K)$ *is uniquely expressible in the form*

(22.6.5.1)
$$\tilde{f} \mapsto \langle f, \omega \rangle = \int_G f(x)\omega(x)\, dm_G(x)$$

where ω *is a function in* $\mathscr{C}(K\backslash G/K)$, *bounded and continuous on* G. *Moreover,* ω *is uniformly continuous with respect to any left-invariant distance on* G, *and* $|\omega(s)| \leqq \omega(e) = 1$ *for all* $s \in G$.

Every character ζ of the commutative subalgebra $L^1_{\mathbf{C}}(K\backslash G/K)$ of $M^1_{\mathbf{C}}(G)$ extends to a character ζ' of the commutative Banach subalgebra $\mathbf{C}\varepsilon_e + L^1_{\mathbf{C}}(K\backslash G/K)$, by defining $\zeta'(\lambda\varepsilon_e + \tilde{f}) = \lambda + \zeta(\tilde{f})$. Since this latter subalgebra has an identity element, we have $|\zeta(\tilde{f})| \leqq N_1(\tilde{f})$ for all $\tilde{f} \in L^1_{\mathbf{C}}(K\backslash G/K)$ **(15.3.1)**. It follows that $\tilde{f} \mapsto \zeta(\tilde{f}^\natural)$ is a *continuous linear form* on the Banach space $L^1_{\mathbf{C}}(G)$, of norm $\leqq 1$ by virtue of **(22.6.4.1)**. Hence, by virtue of **(13.17.1)**, it can be written in the form

(22.6.5.2)
$$\tilde{f} \mapsto \langle \tilde{f}, \tilde{\omega}_0 \rangle = \int_G f(x)\omega_0(x)\, dm_G(x)$$

where $\omega_0 \in \mathscr{L}^\infty_{\mathbf{C}}(G)$ and $N_\infty(\omega_0) \leqq 1$. Moreover, for t and t' in K and $f \in \mathscr{K}(G)$ we have

(22.6.5.3)
$$\int f(txt')\omega_0(x)\, dm_G(x) = \int f(x)\omega_0(x)\, dm_G(x),$$

because if $h_{t,t'}$ denotes the function $x \mapsto f(txt')$, which belongs to $\mathscr{K}(G)$, we have $h^\natural_{t,t'} = f^\natural$ and therefore $\zeta(f^\natural) = \zeta(h^\natural_{t,t'})$. (We have written $\zeta(f) = \zeta(\tilde{f})$ for $f \in \mathscr{L}^1_{\mathbf{C}}(K\backslash G/K)$.)

By definition **(15.3)**, there exists a function $f_0 \in \mathscr{K}(\mathrm{K}\backslash\mathrm{G}/\mathrm{K})$ such that $\zeta(f_0) \neq 0$; since ζ is a character of $\mathrm{L}^1_{\mathbf{C}}(\mathrm{K}\backslash\mathrm{G}/\mathrm{K})$ we have, for all $g \in \mathscr{K}(\mathrm{K}\backslash\mathrm{G}/\mathrm{K})$, by virtue of the Lebesgue–Fubini theorem,

(22.6.5.4)
$$\begin{aligned}\zeta(g) &= \zeta(f_0)^{-1}\zeta(g * f_0)\\ &= \zeta(f_0)^{-1}\iint_{\mathrm{G}\times\mathrm{G}} f_0(s^{-1}x)g(s)\omega_0(s)\,dm_{\mathrm{G}}(s)\,dm_{\mathrm{G}}(x)\\ &= \int_{\mathrm{G}} g(s)\omega(s)\,dm_{\mathrm{G}}(s)\end{aligned}$$

where we have put

(22.6.5.5)
$$\begin{aligned}\omega(s) &= \zeta(f_0)^{-1}\int_{\mathrm{G}} f_0(s^{-1}x)\omega_0(x)\,dm_{\mathrm{G}}(x)\\ &= \zeta(f_0)^{-1}\int_{\mathrm{G}} f_0(x)\omega_0(sx)\,dm_{\mathrm{G}}(x),\end{aligned}$$

the second expression being a consequence of the left-invariance of m_{G}. It follows, first of all **(14.10.6(ii))**, that the function ω is bounded on G and uniformly continuous with respect to any left-invariant distance on G; and, next, that for $t, t' \in \mathrm{K}$ we have

$$\begin{aligned}\omega(tst') &= \zeta(f_0)^{-1}\int_{\mathrm{G}} f_0(t'^{-1}s^{-1}t^{-1}x)\omega_0(x)\,dm_{\mathrm{G}}(x)\\ &= \zeta(f_0)^{-1}\int_{\mathrm{G}} f_0(s^{-1}t^{-1}x)\omega_0(x)\,dm_{\mathrm{G}}(x)\\ &= \zeta(f_0)^{-1}\int_{\mathrm{G}} f_0(s^{-1}x)\omega_0(x)\,dm_{\mathrm{G}}(x)\\ &= \omega(s),\end{aligned}$$

where the first relation follows from the fact that $f_0 \in \mathscr{K}(\mathrm{K}\backslash\mathrm{G}/\mathrm{K})$ and the second follows from **(22.6.5.3)** applied to the function $x \mapsto f_0(s^{-1}x)$, which belongs to $\mathscr{K}(\mathrm{G})$. We have therefore established **(22.6.5.1)**, since $\mathscr{K}(\mathrm{K}\backslash\mathrm{G}/\mathrm{K})$ is dense in $\mathrm{L}^1_{\mathbf{C}}(\mathrm{K}\backslash\mathrm{G}/\mathrm{K})$ **(13.11.6)**. Let us next show that $\tilde{\omega} = \tilde{\omega}_0$, which will prove that $|\omega(s)| \leqq 1$ for all $s \in \mathrm{G}$ by virtue of the continuity of ω. By **(13.17.1)** it will be enough to show that $\langle f, \omega\rangle = \langle f, \omega_0\rangle = \zeta(f^\natural)$ for all $f \in \mathscr{K}(\mathrm{G})$; and this follows from **(22.6.5.4)**, the relation

(22.6.5.6)
$$\langle f^\natural, \psi\rangle = \langle f, \psi^\natural\rangle$$

for $f \in \mathscr{K}(G)$ and $\psi \in \mathscr{C}(G)$, and the fact that $\omega^\natural = \omega$. The relation **(22.6.5.6)** follows from the Lebesgue–Fubini theorem and the two-sided invariance of m_G and m_K, because we have

$$\begin{aligned}\langle f^\natural, \psi\rangle &= \iiint f(tst')\psi(s)\,dm_G(s)\,dm_K(t)\,dm_K(t')\\ &= \iiint f(s)\psi(t^{-1}st'^{-1})\,dm_G(s)\,dm_K(t)\,dm_K(t')\\ &= \iiint f(s)\psi(tst')\,dm_G(s)\,dm_K(t)\,dm_K(t')\\ &= \langle f, \psi^\natural\rangle\end{aligned}$$

by virtue of **(14.3.4.1)**. This argument also establishes the *uniqueness* of the function $\omega \in \mathscr{C}(K\backslash G/K)$ satisfying **(22.6.5.1)**.

Finally, if we set $s = e$ in **(22.6.5.5)**, we obtain $\omega(e) = 1$ by reason of **(22.6.5.2)** and the relation $f_0 = f_0^\natural$, and this completes the proof of **(22.6.5)**.

We remark also that the preceding proof shows that

(22.6.5.7) $$\omega(s) = \langle f_0, \omega\rangle^{-1} \int_G f_0(s^{-1}x)\omega(x)\,dm_G(x)$$

for all functions $f_0 \in \mathscr{K}(K\backslash G/K)$ such that $\langle f_0, \omega\rangle \neq 0$.

(22.6.6) A bounded function $\omega \in \mathscr{C}(K\backslash G/K)$ is said to be a *spherical function* (or *zonal spherical function*) on G relative to K, if $f \mapsto \langle f, \omega\rangle$ is a *nonzero character* of $L^1_{\mathbf{C}}(K\backslash G/K)$. This implies in particular, by virtue of **(22.6.5)**, that $\omega(e) = 1$ and $|\omega(s)| \leq \omega(e)$ for all $s \in G$. If ω is a spherical function, so also are $\bar{\omega}$ and $\check{\omega}$. For if $f \in \mathscr{K}(K\backslash G/K)$, we have $\langle f, \bar{\omega}\rangle = \overline{\langle \bar{f}, \omega\rangle}$ and $\overline{(f * g)} = \bar{f} * \bar{g}$, which proves the first assertion; and since

$$\langle f, \check{\omega}\rangle = \int_G f(s)\omega(s^{-1})\,dm_G(s) = \int_G f(s^{-1})\omega(s)\,dm_G(s) = \langle \check{f}, \omega\rangle$$

because G is unimodular, and $(f * g)\check{} = \check{g} * \check{f} = \check{f} * \check{g}$ for $f, g \in \mathscr{K}(K\backslash G/K)$, the second assertion is established. Together with **(22.6.5)**, this shows that a spherical function is *uniformly continuous* with respect to any left- *or* right-invariant distance on G. It should be noted that in general a spherical function is *not necessarily of compact support*.

(22.6.7) *Let ω be a bounded function belonging to $\mathscr{C}(\mathrm{K}\backslash\mathrm{G}/\mathrm{K})$ and not identically zero. Then the following properties are equivalent:*

(a) *ω is a spherical function on G relative to K.*
(b) *We have*

$$\int_{\mathrm{K}} \omega(xty)\, dm_{\mathrm{K}}(t) = \omega(x)\omega(y) \tag{22.6.7.1}$$

for all $x, y \in \mathrm{G}$.

(c) *We have $\omega(e) = 1$, and for each function $f \in \mathscr{K}(\mathrm{K}\backslash\mathrm{G}/\mathrm{K})$ there exists a number λ_f such that $f * \omega = \lambda_f \omega$ (resp. $\omega * f = \lambda_f \omega$).*

We shall begin by showing that (c) implies (a). Suppose that $f * \omega = \lambda_f \omega$ for all $f \in \mathscr{K}(\mathrm{K}\backslash\mathrm{G}/\mathrm{K})$; since $\omega(e) = 1$ and G is unimodular, it follows that

$$\lambda_{\check{f}} = (\check{f} * \omega)(e) = \int_{\mathrm{G}} f(s)\omega(s)\, dm_{\mathrm{G}}(s) = \langle f, \omega \rangle.$$

If $f, g \in \mathscr{K}(\mathrm{K}\backslash\mathrm{G}/\mathrm{K})$, we have $(f * g) * \omega = f * (g * \omega)$, and consequently $\lambda_{\check{g} * \check{f}} = \lambda_{\check{g}} \lambda_{\check{f}}$, or in other words $\langle f * g, \omega \rangle = \langle f, \omega \rangle \langle g, \omega \rangle$, which proves that ω is a spherical function. The proof is the same if $\omega * f = \lambda_f \omega$ for all $f \in \mathscr{K}(\mathrm{K}\backslash\mathrm{G}/\mathrm{K})$.

Next we shall prove that (a) implies (c). Since the function $f * \omega$ belongs to $\mathscr{C}(\mathrm{K}\backslash\mathrm{G}/\mathrm{K})$, in order to prove that $f * \omega = \langle \check{f}, \omega \rangle \omega$, it is sufficient by virtue of **(22.6.5.6)** to show that for all functions $g \in \mathscr{K}(\mathrm{K}\backslash\mathrm{G}/\mathrm{K})$ we have

$$\langle g, f * \omega \rangle = \langle \check{f}, \omega \rangle \langle g, \omega \rangle.$$

But by virtue of **(14.9.4)** we have $\langle g, f * \omega \rangle = \langle \check{f} * g, \omega \rangle = \langle \check{f}, \omega \rangle \langle g, \omega \rangle$, because ω is a spherical function. Since $\check{\omega}$ is also a spherical function, we have therefore

$$\omega * f = (\check{f} * \check{\omega})^{\vee} = \overline{\langle \overline{\check{f}}, \check{\omega} \rangle} \omega = \langle \check{f}, \omega \rangle \omega,$$

by using the fact that G is unimodular.

Finally, we shall show that (b) and (c) are equivalent. Put

$$h(x, y) = \int_{\mathrm{K}} \omega(xty)\, dm_{\mathrm{K}}(t).$$

It follows from **(14.1.5.5)** that the function $x \mapsto h(x, y)$ is continuous; moreover, for each $t' \in \mathrm{K}$ we have $h(t'x, y) = h(x, y)$ because $\omega(t'xty) = \omega(xty)$, and $h(xt', y) = h(x, y)$ by virtue of the invariance of m_{K}; hence the function $x \mapsto h(x, y)$ belongs to $\mathscr{C}(K\backslash G/K)$. We shall now show that for all $f \in \mathscr{K}(K\backslash G/K)$ we have

(22.6.7.2)
$$\int_G \check{f}(x)h(x, y)\, dm_G(x) = (f * \omega)(y).$$

We calculate:

$$\begin{aligned}\int_G \check{f}(x)h(x, y)\, dm_G(x) &= \iint_{G\times K} \check{f}(x)\omega(xty)\, dm_G(x)\, dm_K(t)\\ &= \iint_{G\times K} f(x)\omega(x^{-1}ty)\, dm_G(x)\, dm_K(t)\\ &= \int_K (f * \omega)(ty)\, dm_K(t)\end{aligned}$$

by virtue of the Lebesgue–Fubini theorem, the fact that G is unimodular, and the relation $f * \omega \in \mathscr{C}(K\backslash G/K)$. This being so, if (b) is true it follows from (22.6.7.1) that $(f * \omega)(y) = \langle \check{f}, \omega\rangle\omega(y)$ for all $y \in G$, in other words $f * \omega = \langle \check{f}, \omega\rangle\omega$ for all $f \in \mathscr{K}(K\backslash G/K)$. Conversely, this relation implies that $h(x, y) = \omega(x)\omega(y)$ for all $y \in G$, by virtue of (22.6.5.6), and hence implies (b).

(22.6.7.3) It should be remarked that if a bounded continuous function ω on G, not identically zero, satisfies the functional equation (22.6.7.1), then it automatically belongs to $\mathscr{C}(K\backslash G/K)$ and hence is a *spherical function*. For the two-sided invariance of m_K shows that $\omega(xt')\omega(y) = \omega(x)\omega(y) = \omega(x)\omega(t'y)$ for all $t \in K$.

(22.6.8) Let **S** (or **S**(G/K)) denote the set of spherical functions on G relative to the compact subgroup K, so that **S** is a subset of $\mathscr{C}(K\backslash G/K) \cap L^{\infty}_{\mathbf{C}}(G)$. In the space $\mathbf{X}(A)$ of characters of the Banach algebra $A = \mathbf{C}\varepsilon_e + L^1_{\mathbf{C}}(K\backslash G/K)$, the set of characters whose restriction to $L^1_{\mathbf{C}}(K\backslash G/K)$ is zero is either empty or consists of one element. The former case occurs only when $\varepsilon_e \in L^1_{\mathbf{C}}(K\backslash G/K)$, which implies that G is discrete and ε_e invariant under translations by elements of K; this can happen only when $K = \{e\}$ and G is *discrete* and *commutative*. We denote by $\mathbf{X}_0(A)$ the set of characters of A whose restriction to $L^1_{\mathbf{C}}(K\backslash G/K)$ is nonzero; this is a subspace of $\mathbf{X}(A)$ which is always *locally compact* (and separable and metrizable). For each spherical function $\omega \in \mathbf{S}$, let ζ_ω denote the character $\check{f} \mapsto \langle f, \omega\rangle$ of $L^1_{\mathbf{C}}(K\backslash G/K)$; then the mapping $\omega \mapsto \zeta_\omega$ is a *bijection* of **S**(G/K) onto $\mathbf{X}_0(A)$. Moreover:

(22.6.9) (i) *The mapping* $\omega \mapsto \zeta_\omega$ *is a homeomorphism of* $\mathbf{S}(G/K)$, *endowed with the topology induced by that of the Fréchet space* $\mathscr{C}_{\mathbf{C}}(G)$, *onto* $\mathbf{X}_0(A)$ *endowed with the topology induced by the weak topology of the dual* A' *of* A **(15.3.2)**.

(ii) *Every compact subset* L *of* $\mathbf{S}(G/K)$ *is equicontinuous* **(7.5)**.

(iii) *The mapping* $(x, \omega) \mapsto \omega(x)$ *of* $G \times \mathbf{S}(G/K)$ *into* $\mathbf{C}$ *is continuous.*

(i) In view of **(22.6.5.6)** it is enough to show that the topology induced on $\mathbf{S}(G/K)$ by that of $\mathscr{C}_{\mathbf{C}}(G)$ is the same as that induced by the weak topology of $L^\infty_{\mathbf{C}}(G)$. We shall apply the criterion of **(22.1.11.2)**, the functions in $\mathbf{S}(G/K)$ being uniformly bounded by virtue of **(22.6.5)**. Let therefore ω_1 be a spherical function, K_1 a compact subset of G, and ε a positive real number. Let $f_1 \in \mathscr{K}(K\backslash G/K)$ be such that $|\langle f_1, \omega_1\rangle| = b > 0$, and take as neighborhood U of ω_1 in $\mathbf{S}$ for the weak topology of $L^\infty_{\mathbf{C}}(G)$ the set of $\omega \in \mathbf{S}$ such that $|\langle f_1, \omega - \omega_1\rangle| \leqq \frac{1}{2}b$, which implies that $|\langle f_1, \omega\rangle| \geqq \frac{1}{2}b$ for all $\omega \in U$. By virtue of **(22.6.5.7)** we have, for each $\omega \in U$,

$$\omega(s) = \langle f_1, \omega\rangle^{-1} \int_G f_1(s^{-1}x)\omega(x)\, dm_G(x). \tag{22.6.9.1}$$

Now let W be any compact neighborhood of e in G, and let $a = m_G(W)$. Then we have

$$|(a^{-1}\varphi_W * \omega)(s) - \omega(s)| \leqq a^{-1} \int_W |\omega(t^{-1}s) - \omega(s)|\, dm_G(t)$$

and therefore it is enough to choose W so that for each $\omega \in U$ we have

$$|\omega(t^{-1}s) - \omega(s)| \leqq \varepsilon \tag{22.6.9.2}$$

for all $s \in K_1$ and $t \in W$. By virtue of **(22.6.9.1)**, this will be the case provided that

$$\int_G |f_1(s^{-1}tx) - f_1(s^{-1}x)|\, dm_G(x) \leqq \tfrac{1}{2}b\varepsilon \tag{22.6.9.3}$$

for all $s \in K_1$ and $t \in W$. Now since f_1 has compact support, the proof of **(14.1.5.5)** shows that it is enough to prove that for each $\eta > 0$ there exists a neighborhood W of e such that $|f_1(s^{-1}tx) - f_1(s^{-1}x)| \leqq \eta$ for all $s \in K_1$, $t \in W$ and $x \in G$. But f_1 is uniformly continuous with respect to a right-invariant distance on G **(3.16.5)**, hence there exists a neighborhood W_1 of e

such that $|f_1(x) - f_1(y)| \leqq \eta$ whenever $yx^{-1} \in W_1$. Hence it is enough to show that there exists a neighborhood W of e such that the relation $t \in W$ implies that $s^{-1}x(s^{-1}tx)^{-1} = s^{-1}ts \in W_1$ for all $s \in K_1$, and this follows from **(21.3.4.1)**.

It follows therefore that the subspace **S**(G/K) of $\mathscr{C}_{\mathbf{C}}(G)$ is *locally compact*.

(ii) Let $x_0 \in G$. For each compact neighborhood V_0 of x_0, the mapping $f \mapsto f|V_0$ of $\mathscr{C}_{\mathbf{C}}(G)$ into $\mathscr{C}_{\mathbf{C}}(V_0)$ is continuous **(12.14.6)**, and hence the image of L under this mapping is compact **(3.17.9)**. By Ascoli's theorem **(7.5.7)**, this image is equicontinuous. Hence for each $\varepsilon > 0$ there exists a neighborhood $V \subset V_0$ of x_0 such that $|\omega(x) - \omega(x_0)| \leqq \varepsilon$ for all $x \in V$ and all $\omega \in L$.

(iii) Let $(x_0, \omega_0) \in G \times \mathbf{S}$. For each $\varepsilon > 0$ there exists, by (ii) above, a compact neighborhood V of x_0 in G and a compact neighborhood W of ω_0 in **S** such that $|\omega(x) - \omega(x_0)| \leqq \varepsilon$ for all $x \in V$ and all $\omega \in W$. Also, by the definition of the topology of $\mathscr{C}_{\mathbf{C}}(G)$ **(12.14.6)**, there exists a neighborhood $U \subset W$ of ω_0 in **S** such that $|\omega(x) - \omega_0(x)| \leqq \varepsilon$ for all $x \in V$ and all $\omega \in U$. It follows now that $|\omega(x) - \omega_0(x_0)| \leqq 3\varepsilon$ for all $x \in V$ and all $\omega \in U$.

Examples

(22.6.10) Suppose first that G is *compact*. If (G, K) is a Gelfand pair, the Hilbert algebra $L^2_{\mathbf{C}}(K\backslash G/K)$, the closure of $\mathscr{K}(K\backslash G/K)$ in $L^2_{\mathbf{C}}(G)$, is *commutative*, and we are therefore in the situation of **(22.5.6)**. Furthermore, every character of $L^1_{\mathbf{C}}(K\backslash G/K)$ restricts to a (continuous) character of $L^2_{\mathbf{C}}(K\backslash G/K)$. The (algebraic) direct sum of the ideals $\mathbf{C}\omega_{\boldsymbol{\rho}}$ for all the classes $\boldsymbol{\rho} \in R(G)$ such that $(\boldsymbol{\rho} : \boldsymbol{\sigma}_0) = 1$ is a dense subalgebra B of $L^2_{\mathbf{C}}(K\backslash G/K)$, and it is immediately seen that the only homomorphisms $\neq 0$ of B into the field **C** are the mappings $f \mapsto \langle f, \bar{\omega}_{\boldsymbol{\rho}} \rangle$; this shows that in this case the spherical functions on G relative to K are the *partial traces* $\omega_{\boldsymbol{\rho}}$ **(22.5.2)** for all the classes $\boldsymbol{\rho} \in R(G)$ *such that* $(\boldsymbol{\rho} : \boldsymbol{\sigma}_0) = 1$. Since $\omega_{\boldsymbol{\rho}} \in L^1_{\mathbf{C}}(G)$, the set of $\tilde{f} \in L^{\infty}_{\mathbf{C}}(G)$ such that $|\langle f - \omega_{\boldsymbol{\rho}}, \bar{\omega}_{\boldsymbol{\rho}} \rangle| \leqq \frac{1}{2}$ is a neighborhood of $\omega_{\boldsymbol{\rho}}$ for the weak topology on $L^{\infty}_{\mathbf{C}}(G)$; since $\langle \omega_{\boldsymbol{\rho}}, \bar{\omega}_{\boldsymbol{\rho}} \rangle = 1$ and $\langle \omega_{\boldsymbol{\rho}'}, \bar{\omega}_{\boldsymbol{\rho}} \rangle = 0$ for all $\boldsymbol{\rho}' \neq \boldsymbol{\rho}$, this neighborhood contains no $\omega_{\boldsymbol{\rho}'}$ other than $\omega_{\boldsymbol{\rho}}$: in other words, the space **S**(G/K) is *discrete*.

(22.6.11) Let G be a noncompact connected semisimple Lie group with finite center, and K its maximal compact subgroup in the description of **(21.18.8)**, so that (G, K) is a Gelfand pair. There exists also in G a solvable closed subgroup S such that the mapping $(x, y) \mapsto xy$ of $K \times S$ into G is a diffeomorphism **(21.21.10)**; and the same is true of the mapping $(x, y) \mapsto yx$, because $(xy)^{-1} = y^{-1}x^{-1}$. Let α be a continuous homomorphism of S into the multiplicative group $\mathbf{C}^*$ (an "exponential" on S). Clearly α can be extended to a continuous mapping (still denoted by α) of G into $\mathbf{C}^*$ by the rule

$$\alpha(st) = \alpha(s) \qquad (s \in \mathrm{S},\ t \in \mathrm{K}). \tag{22.6.11.1}$$

This extended function is such that

$$\alpha(xt) = \alpha(x) \qquad (x \in \mathrm{G},\ t \in \mathrm{K}) \tag{22.6.11.2}$$

$$\alpha(sx) = \alpha(s)\alpha(x) \qquad (s \in \mathrm{S},\ x \in \mathrm{G}). \tag{22.6.11.3}$$

The first of these relations follows directly from **(22.6.11.1)** and the fact that $x = st'$ for some $s \in \mathrm{S}$ and $t' \in \mathrm{K}$. Likewise, to prove **(22.6.11.3)**, we write $x = s't$ with $s' \in \mathrm{S}$ and $t \in \mathrm{K}$, and then we have $\alpha(sx) = \alpha(ss't) = \alpha(ss') = \alpha(s)\alpha(s') = \alpha(s)\alpha(x)$.

We assert now that the function

$$\omega(x) = \int_{\mathrm{K}} \alpha(tx)\, dm_{\mathrm{K}}(t) \tag{22.6.11.4}$$

is a *spherical function* on G relative to K. To prove this it is enough (**(14.1.5.5)** and **(22.6.7.2)**) to verify the functional equation **(22.6.7.1)**. The left-hand side of this equation is here equal to

$$\iint_{\mathrm{K}\times\mathrm{K}} \alpha(txt'y)\, dm_{\mathrm{K}}(t)\, dm_{\mathrm{K}}(t'). \tag{22.6.11.5}$$

Now we can write $tx = s(t)u(t)$ for all $t \in \mathrm{K}$, where $s(t) \in \mathrm{S}$ and $u(t) \in \mathrm{K}$, and s, u are continuous functions of t on K. By the invariance of m_{K} and the Lebesgue–Fubini theorem, the integral **(22.6.11.5)** is also equal to

$$\begin{aligned}\iint_{\mathrm{K}\times\mathrm{K}} \alpha(s(t)u(t)t'y)\, dm_{\mathrm{K}}(t)\, dm_{\mathrm{K}}(t') &= \iint_{\mathrm{K}\times\mathrm{K}} \alpha(s(t)t'y)\, dm_{\mathrm{K}}(t)\, dm_{\mathrm{K}}(t')\\ &= \iint_{\mathrm{K}\times\mathrm{K}} \alpha(s(t))\alpha(t'y)\, dm_{\mathrm{K}}(t)\, dm_{\mathrm{K}}(t')\\ &= \omega(x)\omega(y)\end{aligned}$$

in view of **(22.6.11.2)** and **(22.6.11.3)**.

It can be proved [62] that the formula **(22.6.11.4)** gives *all* the spherical functions on G relative to K.

For example, take for G, S and K the groups of **(22.3.9)**. The image of a double coset KsK in G/K, identified with the half-plane P, is the *orbit* of a point of P under the left action of K on P. The orbit of the point ir (where $0 < r < 1$) is the circle which has the points ir and i/r diametrically opposed, the image of ir under the element

$$\begin{pmatrix} \cos\theta & \sin\theta \\ -\sin\theta & \cos\theta \end{pmatrix}$$

being the point

$$x + iy = \frac{ir\cos\theta + \sin\theta}{-ir\sin\theta + \cos\theta}.$$

This circle is completely determined by its center iv, where $v = \frac{1}{2}(r + r^{-1})$. The matrices in the corresponding double coset are the matrices XU, where $U \in$ K and

$$X = \begin{pmatrix} y^{1/2} & xy^{-1/2} \\ 0 & y^{-1/2} \end{pmatrix},$$

x and y being as above. Since $\operatorname{Tr}({}^tX \cdot X) = y^{-1}(x^2 + y^2 + 1) = 2v$, it follows (since ${}^tU = U^{-1}$) that the matrices

$$\begin{pmatrix} a & b \\ c & d \end{pmatrix} \in \mathrm{G} = \mathbf{SL}(2, \mathbf{R})$$

of this double coset are those for which $a^2 + b^2 + c^2 + d^2 = 2v$; hence the functions belonging to $\mathscr{C}(\mathrm{K}\backslash\mathrm{G}/\mathrm{K})$ are the functions of the form $f(\frac{1}{2}(a^2 + b^2 + c^2 + d^2))$, where f is a continuous complex-valued function defined on the interval $[1, +\infty[$ **(12.10.6)**. Since the value of an "exponential" of the group S at a matrix

$$X = \begin{pmatrix} y^{1/2} & xy^{-1/2} \\ 0 & y^{-1/2} \end{pmatrix}$$

depends only on the coset of X in $\mathrm{S}/\mathscr{D}(\mathrm{S})$, which may be identified with the diagonal matrix

$$\begin{pmatrix} y^{1/2} & 0 \\ 0 & y^{-1/2} \end{pmatrix},$$

it follows from **(4.1.3)** and **(21.3.9)** that every such "exponential" of S is of the form

(22.6.11.6) $$\alpha : \begin{pmatrix} y^{1/2} & xy^{-1/2} \\ 0 & y^{-1/2} \end{pmatrix} \mapsto y^{-\rho} = e^{-\rho \log y}$$

where ρ is *any complex number*. It is easily verified that

$$y^{-1} = r \sin^2 \theta + r^{-1} \cos^2 \theta = v + \sqrt{v^2 - 1} \cos 2\theta,$$

and hence the spherical function on G obtained from the homomorphism (22.6.11.6) by means of the formula (22.6.11.4) may be identified with the continuous function on the interval $[1, +\infty[$:

(22.6.11.7) $$\mathrm{P}_\rho(v) = \frac{1}{2\pi} \int_0^{2\pi} (v + \sqrt{v^2 - 1} \cos \varphi)^\rho \, d\varphi$$

which is the *Legendre function of* (*complex*) *index* ρ. The functional equation (22.6.7.1) here takes the form

(22.6.11.8) $$\frac{1}{2\pi} \int_0^{2\pi} \mathrm{P}_\rho(\mathrm{ch}\, t \,\mathrm{ch}\, u + \mathrm{sh}\, t \,\mathrm{sh}\, u \cos \varphi) \, d\varphi = \mathrm{P}_\rho(\mathrm{ch}\, t)\mathrm{P}_\rho(\mathrm{ch}\, u)$$

for $t, u \in \mathbf{R}$.

(22.6.12) Suppose now that we are in the third case mentioned in (22.6.3), so that G contains a closed commutative normal subgroup A and a compact subgroup K such that the mapping $(t, s) \mapsto ts$ is a homeomorphism of $\mathrm{K} \times \mathrm{A}$ onto G. Let α be a continuous homomorphism of A into the group $\mathbf{C}^*$; we define a function $\omega \in \mathscr{C}(\mathrm{K}\backslash\mathrm{G}/\mathrm{K})$ by the formula

(22.6.12.1) $$\omega(x) = \int_{\mathrm{K}} \alpha(usu^{-1}) \, dm_{\mathrm{K}}(u)$$

where $x = ts$ with $t \in \mathrm{K}$ and $s \in \mathrm{A}$. We shall show that ω is a *spherical function* on G relative to K, again by verifying the functional equation (22.6.7.1). If $x = t_1 s_1$, $y = t_2 s_2$, where $t_1, t_2 \in \mathrm{K}$ and $s_1, s_2 \in \mathrm{A}$, then for $v \in \mathrm{K}$ we have $xvy = (t_1 v t_2)(((vt_2)^{-1} s_1 v t_2) s_2)$, and therefore the left-hand side of the functional equation is the integral

$$\iint_{\mathrm{K} \times \mathrm{K}} \alpha(((vt_2 u^{-1})^{-1} s_1 (vt_2 u^{-1}))(us_2 u^{-1})) \, dm_{\mathrm{K}}(u) \, dm_{\mathrm{K}}(v)$$

$$= \iint_{\mathrm{K} \times \mathrm{K}} \alpha((vt_2 u^{-1})^{-1} s_1 (vt_2 u^{-1}))\alpha(us_2 u^{-1}) \, dm_{\mathrm{K}}(u) \, dm_{\mathrm{K}}(v)$$

and by using the fact that K is unimodular, and the Lebesgue–Fubini theorem, it is clear that this integral is equal to $\omega(x)\omega(y)$. Again it can be shown that *all* the spherical functions on G relative to K are obtained in this way (Section 22.10, Problem 15).

Take for example G to be the group of Euclidean displacements of the plane with determinant $+1$, consisting of the matrices

$$\begin{pmatrix} \cos\theta & \sin\theta & 0 \\ -\sin\theta & \cos\theta & 0 \\ x & y & 1 \end{pmatrix}$$

where θ, x, and y are real. A double coset of K consists of the matrices for which $x^2 + y^2$ has a constant value r^2, so that the functions in $\mathscr{C}(\mathrm{K}\backslash\mathrm{G}/\mathrm{K})$ are those of the form $\psi((x^2 + y^2)^{1/2})$, where ψ is a continuous function on the interval $[0, \infty[$. The group A consists of the matrices for which $\theta = 0$, and may be identified with $\mathbf{R}^2$; hence every continuous homomorphism α of A into $\mathbf{C}^*$ is of the form $(x, y) \mapsto \exp(\lambda x + \mu y)$ where λ and μ are arbitrary *complex* numbers **(16.28.8)**. It now follows easily that the spherical functions on G relative to K, given by **(22.6.12.1)**, may be identified with the continuous functions on the interval $[0, \infty[$ defined by the formula

(22.6.12.2)
$$\psi(r) = \frac{1}{2\pi}\int_0^{2\pi} \exp(r(\lambda\cos\varphi + \mu\sin\varphi))\,d\varphi.$$

In particular, if $\lambda = 0$ and $\mu = i$, the function ψ is the *Bessel function* J_0.†

7. PLANCHEREL AND FOURIER TRANSFORMS

(22.7.1) Let (G, K) be a Gelfand pair **(22.6.2)**. We have seen **(22.6.8)** that the space $\mathbf{S}(\mathrm{G}/\mathrm{K})$ of spherical functions on G relative to K, considered as a subspace of $\mathscr{C}_{\mathbf{C}}(\mathrm{G})$, may be identified with the subspace $\mathbf{X}_0(\mathrm{A})$ of the space $\mathbf{X}(\mathrm{A})$ of characters of the Banach algebra $\mathrm{A} = \mathbf{C}\varepsilon_e + \mathrm{L}^1_{\mathbf{C}}(\mathrm{K}\backslash\mathrm{G}/\mathrm{K})$, with a complement consisting of at most one point. The restriction to $\mathbf{X}_0(\mathrm{A})$ of the *Gelfand transform* **(15.3.3)** of an element $\tilde{f}$ of $\mathrm{L}^1_{\mathbf{C}}(\mathrm{K}\backslash\mathrm{G}/\mathrm{K})$ may therefore be identified with the complex-valued function on $\mathbf{S}(\mathrm{G}/\mathrm{K})$ defined by

(22.7.1.1)
$$\mathscr{F}f : \omega \mapsto \langle f, \omega\rangle = \int_{\mathrm{G}} f(x)\omega(x)\,dm_{\mathrm{G}}(x).$$

† See my book *Infinitesimal Calculus*, Boston (Houghton–Mifflin), 1971, Chapter XV.

This function is called the *Fourier cotransform* of $f \in \mathscr{L}_{\mathrm{C}}^1(\mathrm{K}\backslash\mathrm{G}/\mathrm{K})$ (it depends of course only on the class $\tilde{f}$ of f). The *Fourier transform* of the function f is defined to be the Fourier cotransform of $\check{f}$, that is to say the function

(22.7.1.2) $$\mathscr{F}f : \omega \mapsto \langle \check{f}, \omega \rangle = \langle f, \check{\omega} \rangle = \int_{\mathrm{G}} f(x)\omega(x^{-1})\, dm_{\mathrm{G}}(x)$$

(remember that G is assumed to be unimodular).

The definitions **(22.7.1.1)** and **(22.7.1.2)** apply equally well to *any* function $f \in \mathscr{L}_{\mathrm{C}}^1(\mathrm{G})$; moreover, the same proof as for **(22.6.5.6)** shows that we have

(22.7.1.3) $$\overline{\overline{\mathscr{F}}}f = \overline{\overline{\mathscr{F}}}(f^{\natural}), \qquad \mathscr{F}f = \mathscr{F}(f^{\natural}).$$

It follows from the general properties of the Gelfand transformation **(15.3.4)** that $\mathscr{F}f$ and $\overline{\overline{\mathscr{F}}}f$ are *continuous functions on the locally compact space* $\mathbf{S}(\mathrm{G}/\mathrm{K})$ *and tend to* 0 *at infinity* **(13.20.6)**. Furthermore, for all functions $f, g \in \mathscr{L}_{\mathrm{C}}^1(\mathrm{K}\backslash\mathrm{G}/\mathrm{K})$, we have

(22.7.1.4) $$\mathscr{F}(f * g) = (\mathscr{F}f)(\mathscr{F}g), \qquad \overline{\overline{\mathscr{F}}}(f * g) = (\overline{\overline{\mathscr{F}}}f)(\overline{\overline{\mathscr{F}}}g);$$

and for all $f \in \mathscr{L}_{\mathrm{C}}^1(\mathrm{G})$ we have

(22.7.1.5) $$\|\mathscr{F}f\| \leqq \mathrm{N}_1(f), \qquad \|\overline{\overline{\mathscr{F}}}f\| \leqq \mathrm{N}_1(f)$$

so that $\mathscr{F}$ and $\overline{\overline{\mathscr{F}}}$ are *continuous linear mappings of the Banach space* $\mathrm{L}_{\mathrm{C}}^1(\mathrm{G})$ *into the Banach space* $\mathscr{C}_{\mathrm{C}}^0(\mathbf{S})$ **(13.20.6)**.

It should be noted that, even when f belongs to $\mathscr{K}(\mathrm{G})$, in general the functions $\mathscr{F}f$ and $\overline{\overline{\mathscr{F}}}f$ are *not* compactly supported.

Finally, for each function $f \in \mathscr{L}_{\mathrm{C}}^1(\mathrm{G}/\mathrm{K})$ and each $s \in \mathrm{G}$, we have

(22.7.1.6) $$\overline{\overline{\mathscr{F}}}(\gamma(s)f) = \omega(s)\overline{\overline{\mathscr{F}}}f, \qquad \mathscr{F}(\gamma(s)f) = \omega(s^{-1})\mathscr{F}f.$$

Likewise, for all $f \in \mathscr{L}_{\mathrm{C}}^1(\mathrm{K}\backslash\mathrm{G})$ and all $s \in \mathrm{G}$, we have

(22.7.1.7) $$\overline{\overline{\mathscr{F}}}(\delta(s)f) = \omega(s^{-1})\overline{\overline{\mathscr{F}}}f, \qquad \mathscr{F}(\delta(s)f) = \omega(s)\mathscr{F}f.$$

We shall prove the first of the formulas (22.7.1.6); the others are proved similarly. We have

$$\begin{aligned}\bar{\mathscr{F}}(\gamma(s)f)(\omega) &= \int_G f(s^{-1}x)\omega(x)\,dm_G(x)\\ &= \int_G f(x)\omega(sx)\,dm_G(x)\\ &= \int_G f(x)\omega(stx)\,dm_G(x)\end{aligned}$$

for all $t \in K$, because $f(tx) = f(x)$ for almost all $x \in G$. It follows that

$$\begin{aligned}\bar{\mathscr{F}}(\gamma(s)f)(\omega) &= \iint_{G\times K} f(x)\omega(stx)\,dm_G(x)\,dm_K(t)\\ &= \int_G f(x)\,dm_G(x)\int_K \omega(stx)\,dm_K(t)\\ &= \omega(s)\int_G f(x)\omega(x)\,dm_G(x)\\ &= \omega(s)\bar{\mathscr{F}}f(\omega)\end{aligned}$$

by virtue of the functional equation (22.6.7.1).

(22.7.2) We shall denote by **Z** or **Z**(G/K) the subset of **S**(G/K) consisting of the *spherical functions* on G (relative to K) which are of *positive type* (22.1.3). It follows immediately from the characterization (22.1.3.1) of continuous functions of positive type that **Z** is *closed* in **S**, hence is *locally compact.* We shall see that the spherical functions belonging to **Z** feature in the study of *measures of positive type* (22.2.1) on G.

Recall (22.2.3) that if μ is a *measure of positive type* on G, there is canonically associated with μ a *continuous unitary representation* U_μ of G on a separable Hilbert space E_μ. By *extension* (21.1.7) we obtain a *nondegenerate representation* $\tilde{f} \mapsto U_\mu(f)$ of the involutory algebra $L^1_{\mathbf{C}}(G)$ on E_μ. For any two functions $g, h \in \mathscr{K}(G)$ we have, by (22.2.3.1),

(22.7.2.1) $$(U_\mu(f)\cdot\pi(g)\,|\,\pi(h))_\mu = \iiint f(s)g(s^{-1}u^{-1}t)\overline{h(u^{-1})}\,dm_G(s)\,dm_G(u)\,d\mu(t)$$

or, more simply, when $f \in \mathscr{K}(G)$,

(22.7.2.2) $$U_\mu(f) \cdot \pi(g) = \pi(f * g).$$

The positive Hilbert form $(g, h) \mapsto \mu(\check{\bar{h}} * g)$ on $\mathscr{K}(G)$ therefore satisfies the conditions (U) and (N) of **(15.6)**. The restriction of this form to the *commutative* involutory subalgebra $\mathscr{K}(K\backslash G/K)$ is therefore a *bitrace*, which evidently satisfies condition (U) of **(15.6)**. Moreover, it satisfies condition (N): for if $g \in \mathscr{K}(K\backslash G/K)$, there exists a sequence (h_n) of functions in $\mathscr{K}(G)$, with supports contained in a fixed compact set, such that the functions $h_n * g$ have their supports contained in a fixed compact set and converge uniformly to g **(14.11.2)**; it follows immediately that the functions $(h_n * g)^\natural = h_n^\natural * g$ **(22.6.1.4)** have the same property, and hence **(13.8.4)** the sequence of elements $\pi(h_n^\natural * g) \in \pi(\mathscr{K}(K\backslash G/K))$ converges to $\pi(g)$ in E_μ, which proves our assertion. Let H_μ denote the closure in E_μ of the subspace $\pi(\mathscr{K}(K\backslash G/K))$, then the mapping $f \mapsto U_\mu(f) | H_\mu$ is a *representation* of $\mathscr{K}(K\backslash G/K)$ on the separable Hilbert space H_μ, which we denote by V_μ. For $f, g \in \mathscr{K}(K\backslash G/K)$ we have, therefore, $V_\mu(f) \cdot \pi(g) = \pi(f * g)$, and by virtue of **(21.1.5.1)**

(22.7.2.3) $$\|V_\mu(f)\| \leqq N_1(f),$$

so that V_μ is a *continuous homomorphism* of the algebra $\mathscr{K}(K\backslash G/K)$, endowed with the topology induced by that of $L^1_{\mathbf{C}}(K\backslash G/K)$, into the algebra $\mathscr{L}(H_\mu)$ of continuous operators on H_μ. Since $L^1_{\mathbf{C}}(G)$ is a separable Banach algebra, it follows that the *closure* $\mathscr{A}_\mu$ of the image of $\mathscr{K}(K\backslash G/K)$ under V_μ is a *separable* commutative star subalgebra of $\mathscr{L}(H_\mu)$.

(22.7.3) We have now shown that the bitrace obtained by restricting the form $(g, h) \mapsto \mu(\check{\bar{h}} * g)$ to the *commutative* algebra $\mathscr{K}(K\backslash G/K)$ satisfies the conditions of the Plancherel–Godement theorem **(15.9.2)**. We shall show that that theorem leads in the present context to the following result:

(22.7.4) *Let* (G, K) *be a Gelfand pair and let* μ *be a measure of positive type on* G. *Then there exists, on the locally compact space* **Z**(G/K) *of spherical functions of positive type, one and only one positive measure* μ^Δ *such that, for all functions* $f \in \mathscr{K}(K\backslash G/K)$, *the Fourier cotransform* $\bar{\mathscr{F}}f$ *belongs to* $L^2_{\mathbf{C}}(\mu^\Delta)$, *and such that*

(22.7.4.1) $$\mu(\check{\bar{g}} * f) = \int_{\mathbf{Z}} \bar{\mathscr{F}}f(\omega)\overline{\bar{\mathscr{F}}g(\omega)}\, d\mu^\Delta(\omega)$$

for all $f, g \in \mathscr{K}(\mathrm{K}\backslash\mathrm{G}/\mathrm{K})$. *The mapping* $f \mapsto (\bar{\mathscr{F}}f)^{\sim}$ *of* $\mathscr{K}(\mathrm{K}\backslash\mathrm{G}/\mathrm{K})$ *into* $\mathrm{L}^2_{\mathbf{C}}(\mu^{\Delta})$ *factorizes as* $f \mapsto \pi(f) \overset{T_0}{\mapsto} (\bar{\mathscr{F}}f)^{\sim}$, *where* T_0 *extends to an isomorphism of the Hilbert space* H_μ *onto the Hilbert space* $\mathrm{L}^2_{\mathbf{C}}(\mu^{\Delta})$.

If we put $\mathscr{A}'_\mu = \mathbf{C} \cdot 1_{\mathrm{H}_\mu} + \mathscr{A}_\mu$, the proof of the Plancherel–Godement theorem sets up a correspondence which takes each character ξ' of the commutative star algebra $\mathscr{A}'_\mu$ to the Hermitian character $\zeta = \xi' \circ V_\mu$ of the algebra $\mathscr{K}(\mathrm{K}\backslash\mathrm{G}/\mathrm{K})$. For each function $f \in \mathscr{K}(\mathrm{K}\backslash\mathrm{G}/\mathrm{K})$ we have, by virtue of **(22.7.2.3)** and **(15.3.1)**,

$$|\xi(f)| = |\xi'(V_\mu(f))| \leqq \|V_\mu(f)\| \leqq \mathrm{N}_1(f),$$

so that the linear form ζ on $\mathscr{K}(\mathrm{K}\backslash\mathrm{G}/\mathrm{K})$ is continuous with respect to the topology induced by that of $\mathrm{L}^1_{\mathbf{C}}(\mathrm{G})$, and hence (**(5.5.4)** and **(3.15.2)**) may be extended by continuity to a *character* of $\mathrm{L}^1_{\mathbf{C}}(\mathrm{K}\backslash\mathrm{G}/\mathrm{K})$, the closure of $\mathscr{K}(\mathrm{K}\backslash\mathrm{G}/\mathrm{K})$ in $\mathrm{L}^1_{\mathbf{C}}(\mathrm{G})$. It follows therefore from **(22.6.5)** that ζ is of the form $f \mapsto \langle f, \omega \rangle$, where ω is a *spherical function* on G relative to K. We claim that ω is *of positive type*. For this we have to prove that for each function $f \in \mathscr{K}(\mathrm{G})$ we have $\langle \check{\bar{f}} * f, \omega \rangle \geqq 0$, or equivalently (by virtue of **(22.6.5.6)**) that

(22.7.4.2) $$\langle (\check{\bar{f}} * f)^{\natural}, \omega \rangle \geqq 0.$$

This relation may also be written as $\xi'(V_\mu((\check{\bar{f}} * f)^{\natural})) \geqq 0$, and it is therefore enough to establish that the operator $N = V_\mu((\check{\bar{f}} * f)^{\natural})$ on $\mathscr{L}(\mathrm{H}_\mu)$ is *positive self-adjoint*: for then the positive self-adjoint operator $N^{1/2}$ **(15.11.12)** will belong to the algebra $\mathscr{A}'_\mu$ by virtue of the Gelfand–Naimark theorem **(15.11.1)**; also $\xi'(N^{1/2})$ is *real* **(15.4.12)**, hence $\xi'(N) = (\xi'(N^{1/2}))^2 \geqq 0$.

By definition, to say that $V_\mu((\check{\bar{f}} * f)^{\natural})$ is positive self-adjoint means that for each function $g \in \mathscr{K}(\mathrm{K}\backslash\mathrm{G}/\mathrm{K})$ we have $(\pi((\check{\bar{f}} * f)^{\natural} * g) \mid \pi(g))_\mu \geqq 0$, or equivalently (by virtue of **(22.2.3.1)**) that

$$\langle \mu, \check{\bar{g}} * ((\check{\bar{f}} * f)^{\natural} * g) \rangle \geqq 0.$$

By virtue of **(22.6.1.5)**, we have

$$\check{\bar{g}} * (\check{\bar{f}} * f)^{\natural} * g = \check{\bar{g}} * ((\check{\bar{f}} * f) * g)^{\natural} = (\check{\bar{g}} * \check{\bar{f}} * f * g)^{\natural} = (\check{\bar{h}} * h)^{\natural}$$

where $h = f * g \in \mathscr{K}(\mathrm{G})$. Our assertion is therefore a consequence of the following lemma:

(22.7.4.3) *If* μ *is a measure of positive type on* G, *we have* $\langle \mu, (\check{\bar{h}} * h)^{\natural} \rangle \geqq 0$ *for all* $h \in \mathscr{K}(\mathrm{G})$.

We have

$$\langle \mu, (\check{\tilde{h}} * h)^{\natural} \rangle = \iiiint_{G\times G\times K\times K} h(s^{-1}txt')\overline{h(s^{-1})}\, dm_G(s)\, d\mu(x)\, dm_K(t)\, dm_K(t')$$

$$= \iiiint h(s^{-1}xt')\overline{h(s^{-1}t^{-1})}\, dm_G(s)\, d\mu(x)\, dm_K(t)\, dm_K(t')$$

by the invariance of Haar measure m_G. If we put

$$\psi(s) = \int_K h(st)\, dm_K(t),$$

then $\psi \in \mathscr{K}(G)$ (14.1.5.5), and since K is unimodular we have

$$\langle \mu, (\check{\tilde{h}} * h)^{\natural} \rangle = \iint_{G\times G} \psi(s^{-1}x)\overline{\psi(s^{-1})}\, dm_G(s)\, d\mu(x)$$

$$= \langle \mu, \check{\tilde{\psi}} * \psi \rangle \geqq 0$$

by virtue of the hypotheses.

It follows therefore that we may canonically identify the spectrum of $\mathscr{A}'_\mu$ with a compact metrizable space containing a closed subspace S_μ of **Z**(G/K), and such that the complement of S_μ contains at most one point. Once this identification has been made, the assertions of **(22.7.4)** are simply those of the Plancherel–Godement theorem; observe that the notation $\hat{f}$ in **(15.9.2)** is here replaced by $\mathscr{F}f$. The set S_μ is the *support* of the measure μ^Δ.

The functions in **Z**(G/K) satisfy $\check{\omega} = \omega$ **(22.1.3.5)**; hence for all $\omega \in$ **Z** and all $f \in \mathscr{L}^1_{\mathbf{C}}(G)$ we have

(22.7.4.4) $$\mathscr{F}\bar{f}(\omega) = \overline{\mathscr{F}f}(\omega),$$

because $\mathscr{F}\bar{f}(\omega) = \int \bar{f}(x)\overline{\omega(x^{-1})}\, dm_G(x) = \overline{\int f(x)\omega(x^{-1})\, dm_G(x)}$.

The positive measure μ^Δ on **Z**(G/K) is called the *Plancherel transform* of the measure of positive type μ on G.

For each (complex) measure μ on G, we may define a measure $\mu^\natural$ on G by the condition

(22.7.4.5) $$\langle \mu^\natural, f \rangle = \langle \mu, f^\natural \rangle$$

for all functions $f \in \mathscr{K}(G)$: for it is clear that $\mu^\natural$ is a positive linear form on $\mathscr{K}(G)$ if μ is positive, and the general case then follows by writing μ as a linear combination of four positive measures. For each $t \in K$ we have then, by (14.1.2)

$$\gamma(t)\mu^\natural = \delta(t)\mu^\natural = \mu^\natural,$$

and conversely the relations $\gamma(t)\mu = \delta(t)\mu = \mu$ for all $t \in K$ imply that $\mu = \mu^\natural$. Hence $\mu \mapsto \mu^\natural$ is a *projection* of the vector space $M_C(G)$ onto the subspace of measures which are invariant under left and right translations by all elements of K. We denote this subspace by $M_C(K\backslash G/K)$.

If $\mu \in M_C(K\backslash G/K)$, for each function $g \in \mathscr{C}_C(G)$ we have

(22.7.4.6) $$(g \cdot \mu)^\natural = g^\natural \cdot \mu$$

for if $f \in \mathscr{K}(G)$ we calculate

$$\begin{aligned}\langle (g \cdot \mu)^\natural, f\rangle &= \langle g \cdot \mu, f^\natural\rangle = \langle \mu, gf^\natural\rangle = \langle \mu^\natural, gf^\natural\rangle \\ &= \langle \mu, (gf^\natural)^\natural\rangle = \langle \mu, g^\natural f^\natural\rangle = \langle \mu, (g^\natural f)^\natural\rangle \\ &= \langle \mu^\natural, g^\natural f\rangle = \langle \mu, g^\natural f\rangle = \langle g^\natural \cdot \mu, f\rangle\end{aligned}$$

by (22.6.1.4).

The same argument shows that if $g \in \mathscr{L}_C^p(G)$ ($p = 1$ or 2), then

(22.7.4.7) $$(g \cdot m_G)^\natural = g^\natural \cdot m_G$$

in view of the definition of $g^\natural$ (22.6.4).

The lemma (22.7.4.3) shows that *if μ is of positive type, then so also is $\mu^\natural$.* The relation $\mu^\natural = 0$ means that *μ vanishes on the subspace $\mathscr{K}(K\backslash G/K)$ of $\mathscr{K}(G)$.* From the uniqueness of the measure in the Plancherel–Godement theorem, it follows that

(22.7.4.8) *For every measure μ of positive type on G, we have $(\mu^\natural)^\Delta = \mu^\Delta$. If μ, ν are measures of positive type, we have $\mu^\Delta = \nu^\Delta$ if and only if $\mu^\natural = \nu^\natural$.*

Examples of Plancherel transforms

(22.7.5) We have already remarked (22.2.2) that the Dirac measure ε_e at the identity element of G is of positive type. Its Plancherel transform $m_{\mathbf{Z}} = \varepsilon_e^\Delta$ is called the *canonical measure* on $\mathbf{Z}$. Since

$$\varepsilon_e(\check{g} * f) = (\check{g} * f)(e) = \int_G f(s^{-1})\overline{g(s^{-1})}\, dm_G(s) = \int_G f(s)\overline{g(s)}\, dm_G(s)$$

because G is unimodular, it follows that for $f, g \in \mathscr{K}(\mathrm{K}\backslash\mathrm{G}/\mathrm{K})$ we have

(22.7.5.1)
$$\int_G f(s)\overline{g(s)}\, dm_G(s) = \int_{\mathbf{Z}} (\mathscr{F}f(\omega))\overline{(\mathscr{F}g(\omega))}\, dm_{\mathbf{Z}}(\omega) = \int_{\mathbf{Z}} (\overline{\mathscr{F}}f(\omega))\overline{(\overline{\mathscr{F}}g(\omega))}\, dm_{\mathbf{Z}}(\omega),$$

and the linear mapping $f \mapsto \mathscr{F}f$ (resp. $f \mapsto \overline{\mathscr{F}}f$) therefore extends to an *isomorphism* (6.2) *of the Hilbert space* $\mathrm{L}^2_{\mathbf{C}}(\mathrm{K}\backslash\mathrm{G}/\mathrm{K})$ *onto the Hilbert space* $\mathrm{L}^2_{\mathbf{C}}(\mathbf{Z}(\mathrm{G}/\mathrm{K}), m_{\mathbf{Z}})$, by virtue of (22.7.4). We may extend $\mathscr{F}$ (resp. $\overline{\mathscr{F}}$) to a linear mapping of all $\mathrm{L}^2_{\mathbf{C}}(\mathrm{G})$ onto $\mathrm{L}^2_{\mathbf{C}}(m_{\mathbf{Z}})$ by defining $\mathscr{F}(\tilde{f}) = \mathscr{F}(\tilde{f}^\natural)$ (resp. $\overline{\mathscr{F}}(\tilde{f}) = \overline{\mathscr{F}}(\tilde{f}^\natural)$); by virtue of (22.7.4.4), we have $\overline{\mathscr{F}}(\tilde{f}) = \overline{\mathscr{F}(\overline{\tilde{f}})}$. By abuse of language, if f is a function in $\mathscr{L}^2_{\mathbf{C}}(\mathrm{G})$, we shall denote by $\mathscr{F}f$ (resp. $\overline{\mathscr{F}}f$) a function in the class $\mathscr{F}\tilde{f}$ (resp. $\overline{\mathscr{F}}\tilde{f}$) (in general this class contains no distinguished element). With this notation, the relations (22.7.5.1) hold for *all f, g in* $\mathscr{L}^2_{\mathbf{C}}(\mathrm{K}\backslash\mathrm{G}/\mathrm{K})$.

(22.7.5.2) The formulas (22.7.5.1) are also valid for $f \in \mathscr{L}^1_{\mathbf{C}}(\mathrm{K}\backslash\mathrm{G}/\mathrm{K})$ and $g \in \mathscr{K}(\mathrm{K}\backslash\mathrm{G}/\mathrm{K})$. For is a sequence of functions $f_n \in \mathscr{K}(\mathrm{K}\backslash\mathrm{G}/\mathrm{K})$ converges to $f \in \mathscr{L}^1_{\mathbf{C}}(\mathrm{G})$, then the sequence $(\mathscr{F}f_n)$ converges *uniformly* to $\mathscr{F}f$ in $\mathbf{Z}$, by virtue of (22.7.1.5); the result now follows by applying (13.10.3) and (13.12.5), since g and $\mathscr{F}g$ are bounded.

(22.7.6) For every spherical function of positive type $\omega \in \mathbf{Z}(\mathrm{G}/\mathrm{K})$, the measure $\omega \cdot m_G$ with density ω is a measure of positive type (22.2.1), and its Plancherel transform is given by

(22.7.6.1)
$$(\omega \cdot m_G)^\Delta = \varepsilon_\omega .$$

For if $f, g \in \mathscr{K}(\mathrm{K}\backslash\mathrm{G}/\mathrm{K})$ we have

$$\langle \check{g} * f, \omega \cdot m_G \rangle = \overline{\mathscr{F}}(\check{g} * f)(\omega) = \overline{\mathscr{F}}f(\omega)\overline{\overline{\mathscr{F}}g(\omega)} = \langle (\overline{\mathscr{F}}f)\overline{(\overline{\mathscr{F}}g)}, \varepsilon_\omega \rangle$$

and the result follows from (22.7.4).

(22.7.7) Let μ be a measure of positive type on G which is invariant under left and right translations by the elements of K (i.e., $\mu^\natural = \mu$). For each function $f \in \mathscr{K}(\mathrm{K}\backslash\mathrm{G}/\mathrm{K})$, the measure $\mu * \check{\bar{f}} * f$ is of positive type and belongs to $\mathrm{M}_{\mathbf{C}}(\mathrm{K}\backslash\mathrm{G}/\mathrm{K})$, and we have

$$(22.7.7.1) \qquad (\mu * \check{\bar{f}} * f)^\Delta = |\bar{\mathscr{F}} f|^2 \cdot \mu^\Delta.$$

For if $g \in \mathscr{K}(\mathrm{G})$, we have by **(14.10.9)**

$$\begin{aligned}\langle \mu * \check{\bar{f}} * f, \check{\bar{g}} * g\rangle &= \langle \check{\mu}, \check{\bar{f}} * f * \check{g} * \bar{g}\rangle \\ &= \langle \mu, \check{\bar{g}} * g * \check{f} * \bar{f}\rangle \\ &= \langle \mu, (\check{\bar{g}} * g * \check{f} * \bar{f})^\natural\rangle.\end{aligned}$$

But since $f \in \mathscr{K}(\mathrm{K}\backslash\mathrm{G}/\mathrm{K})$, we have

$$(\check{\bar{g}} * g * \check{f} * \bar{f})^\natural = (\check{\bar{g}} * g)^\natural * \check{f} * \bar{f} = \check{f} * (\check{\bar{g}} * g)^\natural * \bar{f} = (\check{f} * \check{\bar{g}} * g * \bar{f})^\natural$$

by virtue of **(22.6.1.4)** and the commutativity of $\mathscr{K}(\mathrm{K}\backslash\mathrm{G}/\mathrm{K})$. Putting $h = g * \bar{f}$, we see that

$$\langle \mu * \check{\bar{f}} * f, \check{\bar{g}} * g\rangle = \langle \mu, (\check{\bar{h}} * h)^\natural\rangle$$

which is $\geqq 0$ by **(22.7.4.3)**. If now $g, h \in \mathscr{K}(\mathrm{K}\backslash\mathrm{G}/\mathrm{K})$ we have likewise

$$\begin{aligned}\langle \mu * \check{\bar{f}} * f, \check{\bar{g}} * h\rangle &= \langle \mu, \check{\bar{g}} * h * \check{f} * \bar{f}\rangle \\ &= \int_{\mathbf{Z}} (\bar{\mathscr{F}} h)\overline{(\bar{\mathscr{F}} g)}(\bar{\mathscr{F}} f)\overline{(\bar{\mathscr{F}} f)}\, d\mu^\Delta\end{aligned}$$

by **(22.7.4)**, which suffices to prove **(22.7.7.1)**.

(22.7.8) *Let μ be a measure of positive type on* G, *and μ^Δ its Plancherel transform. For each function $f \in \mathscr{K}(\mathrm{K}\backslash\mathrm{G}/\mathrm{K})$ such that its Fourier cotransform $\bar{\mathscr{F}} f$ belongs to $\mathrm{L}^1_{\mathbf{C}}(\mu^\Delta)$, we have*

$$(22.7.8.1) \qquad \mu(f) = \int_{\mathbf{Z}} \bar{\mathscr{F}} f(\omega)\, d\mu^\Delta(\omega).$$

Since $\mu^\natural(f) = \mu(f)$ and $(\mu^\natural)^\Delta = \mu^\Delta$, we may assume without loss of generality that $\mu = \mu^\natural$. There exists a sequence (g_n) of functions $\geqq 0$ in $\mathscr{K}(\mathrm{G})$, with supports contained in a fixed compact set, such that $\int g_n\, dm_{\mathrm{G}} = 1$ for each n and such that the sequence $(f * g_n)$ converges uniformly to f in G **(14.11.2)**; since $(f * g_n)^\natural = f * g_n^\natural$ **(22.6.4.1)** and since the sequence $(g_n^\natural)$ has the same properties as (g_n), we may assume that $g_n \in \mathscr{K}(\mathrm{K}\backslash\mathrm{G}/\mathrm{K})$. By the definition of a measure, $\mu(f * g_n) \to \mu(f)$ as $n \to \infty$. By **(22.7.4)** we have

$$\mu(f * g_n) = \int_{\mathbf{Z}} \bar{\mathscr{F}}f(\omega)\bar{\mathscr{F}}g_n(\omega)\, d\mu^{\Delta}(\omega).$$

Also the sequence $(f * g_n)$ converges to f in $\mathscr{L}^1_{\mathbf{C}}(\mathrm{G})$ **(14.11.1)**, hence by **(22.7.1.5)** the sequence of functions $\bar{\mathscr{F}}(f * g_n)$ converges uniformly to $\bar{\mathscr{F}}f$ in $\mathbf{Z}$. Furthermore **(22.7.1.5)**, we have $|\bar{\mathscr{F}}g_n(\omega)| \leqq \mathrm{N}_1(g_n) = 1$ for all $\omega \in \mathbf{Z}$, and therefore

$$|\bar{\mathscr{F}}(f * g_n)(\omega)| = |\bar{\mathscr{F}}f(\omega)\bar{\mathscr{F}}g_n(\omega)| \leqq |\bar{\mathscr{F}}f(\omega)|$$

for all $\omega \in \mathbf{Z}$. Since by hypothesis $\bar{\mathscr{F}}f \in \mathscr{L}^1_{\mathbf{C}}(\mu^{\Delta})$, we may apply the dominated convergence theorem **(13.8.4)** to conclude that $\mu(f * g_n)$ tends to $\int_{\mathbf{Z}} \bar{\mathscr{F}}f(\omega)\, d\mu^{\Delta}(\omega)$, thereby completing the proof.

(22.7.9) It should be noted that for a function $f \in \mathscr{K}(\mathrm{K}\backslash\mathrm{G}/\mathrm{K})$, its Fourier cotransform $\bar{\mathscr{F}}f$ always belongs to $\mathscr{L}^2_{\mathbf{C}}(\mu^{\Delta})$ **(22.7.4)**, but does not necessarily belong to $\mathscr{L}^1_{\mathbf{C}}(\mu^{\Delta})$ (Section **22.19**, Problem 8). The proof of **(22.7.8)** shows, more generally, that for $f \in \mathscr{K}(\mathrm{K}\backslash\mathrm{G}/\mathrm{K})$, without requiring $\bar{\mathscr{F}}f$ to belong to $\mathscr{L}^1_{\mathbf{C}}(\mu^{\Delta})$, we may write

(22.7.9.1)
$$\mu(f) = \lim_{n\to\infty} \int_{\mathbf{Z}} \bar{\mathscr{F}}f(\omega)\bar{\mathscr{F}}g_n(\omega)\, d\mu^{\Delta}(\omega)$$

for any sequence (g_n) of functions in $\mathscr{K}(\mathrm{K}\backslash\mathrm{G}/\mathrm{K})$ which satisfy the conditions of **(14.11.2)** ("summation method" for $\mu(f)$).

For example, if $\mu = \varepsilon_e$, we have $\varepsilon_e(\gamma(s)f) = f(s^{-1})$ for $s \in \mathrm{G}$ and $f \in \mathscr{K}(\mathrm{K}\backslash\mathrm{G}/\mathrm{K})$, and hence by virtue of **(22.7.1.6)**

(22.7.9.2)
$$f(s) = \lim_{n\to\infty} \int_{\mathbf{Z}} \omega(s^{-1})\bar{\mathscr{F}}f(\omega)\bar{\mathscr{F}}g_n(\omega)\, dm_{\mathbf{Z}}(\omega).$$

(22.7.10) *Let* $\mathscr{P}_+(\mathrm{K}\backslash\mathrm{G}/\mathrm{K})$ *denote the set of functions in* $\mathscr{C}(\mathrm{K}\backslash\mathrm{G}/\mathrm{K})$ *which are of positive type. Then the mapping* $p \mapsto (p \cdot m_{\mathrm{G}})^{\Delta}$ *is a bijection of* $\mathscr{P}_+(\mathrm{K}\backslash\mathrm{G}/\mathrm{K})$ *onto the set* $\mathrm{M}^1_+(\mathbf{Z})$ *of bounded positive measures on* $\mathbf{Z}$, *and the inverse mapping* $\bar{\mathscr{F}}'$ *is defined by the formula*

(22.7.10.1)
$$\bar{\mathscr{F}}'\mu'(x) = \int_{\mathbf{Z}} \omega(x)\, d\mu'(\omega).$$

We have then, for each function $f \in \mathscr{L}^1_{\mathbf{C}}(G)$,

(22.7.10.2)
$$\int_G f(x)\overline{\mathscr{F}}'\mu'(x)\,dm_G(x) = \int_{\mathbf{Z}} \overline{\mathscr{F}}f(\omega)\,d\mu'(\omega).$$

To show that $(p \cdot m_G)^{\wedge}$ is bounded we shall apply the criterion of the Bochner–Godement theorem **(15.9.4)** by showing that

(22.7.10.3)
$$|\langle p, f\rangle|^2 \leqq p(e)\langle p, \check{\tilde{f}} * f\rangle$$

for all $f \in \mathscr{K}(G)$. If (g_n) is a sequence satisfying the conditions of **(14.11.2)**, the inequality **(15.6.2.2)** gives

(22.7.10.4)
$$|\langle p, \check{\tilde{g}}_n * f\rangle|^2 \leqq \langle p, \check{\tilde{g}}_n * g_n\rangle\langle p, \check{\tilde{f}} * f\rangle$$

and we have

$$\begin{aligned}\langle p, \check{\tilde{g}}_n * g_n\rangle &= \iint p(y^{-1}z)\overline{g_n(y)}g_n(z)\,dm_G(y)\,dm_G(z)\\ &\leqq p(e)\end{aligned}$$

because $|p(x)| \leqq p(e)$ **(22.1.3.5)** for all $x \in G$, and $\int g_n(x)\,dm_G(x) = 1$. Substituting this inequality in **(22.7.10.4)** and passing to the limit, we obtain **(22.7.10.3)**.

Since the function $\omega \mapsto \omega(x)$ is continuous and bounded on $\mathbf{Z}$ **(22.6.9(iii))**, the right-hand side of **(22.7.10.1)** makes sense; moreover, the function $\overline{\mathscr{F}}'\mu'$ so defined is *continuous* on G. To prove this, let $x_0 \in G$ and let ε be a positive real number. Since $|\omega(x)| \leqq 1$ for all $x \in G$ and $\omega \in \mathbf{Z}$, there exists a compact subset L of $\mathbf{Z}$ such that $\int_{\mathbf{Z}-L} |\omega(x)|\,d\mu'(\omega) \leqq \frac{1}{4}\varepsilon$ for all $x \in G$ **(13.9.14)**; also there exists a neighborhood V of x_0 such that

$$|\omega(x) - \omega(x_0)| \leqq \frac{\varepsilon}{2\|\mu'\|}$$

for all $x \in V$ and all $\omega \in L$ **(22.6.9(ii))**; this implies that for all $x \in V$

$$\begin{aligned}|\overline{\mathscr{F}}'\mu'(x) - \overline{\mathscr{F}}'\mu'(x_0)| \leqq \int_{\mathbf{Z}-L} |\omega(x)|\,d\mu'(\omega) &+ \int_{\mathbf{Z}-L} |\omega(x_0)|\,d\mu'(\omega)\\ &+ \int_L |\omega(x) - \omega(x_0)|\,d\mu'(\omega) \leqq \varepsilon.\end{aligned}$$

The fact that $\bar{\mathscr{F}}'\mu'$ belongs to $\mathscr{C}(\mathrm{K}\backslash\mathrm{G}/\mathrm{K})$ follows directly from the invariance of ω under left and right translations by elements of K. Next, we have to verify that $\bar{\mathscr{F}}'\mu'$ is a function of positive type. The integral

$$\iint \bar{\mathscr{F}}'\mu'(y^{-1}z)f(y)\overline{f(z)}\,dm_{\mathrm{G}}(y)\,dm_{\mathrm{G}}(z)$$

can be rewritten in the form

$$\int_{\mathbf{Z}} d\mu'(\omega) \iint \omega(y^{-1}z)f(y)\overline{f(z)}\,dm_{\mathrm{G}}(y)\,dm_{\mathrm{G}}(z)$$

and is therefore $\geqq 0$ for all $f \in \mathscr{K}(\mathrm{G})$, because μ' is a positive measure and each $\omega \in \mathbf{Z}$ is of positive type. Again, for $f, g \in \mathscr{K}(\mathrm{K}\backslash\mathrm{G}/\mathrm{K})$ we have

$$\begin{aligned}\langle \check{\bar{g}} * f, \bar{\mathscr{F}}'\mu' \rangle &= \int_{\mathbf{Z}} \langle \check{\bar{g}} * f, \omega \rangle \, d\mu'(\omega) \\ &= \int_{\mathbf{Z}} \bar{\mathscr{F}}f(\omega)\overline{\bar{\mathscr{F}}g(\omega)}\,d\mu'(\omega)\end{aligned}$$

which shows, by virtue of the uniqueness property of **(22.7.4)**, that $\mu' = (\bar{\mathscr{F}}'\mu')^{\wedge}$.

Finally, for $f \in \mathscr{L}^1_{\mathbf{C}}(\mathrm{G})$, the two sides of **(22.7.10.2)** are equal by virtue of the Lebesgue–Fubini theorem, because the mapping $(x, \omega) \mapsto \omega(x)$ of $\mathrm{G} \times \mathrm{L}$ into $\mathbf{C}$ is continuous **(22.6.9**(iii)**)** and therefore universally measurable.

(22.7.11) *Remarks.* (i) If we replace the Haar measure m_{G} by am_{G}, where a is a constant > 0, the set of spherical functions on G relative to K is unchanged **(14.10)**. The Fourier transform and cotransform are multiplied by a, and the Plancherel transform is multiplied by $1/a$.

(ii) We have seen **(22.5.6)** that when G is compact, *all* the spherical functions are of positive type (cf. also Section **22.10**). This is not the case in general: in the example **(22.6.11)**, the spherical functions constructed from the "exponentials" **(22.6.11.6)** satisfy the relation $\check{\bar{\omega}} = \omega$ only when $\mathscr{R}(\rho) = -\frac{1}{2}$. It can be proved that this condition is sufficient for the corresponding spherical functions to be of positive type (Section **22.9**, Problem 1).

(iii) When G is *compact*, the (discrete) space $\mathbf{S}(\mathrm{G}/\mathrm{K})$ **(22.6.10)** may be identified with the subset of R(G) consisting of the classes $\boldsymbol{\rho}$ such that $(\boldsymbol{\rho} : \boldsymbol{\sigma}_0) = 1$, with the discrete topology. The Fourier transform of a function $f \in \mathscr{L}^1_{\mathbf{C}}(\mathrm{G})$ is the family $\boldsymbol{\rho} \mapsto c_{\boldsymbol{\rho}} = n_{\boldsymbol{\rho}}^{-1}(f \mid m_{11}^{(\boldsymbol{\rho})})$, where $\boldsymbol{\rho} \in \mathbf{S}(\mathrm{G}/\mathrm{K})$.

PROBLEMS

1. Let (G, K) be a Gelfand pair and let g be a function belonging to $\mathscr{C}(K\backslash G)$ (22.6.1) such that for each $f \in \mathscr{K}(K\backslash G/K)$ there exists a complex number λ_f such that $f * g = \lambda_f g$. If $g(s_0) \neq 0$ for some $s_0 \in G$, show that the function

$$\omega(s) = g(s_0)^{-1} \int g(sts_0)\, dm_K(t)$$

is a spherical function.

2. Let G be a (separable, metrizable) locally compact group, K a compact subgroup of G, and Γ a *discrete* subgroup of G such that the homogeneous space G/Γ is *compact*. If f is a complex-valued function defined in G and constant on each double coset $Ks\Gamma$ (12.10.2), and if $p: G \to K\backslash G$ and $\pi: G \to G/\Gamma$ are the canonical mappings, then we may write $f(s) = g(p(s))$ and $f(s) = h(\pi(s))$, where g is a function defined on $K\backslash G$ such that $g(x \cdot \xi) = g(x)$ for all $x \in K\backslash G$ and all $\xi \in \Gamma$, and h is a function defined on G/Γ such that $h(t \cdot y) = h(y)$ for all $y \in G/\Gamma$ and all $t \in K$. If we consider Γ as acting on the right on $X = K\backslash G$, the function g is an *automorphic form* for the factor of automorphy equal to the constant 1 (Section 22.3, Problem 13), and g is also said to be an *automorphic function* relative to the action of Γ.

Suppose that G is unimodular and that (G, K) is a *Gelfand pair*, and retain the notation of Section 22.3, Problem 9. Denote by $L^2_{\mathbf{C}}(K\backslash G/\Gamma)$ the closed subspace of $L^2_{\mathbf{C}}(m_{G/\Gamma})$ consisting of the classes of functions $f \in \mathscr{L}^2_{\mathbf{C}}(m_{G/\Gamma})$ such that $f(t \cdot \dot{x}) = f(\dot{x})$ for all $\dot{x} \in G/\Gamma$ and all $t \in K$.

(a) Show that the space $L^2_{\mathbf{C}}(K\backslash G/\Gamma)$ is the Hilbert sum of a sequence (H_n) of finite-dimensional closed subspaces consisting of classes of *continuous* functions on G/Γ, and that corresponding to each of these subspaces there exists a character ζ_n of the commutative Banach algebra $L^1_{\mathbf{C}}(K\backslash G/K)$ such that, for each continuous function $g \in H_n$ and each $f \in \mathscr{L}^1_{\mathbf{C}}(K\backslash G/K)$, we have $V(f) \cdot g = f * g = \zeta_n(f)g$. (Use the fact that the $V(f)$, for $f \in \mathscr{L}^1_{\mathbf{C}}(K\backslash G/K)$, are compact normal operators on $L^2_{\mathbf{C}}(m_{G/\Gamma})$ which commute with each other, and also the fact that for $f \in \mathscr{K}(K\backslash G/K)$ the function $V(f) \cdot g$ is continuous, by virtue of the formula (*) of Section 22.3, Problem 9.)

(b) Let H'_n be the set of functions $g \circ \pi$, where $g \in H_n$; these are therefore bounded continuous functions f on G such that $f(ts\xi) = f(s)$ for $s \in G$, $t \in K$ and $\xi \in \Gamma$. Show that for each such function $g \circ \pi \neq 0$, the function ω_n on G defined by

$$\omega_n(s) = (N_2(g))^{-2} \int_{G/\Gamma} g(s \cdot \dot{x})\overline{g(\dot{x})}\, dm_{G/\Gamma}(\dot{x})$$

is a spherical function of positive type such that the character ζ_n is the mapping $f \mapsto \langle f, \omega_n \rangle$; furthermore, we have

$$\omega_n(s) = (g(\pi(s_0)))^{-1} \int_K g(\pi(sts_0))\, dm_K(t)$$

if $g(\pi(s_0)) \neq 0$ (Problem 1). (Use (22.6.7(c)).) Show that the set of functions ω_n is closed and discrete in $\mathbf{Z}(G/K)$.

8. THE SPACES P(G) AND P′(Z)

(22.8.1) Given a unimodular locally compact group G (separable and metrizable) we shall denote by $P_+(G)$ the set of *measures of positive type* on G, and by P(G) the vector subspace of $M_C(G)$ *generated* by $P_+(G)$, so that the elements of P(G) are of the form $\mu_1 - \mu_2 + i(\mu_3 - \mu_4)$, where $\mu_j (1 \leqq j \leqq 4)$ are measures of positive type. It is clear that if $\mu \in P(G)$, then also $\mu^\natural \in P(G)$ (22.7.4.3), and the image of P(G) under the mapping $\mu \mapsto \mu^\natural$ is $P(G) \cap M_C(K\backslash G/K)$.

Suppose now that (G, K) is a *Gelfand pair*. With the same notation as in (22.7), the Plancherel transformation $\mu \mapsto \mu^\Delta$ is a mapping of $P_+(G)$ into the set $M_+(\mathbf{Z})$ of *positive measures* on $\mathbf{Z} = \mathbf{Z}(G/K)$, such that $(\mu + \nu)^\Delta = \mu^\Delta + \nu^\Delta$ and $(c\mu)^\Delta = c\mu^\Delta$ for all real numbers $c > 0$. It follows immediately (as in the proof of (13.3.2)) that for every combination $\mu = \mu_1 - \mu_2 + i(\mu_3 - \mu_4)$ of measures in $P_+(G)$, the sum $\mu_1^\Delta - \mu_2^\Delta + i(\mu_3^\Delta - \mu_4^\Delta)$ is a measure on $\mathbf{Z}$ which depends only on μ and not on the particular decomposition chosen, and that if we denote by μ^Δ this complex measure on $\mathbf{Z}$, then the mapping $\mu \mapsto \mu^\Delta$ so defined, from P(G) to $M(\mathbf{Z}) = M_C(\mathbf{Z})$, is **C**-linear. We call μ^Δ the *Plancherel transform* of μ.

It follows from (22.7.4.1) that for each measure $\mu \in P(G)$ and each pair of functions $f, g \in \mathscr{K}(K\backslash G/K)$ we have

(22.8.1.1)
$$\mu(f * g) = \int_{\mathbf{Z}} \mathscr{F}f(\omega)\overline{\mathscr{F}g(\omega)}\, d\mu^\Delta(\omega).$$

By regularization (14.11.2) we deduce from (22.8.1.1) that if μ and ν are two measures belonging to $P(G) \cap M_C(K\backslash G/K)$ such that $\mu^\Delta = \nu^\Delta$, then we have $\langle \mu, f \rangle = \langle \nu, f \rangle$ for all functions $f \in \mathscr{K}(K\backslash G/K)$; and since $\mu = \mu^\natural$ and $\nu = \nu^\natural$, it follows that $\langle \mu, f \rangle = \langle \nu, f \rangle$ for all $f \in \mathscr{K}(G)$ (22.7.4.5), and hence $\mu = \nu$. In other words, the *kernel* of the Plancherel transformation is the set of measures $\mu \in P(G)$ such that $\mu^\natural = 0$. We shall denote by $P'(\mathbf{Z})$ the *image* of P(G) (or of $P(G) \cap M_C(K\backslash G/K)$) under the Plancherel transformation.

Let $\mathscr{P}(G)$ denote the space of all complex linear combinations of *continuous functions of positive type*, which may be canonically identified with a vector subspace of $L_C^\infty(G)$ and also with a vector subspace of P(G) (the function $f \in \mathscr{P}(G)$ being identified with the measure $f \cdot m_G$). It follows from (22.7.10) that *the image of* $\mathscr{P}(G)$ *under the Plancherel transformation is the space* $M_C^1(\mathbf{Z})$ *of bounded measures on* $\mathbf{Z}$. We denote by $\mathscr{P}(K\backslash G/K)$ the subspace of $\mathscr{P}(G)$ consisting of the functions which are invariant under left and right translations by elements of K. From the remarks above it follows that the mapping $f \mapsto (f \cdot m_G)^\Delta$ is a linear *bijection* of $\mathscr{P}(K\backslash G/K)$ onto $M_C^1(\mathbf{Z})$; the inverse bijection is denoted again by $\bar{\mathscr{F}}'$ and is given by the formula (22.7.10.1).

We remark that

(22.8.1.2)
$$\|\bar{\mathscr{F}}'\mu'\| \leqq \|\mu'\|$$

because $|\omega(x)| \leqq 1$ for all $x \in G$ and $\omega \in \mathbf{S}$; hence the linear mapping $\bar{\mathscr{F}}'$ of $M_C^1(\mathbf{Z})$ onto $\mathscr{P}(K\backslash G/K)$ is *continuous* with respect to the topology of the Banach space $\mathscr{C}_C^\infty(G)$. It should be noted, however, that the image space $\mathscr{P}(K\backslash G/K)$ is not in general closed in $\mathscr{C}_C^\infty(G)$ (22.8.11).

(22.8.2) No usable criteria are known which *characterize* the measures belonging to P(G) or to P′(**Z**). A *necessary* condition for a measure μ' on **Z** to belong to P′(**Z**) is that the Fourier cotransforms $\bar{\mathscr{F}}f$ of all $f \in \mathscr{K}(K\backslash G/K)$ should belong to $\mathscr{L}_C^2(|\mu'|)$ (22.7.4), but this condition is not sufficient (Section 22.11, Problem 10). Again, we have just mentioned that $M_C^1(\mathbf{Z}) \subset P'(\mathbf{Z})$, but there exist unbounded measures in P′(**Z**) (22.10.5). In particular, we have the following result:

(22.8.3) *For all measures* $\mu' \in P'(\mathbf{Z})$ *and all functions* $g' \in \mathscr{L}_C^2(|\mu'|)$, *the measure* $g' \cdot \mu'$ *belongs to* P′(**Z**). *If* $\nu \in P(G)$ *is a measure such that* $\nu^\Delta = g' \cdot \mu'$, *then for all functions* $f \in \mathscr{K}(K\backslash G/K)$ *we have*

(22.8.3.1)
$$\nu(f) = \int_{\mathbf{Z}} \bar{\mathscr{F}}f(\omega) g'(\omega)\, d\mu'(\omega).$$

Clearly it is enough to prove the proposition when $\mu' = \mu^\Delta \geqq 0$, μ being a measure of positive type on G, which we may assume is equal to $\mu^\natural$ (22.7.4.7); likewise, we may limit ourselves to the case where $g' \geqq 0$. Then there exists an increasing sequence of functions $u_n' \geqq 0$ on **Z** which converge simply to g' and belong to $\mathscr{L}_C^1(|\mu'|) \cap \mathscr{L}_C^2(|\mu'|)$: consider, for example, an increasing sequence (A_n) of compact subsets of **Z** whose union is **Z**, and take $u_n' = g'\varphi_{A_n}$ (13.11.7). Since the decreasing sequence of μ'-integrable functions $(g' - u_n')^2$ converges to 0, their integrals converge to 0 (13.8.4): in other words, the sequence (u_n') converges to g' in $\mathscr{L}_C^2(|\mu'|)$. The measure $u_n' \cdot \mu'$ on **Z**, being positive and bounded, is of the form ν_n^Δ, where ν_n is a measure of positive type on G (22.7.10), and for every function $f \in \mathscr{K}(G)$ we have (22.7.10.2)

$$\nu_n(f) = \int_{\mathbf{Z}} \bar{\mathscr{F}}f(\omega) u_n'(\omega)\, d\mu'(\omega).$$

Since $\overline{\mathscr{F}}f$ and g' belong to $\mathscr{L}^2_{\mathbf{C}}(|\mu'|)$, and $0 \leqq u'_n \leqq g$, it follows from **(13.11.7)** and the dominated convergence theorem **(13.8.4)** that $(v_n(f))$ converges to $\int_{\mathbf{Z}} \overline{\mathscr{F}}f(\omega)g'(\omega)\,d\mu'(\omega)$. Hence the sequence (v_n) converges *vaguely* **(13.4.1)** to a measure v on G which is clearly of positive type, and for each function $f \in \mathscr{K}(\mathrm{G})$ the formula **(22.8.3.1)** holds, since the function $\overline{\mathscr{F}}f$ is integrable with respect to the measure $g' \cdot \mu'$. If we now replace f by $g * h$, where g and h are in $\mathscr{K}(\mathrm{K}\backslash\mathrm{G}/\mathrm{K})$, and use the uniqueness property of (22.7.4), we see that $v^{\Delta} = g' \cdot \mu'$.

In particular:

(22.8.4) *The space* $\mathrm{L}^2_{\mathbf{C}}(m_{\mathbf{Z}})$ *(regarded as a subspace of* $\mathrm{M}_{\mathbf{C}}(\mathbf{Z})$*) is contained in* $\mathrm{P}'(\mathbf{Z})$. *The space* $\mathrm{L}^2_{\mathbf{C}}(\mathrm{K}\backslash\mathrm{G}/\mathrm{K})$ *(regarded as a subspace of* $\mathrm{M}_{\mathbf{C}}(\mathrm{K}\backslash\mathrm{G}/\mathrm{K})$*) is contained in* $\mathrm{P}(\mathrm{G})$. *The restriction of the Plancherel transformation to* $\mathrm{L}^2_{\mathbf{C}}(\mathrm{K}\backslash\mathrm{G}/\mathrm{K})$ *is the isomorphism* $\mathscr{F}$ *of* $\mathrm{L}^2_{\mathbf{C}}(\mathrm{K}\backslash\mathrm{G}/\mathrm{K})$ *onto* $\mathrm{L}^2_{\mathbf{C}}(m_{\mathbf{Z}})$ *which extends the Fourier transformation* (22.7.5).

The first assertion is a particular case of **(22.8.3)**, with $\mu' = m_{\mathbf{Z}} = \varepsilon_e^{\Delta}$. Now let $\tilde{g} \in \mathrm{L}^2_{\mathbf{C}}(\mathrm{K}\backslash\mathrm{G}/\mathrm{K})$ and $\tilde{g}' = \mathscr{F}\tilde{g} \in \mathrm{L}^2_{\mathbf{C}}(m_{\mathbf{Z}})$, and assume first that $g' \geqq 0$. Let (g_n) be a sequence of functions in $\mathscr{K}(\mathrm{K}\backslash\mathrm{G}/\mathrm{K})$ which converges to g in $\mathscr{L}^2_{\mathbf{C}}(\mathrm{G})$. By virtue of (22.7.4.1) applied to $\mu = \varepsilon_e$, for all $f \in \mathscr{K}(\mathrm{K}\backslash\mathrm{G}/\mathrm{K})$ we have

$$\text{(22.8.4.1)} \qquad \int f(s)g_n(s)\,dm_{\mathrm{G}}(s) = \int_{\mathbf{Z}} \overline{\mathscr{F}}f(\omega)\mathscr{F}g_n(\omega)\,dm_{\mathbf{Z}}(\omega).$$

Since $\overline{\mathscr{F}}f$ belongs to $\mathrm{L}^2_{\mathbf{C}}(m_{\mathbf{Z}})$ (22.7.4) and $\mathscr{F}\tilde{g}_n$ tends to $\tilde{g}'$ in $\mathrm{L}^2_{\mathbf{C}}(m_{\mathbf{Z}})$ by virtue of (22.7.5), it follows from the continuity of the scalar product in a Hilbert space that

$$\text{(22.8.4.2)} \qquad \langle f, g \cdot m_{\mathrm{G}} \rangle = \langle \overline{\mathscr{F}}f, g' \cdot m_{\mathbf{Z}} \rangle$$

for all $f \in \mathscr{K}(\mathrm{K}\backslash\mathrm{G}/\mathrm{K})$. Now replace f in this relation by $(\check{\bar{h}} * h)^{\natural}$, where $h \in \mathscr{K}(\mathrm{G})$; it follows from (22.7.4.3) that $\langle (\check{\bar{h}} * h)^{\natural}, \omega \rangle \geqq 0$ for all $\omega \in \mathbf{Z}$, because ω is a function of positive type: in other words, $\mathscr{F}((\check{\bar{h}} * h)^{\natural})(\omega) \geqq 0$ for all $\omega \in \mathbf{Z}$. But since $(g \cdot m_{\mathrm{G}})^{\natural} = g \cdot m_{\mathrm{G}}$, we deduce from **(22.8.4.2)** that $\langle \check{\bar{h}} * h, g \cdot m_{\mathrm{G}} \rangle = \langle (\check{\bar{h}} * h)^{\natural}, g \cdot m_{\mathrm{G}} \rangle \geqq 0$, hence that $g \cdot m_{\mathrm{G}}$ is a measure of positive type. Furthermore, by replacing f by $u * v$ with $u, v \in \mathscr{K}(\mathrm{K}\backslash\mathrm{G}/\mathrm{K})$, it follows from (22.7.4) that $(g \cdot m_{\mathrm{G}})^{\Delta} = g' \cdot m_{\mathbf{Z}}$, which completes the proof when $g' \geqq 0$. The general case follows from this by decomposing a function belonging to $\mathscr{L}^2_{\mathbf{C}}(m_{\mathbf{Z}})$ as a linear combination of four functions $\geqq 0$ belonging to this space.

(22.8.4.3) It should be noted that it can happen that a positive measure μ' on **Z** belongs to $P'(\mathbf{Z})$ and yet there exist positive measures ν' such that $0 \leqq \nu' \leqq \mu'$ which do not belong to $P'(\mathbf{Z})$ (Section **22.11**, Problem 10).

(22.8.5) *The space* $M^1_{\mathbf{C}}(G)$ *of bounded measures on* G *is contained in* P(G). *For each bounded measure* μ *on* G *we have*

$$\mu^{\Delta} = (\mathscr{F}\mu) \cdot m_{\mathbf{Z}} \tag{22.8.5.1}$$

where $\mathscr{F}\mu$ *is a bounded continuous function on* **S**, *defined by*

$$(\mathscr{F}\mu)(\omega) = \int_G \omega(x^{-1})\, d\mu(x). \tag{22.8.5.2}$$

Furthermore, for each $f \in \mathscr{L}^1_{\mathbf{C}}(K\backslash G/K)$ (resp. $f \in \mathscr{L}^2_{\mathbf{C}}(K\backslash G/K)$), *the functions* $\mu * f$ *and* $f * \mu$, *which belong to* $\mathscr{L}^1_{\mathbf{C}}(G)$ (resp. $\mathscr{L}^2_{\mathbf{C}}(G)$) **(14.9.2)** *satisfy the relations*

$$\mathscr{F}(\mu * f) = \mathscr{F}(f * \mu) = (\mathscr{F}\mu)(\mathscr{F}f) \tag{22.8.5.3}$$

(resp. *satisfy these relations almost everywhere with respect to* $m_{\mathbf{Z}}$).

For every bounded measure μ on G, the measure $\check{\mu} * \mu$ is of positive type **(22.2.2)**. By virtue of **(22.2.1.3)**, it follows that $\mu * \nu$ belongs to P(G) whenever μ and ν are bounded measures; in particular, $\mu = \mu * \varepsilon_e \in P(G)$, and hence $M^1_{\mathbf{C}}(G) \subset P(G)$. (We have in fact proved more, namely that μ is a linear combination of *bounded* measures of positive type.)

The continuity of the function $\mathscr{F}\mu$ on **S** follows from the continuity of the function $(x, \omega) \mapsto \omega(x^{-1})$ on G × **S** **(22.6.9)**. For if ω_0 is any point of **S**, then for each $\varepsilon > 0$ there exists a compact subset L of G such that $|\mu|(G - L) \leqq \frac{1}{4}\varepsilon$ **(13.9.18)** and a compact neighborhood V of ω_0 in **S** such that

$$|\omega(x^{-1}) - \omega_0(x^{-1})| \leqq \frac{\varepsilon}{2|\mu|(G)}$$

for all $\omega \in V$ and $x \in L$ (3.16.5). Since $|\omega(x^{-1}) - \omega_0(x^{-1})| \leqq 2$ for all $x \in G$, it follows that

$$\left| \int_G (\omega(x^{-1}) - \omega_0(x^{-1}))\, d\mu(x) \right|$$

$$\leqq \left| \int_L (\omega(x^{-1}) - \omega_0(x^{-1}))\, d\mu(x) \right| + \left| \int_{G-L} (\omega(x^{-1}) - \omega_0(x^{-1}))\, d\mu(x) \right|$$

$$\leqq \tfrac{1}{2}\varepsilon + \tfrac{1}{2}\varepsilon = \varepsilon$$

for all $\omega \in V$, which establishes our assertion.

We shall next prove (22.8.5.3) for $f \in \mathscr{L}^1_{\mathbf{C}}(K\backslash G/K)$. By the Lebesgue–Fubini theorem we have

$$\begin{aligned} \mathscr{F}(\mu * f)(\omega) &= \iint f(s^{-1}x)\omega(x^{-1})\, d\mu(s)\, dm_G(x) \\ &= \iint f(x)\omega(x^{-1}s^{-1})\, d\mu(s)\, dm_G(x) \\ &= \iint f(t^{-1}x)\omega(x^{-1}ts^{-1})\, d\mu(s)\, dm_G(x) \end{aligned}$$

for all $t \in K$, by the invariance of m_G. Now by definition, for almost all $x \in G$ we have $f(t^{-1}x) = f(x)$ for all $t \in K$; hence by integrating with respect to t we obtain

$$\begin{aligned} \mathscr{F}(\mu * f)(\omega) &= \iiint_{G \times G \times K} f(x)\omega(x^{-1}ts^{-1})\, d\mu(s)\, dm_G(x)\, dm_K(t) \\ &= \iint_{G \times G} f(x)\omega(x^{-1})\omega(s^{-1})\, d\mu(s)\, dm_G(x) \\ &= ((\mathscr{F}\mu)(\omega))((\mathscr{F}f)(\omega)) \end{aligned}$$

by the Lebesgue–Fubini theorem and the functional equation (22.6.7.1). The proof is the same for $f * \mu$.

Since we have $\mathscr{F}(\mu^{\natural}) = \mathscr{F}\mu$ by virtue of (22.8.5.1), in proving (22.8.5.1) we may limit ourselves to the case where $\mu \in M^1_{\mathbf{C}}(K\backslash G/K)$. Now, for two functions $f, g \in \mathscr{K}(K\backslash G/K)$, the function $\mu * f$ belongs to $\mathscr{L}^1_{\mathbf{C}}(K\backslash G/K)$, and by (14.10.9) we have $\langle \mu, f * g \rangle = \langle \check{\mu} * f, \check{g} \rangle$; hence, by (22.7.5.2) and (22.8.5.3), we have

$$\langle \mu, f * g \rangle = \int (\overline{\mathscr{F}}(\check{\mu} * f))\overline{(\mathscr{F}\check{g})}\, dm_{\mathbf{Z}}$$

$$= \int (\mathscr{F}(\check{f} * \mu))(\mathscr{F} g)\, dm_{\mathbf{Z}}$$

$$= \int (\mathscr{F}\mu)(\overline{\mathscr{F}} f)(\overline{\mathscr{F}} g)\, dm_{\mathbf{Z}}$$

in view of (22.7.4.4). By virtue of (22.7.4), this proves (22.8.5.1).

Finally, we have to prove (22.8.5.3) when $f \in \mathscr{L}^2_{\mathbf{C}}(K\backslash G/K)$. There exists a sequence of functions (f_n) in $\mathscr{K}(K\backslash G/K)$ which converges to f in $\mathscr{L}^2_{\mathbf{C}}(K\backslash G/K)$. Since $f_n \in \mathscr{L}^1_{\mathbf{C}}(K\backslash G/K)$, we have $\mathscr{F}(\mu * f_n) = (\mathscr{F}\mu)(\mathscr{F} f_n)$; also $\mu * f_n$ tends to $\mu * f$ in $\mathscr{L}^2_{\mathbf{C}}(G)$ (14.9.2), hence (22.7.5) $\mathscr{F}(\mu * f_n)$ tends to $\mathscr{F}(\mu * f)$ in $\mathscr{L}^2_{\mathbf{C}}(m_{\mathbf{Z}})$; on the other hand, by (22.7.5) $(\mathscr{F}\mu)(\mathscr{F} f_n)$ tends to $(\mathscr{F}\mu)(\mathscr{F} f)$ in $L^2_{\mathbf{C}}(m_{\mathbf{Z}})$, and the proof is complete.

(22.8.6) The function $\mathscr{F}\mu$ (defined on **S**) is called the *Fourier transform* of the bounded measure μ on G. This terminology is justified by the fact that $\mathscr{F} f = \mathscr{F}(f \cdot m_G)$ for all functions $f \in \mathscr{L}^1_{\mathbf{C}}(G)$. Likewise, the function $\mathscr{F}(\check{\mu})$ is called the *Fourier cotransform* of μ, and is denoted by $\overline{\mathscr{F}}\mu$.

We have in particular

(22.8.6.1) $$(\mathscr{F}\varepsilon_x)(\omega) = \omega(x^{-1}), \qquad (\overline{\mathscr{F}}\varepsilon_x)(\omega) = \omega(x).$$

It is clear that $\|\mathscr{F}\mu\| \leq \|\mu\|$, hence that $\mathscr{F}$ is a continuous linear mapping of the Banach space $M^1_{\mathbf{C}}(G)$ into the Banach space $\mathscr{C}^{\infty}_{\mathbf{C}}(\mathbf{S})$ (7.2). Observe that the bounded continuous function $\mathscr{F}\mu$ *does not necessarily tend to* 0 at infinity, as is shown for example by the case of the measure $\mu = \varepsilon_e$, for which $\mathscr{F}(\varepsilon_e)$ is the constant function 1.

(22.8.7) It follows from the characterizations of the Plancherel transformation on $L^2_{\mathbf{C}}(G)$ (22.8.4) and on $M^1_{\mathbf{C}}(G)$ (22.8.5) that, on the space $L^1_{\mathbf{C}}(G) \cap L^2_{\mathbf{C}}(G) = M^1_{\mathbf{C}}(G) \cap L^2_{\mathbf{C}}(G)$, the class of the Fourier transform $\mathscr{F} f$ of a function $f \in \mathscr{L}^1_{\mathbf{C}}(G) \cap \mathscr{L}^2_{\mathbf{C}}(G)$ is the same as the class $\mathscr{F}\tilde{f}$ defined in (22.7.5).

(22.8.8) *If $f, g \in \mathscr{L}^2_{\mathbf{C}}(\mathrm{K}\backslash\mathrm{G}/\mathrm{K})$, then the bounded continuous function $f * g$* **(14.10.7)** *belongs to $\mathscr{P}(\mathrm{K}\backslash\mathrm{G}/\mathrm{K})$, and we have*

(22.8.8.1)
$$((f * g) \cdot m_{\mathrm{G}})^{\wedge} = (\mathscr{F}f)(\mathscr{F}g) \cdot m_{\mathbf{Z}}.$$

By using **(22.2.1.3)**, we may limit ourselves to the case where $g = \overset{*}{f}$. The bounded continuous function $f * \overset{*}{f}$ is of positive type **(22.1.4)**. To prove **(22.8.8.1)**, consider first the case where $f \in \mathscr{K}(\mathrm{K}\backslash\mathrm{G}/\mathrm{K})$; then for $u, v \in \mathscr{K}(\mathrm{K}\backslash\mathrm{G}/\mathrm{K})$ we have

$$\begin{aligned}\langle f * \overset{*}{f}, \check{v} * u\rangle &= \iiint f(s)\overline{\check{f}(s^{-1}x)u(t^{-1})\check{v}(t)}\, dm_{\mathrm{G}}(s)\, dm_{\mathrm{G}}(t)\, dm_{\mathrm{G}}(x) \\ &= \iint f(s)\overline{\check{v}(t)}\, dm_{\mathrm{G}}(s)\, dm_{\mathrm{G}}(t) \int \overline{\check{f}(s^{-1}x)u(t^{-1}x)}\, dm_{\mathrm{G}}(x)\end{aligned}$$

But by **(22.7.5.1)** and **(22.7.1.6)** we have

$$\int_{\mathrm{G}} \overline{\check{f}(s^{-1}x)u(t^{-1}x)}\, dm_{\mathrm{G}}(x) = \int_{\mathbf{Z}} \omega(s^{-1})\omega(t^{-1})\overline{(\mathscr{F}f(\omega))(\overline{\mathscr{F}}u(\omega))}\, dm_{\mathbf{Z}}(\omega);$$

hence

(22.8.8.2)

$$\begin{aligned}&\langle f * \overset{*}{f}, \check{v} * u\rangle \\ &= \int_{\mathbf{Z}} \overline{(\mathscr{F}f(\omega))}(\overline{\mathscr{F}}u(\omega))\, dm_{\mathbf{Z}}(\omega) \iint_{\mathrm{G}\times\mathrm{G}} \omega(s^{-1})f(s)\omega(t^{-1})\overline{\check{v}(t)}\, dm_{\mathrm{G}}(s)\, dm_{\mathrm{G}}(t) \\ &= \int_{\mathbf{Z}} |\mathscr{F}f(\omega)|^2(\overline{\mathscr{F}}u(\omega))\overline{(\overline{\mathscr{F}}v(\omega))}\, dm_{\mathbf{Z}}(\omega)\end{aligned}$$

which shows that in this case **(22.7.4)** we have

(22.8.8.3)
$$((f * \overset{*}{f}) \cdot m_{\mathrm{G}})^{\wedge} = |\mathscr{F}f|^2 \cdot m_{\mathbf{Z}}.$$

To prove **(22.8.8.2)** (and hence also **(22.8.8.3)**) for any $f \in \mathscr{L}^2_{\mathbf{C}}(\mathrm{K}\backslash\mathrm{G}/\mathrm{K})$, we consider a sequence (f_n) of functions in $\mathscr{K}(\mathrm{K}\backslash\mathrm{G}/\mathrm{K})$ which converges to f in $\mathscr{L}^2_{\mathbf{C}}(\mathrm{G})$; then by **(14.10.7)** the sequence $(f_n * \overset{*}{f}_n)$ converges uniformly to $f * \overset{*}{f}$, and $\mathscr{F}f_n$ tends to $\mathscr{F}f$ in $\mathscr{L}^2_{\mathbf{C}}(m_{\mathbf{Z}})$ **(22.7.5)**, hence $|\mathscr{F}f_n|^2 \to |\mathscr{F}f|^2$ in $\mathscr{L}^1_{\mathbf{C}}(m_{\mathbf{Z}})$. If we now replace f by f_n in **(22.8.8.2)** and let $n \to \infty$, we obtain the desired result.

(22.8.9) We denote by $\mathscr{P}^2(K\backslash G/K)$ (resp. $\mathscr{P}^1(K\backslash G/K)$) the intersection of the vector spaces $\mathscr{P}(G)$ and $L^2_{\mathbf{C}}(K\backslash G/K)$ (resp. $\mathscr{P}(G)$ and $L^1_{\mathbf{C}}(K\backslash G/K)$). For $f \in \mathscr{L}^{\infty}_{\mathbf{C}}(G) \cap \mathscr{L}^1_{\mathbf{C}}(G)$, we have $|f(x)|^2 \leqq N_{\infty}(f)|f(x)|$ almost everywhere, hence $f \in \mathscr{L}^2_{\mathbf{C}}(G)$ by virtue of the mean-value theorem; consequently $\mathscr{P}^1(K\backslash G/K) \subset \mathscr{P}^2(K\backslash G/K)$, the functions in $\mathscr{P}(G)$ being bounded.

(22.8.10) *The image of* $\mathscr{P}^2(K\backslash G/K)$ (*identified with a subset of* $M_{\mathbf{C}}(K\backslash G/K)$) *under the Plancherel transformation*) *is the subspace* $L^1_{\mathbf{C}}(m_{\mathbf{Z}}) \cap L^2_{\mathbf{C}}(m_{\mathbf{Z}})$ *of* $M^1_{\mathbf{C}}(\mathbf{Z})$, *and for all* $f \in \mathscr{P}^2(K\backslash G/K)$ *we have*

(22.8.10.1)
$$f = \bar{\mathscr{F}}'((\mathscr{F}f) \cdot m_{\mathbf{Z}}).$$

If in addition $f \in \mathscr{P}^1(K\backslash G/K)$, *then for all* $x \in G$ *we have*

(22.8.10.2)
$$f(x) = \int_{\mathbf{Z}} \omega(x)\, dm_{\mathbf{Z}}(\omega) \int_{G} f(y)\omega(y^{-1})\, dm_G(y)$$

(*Fourier's reciprocity formulas*).

The first assertion follows directly from **(22.7.10)** and **(22.8.4)**, in view of the definition of $\mathscr{P}^2(K\backslash G/K)$ and the fact that $M^1_{\mathbf{C}}(\mathbf{Z}) \cap L^2_{\mathbf{C}}(m_{\mathbf{Z}}) = L^1_{\mathbf{C}}(m_{\mathbf{Z}}) \cap L^2_{\mathbf{C}}(m_{\mathbf{Z}})$ **(13.14.4)**. The relation **(22.8.10.1)** is a consequence of **(22.7.10.1)** and the fact that the Plancherel transform of $f \cdot m_G$ is $(\mathscr{F}f) \cdot m_{\mathbf{Z}}$, because $f \in \mathscr{L}^2_{\mathbf{C}}(K\backslash G/K)$ **(22.8.4)**.

It should be carefully noted that on the right-hand side of **(22.8.10.2)** it is *not* in general possible to replace the repeated integrals by the double integral $\iint_{G \times \mathbf{Z}} \omega(x)\omega(y^{-1})f(y)\, dm_G(y)\, dm_{\mathbf{Z}}(\omega)$, since the latter *is not defined* in general for a function $f \in \mathscr{P}^1(K\backslash G/K)$ (cf. Chapter XXIII).

(22.8.11) *Remark.* If $f, g \in \mathscr{K}(K\backslash G/K)$, the convolution product $f * g$ belongs to $\mathscr{K}(K\backslash G/K) \cap \mathscr{P}(K\backslash G/K) \subset \mathscr{K}(K\backslash G/K) \cap \mathscr{P}^1(K\backslash G/K)$ by virtue of **(22.8.8)**, and therefore $\mathscr{F}(f * g)$ is $m_{\mathbf{Z}}$-*integrable*. On the other hand, one can give examples of functions $f \in \mathscr{K}(K\backslash G/K)$ whose Fourier transform $\mathscr{F}f$ is not $m_{\mathbf{Z}}$-integrable (Section **22.10**, Problem 13), and therefore $\mathscr{K}(K\backslash G/K) \not\subset \mathscr{P}(K\backslash G/K)$.

By virtue of regularization **(14.11.2)**, every function $f \in \mathscr{K}(K\backslash G/K)$ is the uniform limit of a sequence of convolution products $f * g_n$ of f with functions $g_n \in \mathscr{K}(K\backslash G/K)$ whose supports are contained in a fixed compact set. This shows that the spaces $\mathscr{P}(K\backslash G/K)$, $\mathscr{P}^1(K\backslash G/K)$ and $\mathscr{P}^2(K\backslash G/K)$ *are not* *necessarily closed* in the Banach space $\mathscr{C}_{\mathbf{C}}^{\infty}(G)$. It follows immediately from the preceding results that $\mathscr{K}(K\backslash G/K) \cap \mathscr{P}^1(K\backslash G/K)$ is *dense* in $L_{\mathbf{C}}^1(K\backslash G/K)$ and in $L_{\mathbf{C}}^2(K\backslash G/K)$; *a fortiori*, $\mathscr{P}^1(K\backslash G/K)$ is *dense* in $L_{\mathbf{C}}^1(K\backslash G/K)$, and $\mathscr{P}^2(K\backslash G/K)$ is *dense* in $L_{\mathbf{C}}^2(K\backslash G/K)$.

9. SPHERICAL FUNCTIONS OF POSITIVE TYPE AND IRREDUCIBLE REPRESENTATIONS

(22.9.1) Let (G, K) be a Gelfand pair. Recall that to each spherical function *of positive type* $\omega \in \mathbf{Z}(G/K)$ there corresponds canonically a *topologically cyclic continuous unitary representation* of G (22.1.3). We denote by E_ω the Hilbert space of this representation, and by π the canonical mapping of the subalgebra $A = L_{\mathbf{C}}^1(G) + \mathbf{C}\varepsilon_e \subset M_{\mathbf{C}}^1(G)$ into E_ω. The image $\pi(A)$ is dense in E_ω. The scalar product on E_ω satisfies the relation

$$\text{(22.9.1.1)} \quad (\pi(\mu) \mid \pi(\nu)) = \langle \omega, \check{\bar{\nu}} * \mu \rangle = \iint \omega(y^{-1}z)\, d\bar{\nu}(y)\, d\mu(z)$$

for all μ, ν in A.

We shall denote by U_ω the linear representation corresponding to ω; it is defined by the formula

$$\text{(22.9.1.2)} \qquad U_\omega(s) \cdot \pi(\tilde{f}) = \pi(\varepsilon_s * \tilde{f})$$

for $s \in G$ and $\tilde{f} \in L_{\mathbf{C}}^1(G)$. The corresponding representation of $M_{\mathbf{C}}^1(G)$ on E_ω (21.1.4) satisfies

$$\text{(22.9.1.3)} \qquad U_\omega(\nu) \cdot \pi(\mu) = \pi(\nu * \mu)$$

for μ, ν in A. We shall denote the element $U_\omega(s) \cdot \pi(\varepsilon_e)$ by $\pi(\varepsilon_s)$.

The following proposition generalizes the results of (22.5.6) to the situation where the group G is no longer assumed to be compact:

(22.9.2) (i) *Let* (G, K) *be a Gelfand pair and* $\omega \in \mathbf{Z}(G/K)$ *a spherical function of positive type on* G. *Then the corresponding linear representation* U_ω *is irreducible, and its restriction to* K *contains the trivial representation of* K.

(ii) *Conversely, every irreducible unitary representation U of* G *whose restriction to* K *contains the trivial representation of* K *is equivalent to exactly one of the representations U_ω, and the multiplicity of the trivial representation of* K *in U is equal to* 1.

(i) Since $U_\omega(\mu) \cdot \pi(\varepsilon_e) = \pi(\mu)$ for all $\mu \in \mathrm{A}$, the vector $x_0 = \pi(\varepsilon_e)$ is a *totalizer* for U_ω. We shall show that the operator $P = U_\omega(m_\mathrm{K})$ (where the Haar measure m_K on K is canonically identified with a bounded measure on G **(13.1.7)**) is the *orthogonal projection* of E_ω onto the one-dimensional subspace $\mathbf{C}x_0$. This will establish that U_ω is irreducible: for if a closed subspace F of E_ω is stable under U_ω and is not orthogonal to $\mathbf{C}x_0$, then $P(\mathrm{F})$ is contained in F and contains x_0, whence $\mathrm{F} = \mathrm{E}_\omega$ because x_0 is a totalizing vector; if on the other hand F is orthogonal to $\mathbf{C}x_0$, its orthogonal supplement $\mathrm{F}^\perp$ is also stable under U_ω and contains x_0, whence $\mathrm{F}^\perp = \mathrm{E}_\omega$ and therefore $\mathrm{F} = \{0\}$.

To show that P is an orthogonal projection on E, it is enough to prove that $P^2 = P$ and $P^* = P$ **(11.5.3)**. Since U_ω is a representation of $\mathrm{M}^1_\mathbf{C}(\mathrm{G})$, these relations are consequences of the relations $m_\mathrm{K} * m_\mathrm{K} = m_\mathrm{K}$ and $\check{m}_\mathrm{K} = m_\mathrm{K}$, as one verifies immediately. To show that $P(\mathrm{E}_\omega) = \mathbf{C}x_0$, we calculate the scalar product $(P \cdot \pi(\varepsilon_s) \mid \pi(f))$, where $s \in \mathrm{G}$ and $f \in \mathscr{K}(\mathrm{G})$. By definition, it is equal to

$$\begin{aligned}
\langle \omega, \tilde{\bar{f}} * m_\mathrm{K} * \varepsilon_s \rangle &= \iint_{\mathrm{G}\times\mathrm{K}} \omega(x)\overline{f(tsx^{-1})}\, dm_\mathrm{G}(x)\, dm_\mathrm{K}(t) \\
&= \iint_{\mathrm{G}\times\mathrm{K}} \omega(xts)\overline{f(x^{-1})}\, dm_\mathrm{G}(x)\, dm_\mathrm{K}(t) \\
&= \int_\mathrm{G} \overline{f(x^{-1})}\, dm_\mathrm{G}(x) \int_\mathrm{K} \omega(xts)\, dm_\mathrm{K}(t) \\
&= \omega(s) \int_\mathrm{G} \omega(x)\overline{f(x^{-1})}\, dm_\mathrm{G}(x) \\
&= (\omega(s)\pi(\varepsilon_e) \mid \pi(f)),
\end{aligned}$$

by virtue of the invariance of m_G and the functional equation **(22.6.7.1)** of the spherical functions. Since the $\pi(f)$ form a total set in E_ω, we have $P \cdot \pi(\varepsilon_s) = \omega(s)x_0$ for all $s \in \mathrm{G}$, and since the $\pi(\varepsilon_s) = U_\omega(s) \cdot x_0$ form a total set in E_ω, it follows that $P(\mathrm{E}_\omega) = \mathbf{C}x_0$. The proof shows also that

(22.9.2.1) $$\omega(s) = (\pi(\varepsilon_s) \mid x_0) = (U_\omega(s) \cdot x_0 \mid x_0).$$

Finally, with the same notation, for $t \in \mathrm{K}$ we have

$$
\begin{aligned}
(U_\omega(t) \cdot x_0 \mid \pi(f)) &= (\pi(\varepsilon_t) \mid \pi(f)) \\
&= \langle \omega, \check{f} * \varepsilon_t \rangle \\
&= \int_{\mathrm{G}} \omega(x)\overline{f(tx^{-1})}\, dm_{\mathrm{G}}(x) \\
&= \int_{\mathrm{G}} \omega(xt)\overline{f(x^{-1})}\, dm_{\mathrm{G}}(x) \\
&= (x_0 \mid \pi(f))
\end{aligned}
$$

because $\omega(xt) = \omega(x)$; consequently $U_\omega(t) \cdot x_0 = x_0$ for each $t \in \mathrm{K}$, and the restriction of U_ω to K contains the trivial representation.

(ii) Conversely, let U be a unitary representation of G on a separable Hilbert space E, and suppose that the restriction of U to K contains the trivial representation of K; in other words, the closed vector subspace F of E formed by the vectors $x \in \mathrm{E}$ such that $U(t) \cdot x = x$ for all $t \in \mathrm{K}$ is *nonzero*. The subspace F is *stable* under the operators $U(f)$, where f belongs to the algebra $\mathscr{K}(\mathrm{K}\backslash\mathrm{G}/\mathrm{K})$. For if $x \in \mathrm{F}$, $y \in \mathrm{E}$ and $t \in \mathrm{K}$ we have

$$
\begin{aligned}
((U(t)U(f)) \cdot x \mid y) &= \int f(s)((U(t)U(s)) \cdot x \mid y)\, dm_{\mathrm{G}}(s) \\
&= \int f(s)(U(ts) \cdot x \mid y)\, dm_{\mathrm{G}}(s) \\
&= \int f(t^{-1}s)(U(s) \cdot x \mid y)\, dm_{\mathrm{G}}(s) \\
&= \int f(s)(U(s) \cdot x \mid y)\, dm_{\mathrm{G}}(s) \\
&= (U(f) \cdot x \mid y)
\end{aligned}
$$

by virtue of the definition of $\mathscr{K}(\mathrm{K}\backslash\mathrm{G}/\mathrm{K})$ and the invariance of m_{G}; whence $U(t) \cdot (U(f) \cdot x) = U(f) \cdot x$.

Now let V be the representation $f \mapsto U(f) \mid \mathrm{F}$ of the algebra $\mathscr{K}(\mathrm{K}\backslash\mathrm{G}/\mathrm{K})$ on the Hilbert space F. We shall show that *if U is irreducible then so also is V*. Suppose, then, that F contains a closed subspace F_1, stable under V and distinct from $\{0\}$ and F; then the orthogonal supplement F_2 of F_1 in F is also stable under V (15.5) and distinct from $\{0\}$ and F. Let x_1 be a nonzero vector in F_1, and consider the smallest closed vector subspace E_1 of E which contains all the vectors $U(g) \cdot x_1$, where g runs through $\mathscr{K}(\mathrm{G})$. This subspace is *stable* under U (21.1.7); we shall show that E_1 is *orthogonal* to F_2, hence distinct from E, and it is nonzero because it contains x_1; this will therefore show that if V is reducible then so also is U. Now, for $x_2 \in F_2$ we have $U(t) \cdot x_1 = x_1$ and $U(t') \cdot x_2 = x_2$ for all $t, t' \in K$; we may therefore write

$$(U(g) \cdot x_1 | x_2) = \iint_{K \times K} (U(g)U(t) \cdot x_1 | U(t') \cdot x_2) \, dm_K(t) \, dm_K(t')$$

$$= \iint_{K \times K} (((U(t'))^* U(g) U(t)) \cdot x_1 | x_2) \, dm_K(t) \, dm_K(t').$$

But $(U(t'))^* U(g) U(t) = U(\check{\varepsilon}_{t'} * g * \varepsilon_t)$, hence by definition **(22.6.1.3)** we have

$$(U(g) \cdot x_1 | x_2) = (U(g^{\natural}) \cdot x_1 | x_2) = 0$$

because by hypothesis $U(g^{\natural}) \cdot x_1 \in F_1$.

We can now show that if U is irreducible and if $U | K$ contains the trivial representation of K, then it contains it *exactly once*; this means that $\dim(F) = 1$, and is a consequence of the results established above, together with the following general lemma:

(22.9.2.2) *Every irreducible representation V, not identically zero, of a commutative algebra with involution* A *on a separable Hilbert space* E, *such that $V(A)$ is separable, is a representation on a space of dimension* 1.

Let $\mathscr{A}'$ be the closure in $\mathscr{L}(E)$ of the subalgebra generated by the identity 1_E and $V(A)$; then $\mathscr{A}'$ is a separable commutative star algebra, and the representation of $\mathscr{A}'$ on E defined by the canonical injection $\mathscr{A}' \to \mathscr{L}(E)$ is evidently irreducible. By virtue of the Gelfand–Naimark theorem **(15.4.14)**, we may restrict ourselves to the case where $A = \mathscr{C}_C(L)$, where L is a compact space. The representation V, being irreducible, is *a fortiori* topologically cyclic, and we may therefore apply the fundamental theorem **(15.10.1)**: with the notation of this theorem, it has to be shown that if $u \mapsto M_\mu(u)$ is irreducible, the support of the measure μ on L consists of a *single point*. Indeed, if this were not the case, there would exist two distinct points z_1, z_2 in the support of μ, hence an open neighborhood L_1 of z_1 not containing z_2; the sets L_1 and $L_2 = L - L_1$ would both be integrable and nonnegligible with respect to μ. Consequently **(15.10.6)**, the operator $M_\mu(\varphi_{L_1})$ would be an orthogonal projection, distinct from 0 and the identity, hence its image would be a closed subspace of $L^2_C(L, \mu)$, stable under M_μ and distinct from $\{0\}$ and $L^2_C(L, \mu)$; consequently M_μ would not be irreducible.

Returning to the proof of (22.9.2(ii)), we see that the assumptions allow us to write $F = \mathbf{C}x_0$ for some $x_0 \neq 0$ in E. It follows that $U(f) \cdot x_0 = \zeta(f)x_0$ for each function $f \in \mathscr{K}(K\backslash G/K)$, where ζ is a *character* of the Banach algebra $L^1_{\mathbf{C}}(K\backslash G/K)$. Moreover, we may assume that $\|x_0\| = 1$, and then we have

$$\zeta(f) = (U(f) \cdot x_0 | x_0) = \int_G f(s)(U(s) \cdot x_0 | x_0) \, dm_G(s),$$

and the function $\omega(s) = (U(s) \cdot x_0 | x_0)$ is evidently continuous and bounded on G; moreover, for each $t \in K$ we have

$$\omega(st) = (U(s) \cdot (U(t) \cdot x_0) | x_0) = \omega(s)$$

and

$$\omega(ts) = (U(t) \cdot (U(s) \cdot x_0) | x_0) = (U(s) \cdot x_0 | U(t^{-1}) \cdot x_0) = \omega(s)$$

by virtue of the definition of x_0, so that ω is a *spherical function* on G (22.6.6). Also we know (22.1.3) that ω is of positive type. The representation U is therefore equivalent to U_ω, by virtue of (15.6.7).

If $U' = TUT^{-1}$ is a continuous unitary representation of G on a Hilbert space E′, equivalent to U (so that T is an isomorphism of E onto E′), then a vector $x \in E$ is invariant under the operators $U(t)$, $t \in K$ if and only if $T(x) \in E'$ is invariant under the operators $U'(t)$, $t \in K$. Consequently, if U is equivalent to a representation $U_{\omega'}$, we must have $\omega' = \omega$. Q.E.D.

(22.9.3) By contrast with the case of Gelfand pairs (G, K) with G compact (22.5), in general the irreducible representations U_ω *are not "contained"* in the canonical representation of G on $L^2_{\mathbf{C}}(G/K)$ induced by the trivial representation of K on **C** (22.3.7); in other words, there exists in general no closed subspace of $L^2_{\mathbf{C}}(G/K)$, stable under the canonical representation, such that the subrepresentation of the latter on the subspace in question is equivalent to U_ω (Section 22.10, Problem 12). Nevertheless, we shall see that the decomposition of the canonical representation into irreducible representations, described in (22.5), has analogues in the general case.

(22.9.4) *With the notation of* (22.9.1):

(i) *For each function* $f \in \mathscr{L}^1_{\mathbf{C}}(G/K)$, *the operator* $U_\omega(f)$ *on* E_ω *has rank* ≤ 1; *more precisely, for each* $z \in E_\omega$ *we have*

(22.9.4.1) $$U_\omega(f) \cdot z = (z | x_0) U_\omega(f) \cdot x_0$$

where $x_0 = \pi(\varepsilon_e)$, *and the trace*† *of this operator is given by*

$$\text{(22.9.4.2)} \qquad \operatorname{Tr}(U_\omega(f)) = \bar{\mathscr{F}}f(\omega).$$

(ii) *If* f, g *are two functions belonging to* $\mathscr{L}^1_{\mathbf{C}}(G/K) \cap \mathscr{L}^2_{\mathbf{C}}(G/K)$, *we have*

$$\text{(22.9.4.3)} \qquad \operatorname{Tr}(U_\omega(f)U_\omega(g)^*) = \bar{\mathscr{F}}(\check{\bar{g}} * f)(\omega)$$

which as a function of ω *is integrable with respect to the canonical measure* $m_{\mathbf{Z}}$, *and*

$$\text{(22.9.4.4)} \qquad \int_G f(s)\overline{g(s)}\,dm_G(s) = \int_{\mathbf{Z}} \operatorname{Tr}(U_\omega(f)U_\omega(g)^*)\,dm_{\mathbf{Z}}(\omega).$$

(iii) *For each continuous bounded function* $f \in \mathscr{L}^1_{\mathbf{C}}(G/K)$ *such that, for each* $s \in G$, *the function* $\bar{\mathscr{F}}(\varepsilon_s * f)$ *is integrable with respect to* $m_{\mathbf{Z}}$, *we have*

$$\text{(22.9.4.5)} \qquad f(s) = \int_{\mathbf{Z}} \bar{\mathscr{F}}(\varepsilon_s * f)(\omega)\,dm_{\mathbf{Z}}(\omega) = \int_{\mathbf{Z}} \operatorname{Tr}(U_\omega(s)U_\omega(f))\,dm_{\mathbf{Z}}(\omega).$$

(i) For $s \in G$ and $g \in \mathscr{K}(G)$ we calculate

$$\begin{aligned}(U_\omega(f)\cdot\pi(\varepsilon_s)\,|\,\pi(\bar{g})) &= (\pi(f * \varepsilon_s)\,|\,\pi(\bar{g}))\\ &= \iint \omega(x)f(y^{-1}xs^{-1})g(y)\,dm_G(x)\,dm_G(y)\\ &= \iint \omega(yxs)g(y)f(x)\,dm_G(x)\,dm_G(y).\end{aligned}$$

† An endomorphism u of rank 1 (A.4.16) of a vector space E (of *arbitrary* dimension) may be written in the form $x \mapsto f(x)a$, where $a \neq 0$ in E and f is a linear form on E; the vector a and the form f are not uniquely determined, but if a is replaced by λa, where λ is a nonzero scalar, then f is replaced by $\lambda^{-1}f$. We define the *trace* of u to be $\operatorname{Tr}(u) = f(a)$, a scalar which is independent of the choice of a. If v is any endomorphism of E, then $u \circ v$ and $v \circ u$ are endomorphisms of rank $\leqq 1$, and it is easy to verify that $\operatorname{Tr}(u \circ v) = \operatorname{Tr}(v \circ u)$.

When the vector space E is finite-dimensional, the trace so defined agrees with the trace as defined in (A.11.3).

Since for almost all x we have $f(xt) = f(x)$ for all $t \in \mathrm{K}$, the above integral is also equal to

$$\iint \omega(yxts)g(y)f(x)\, dm_{\mathrm{G}}(x)\, dm_{\mathrm{G}}(y),$$

and therefore to

$$\iiint_{\mathrm{G}\times\mathrm{G}\times\mathrm{K}} \omega(yxts)g(y)f(x)\, dm_{\mathrm{G}}(x)\, dm_{\mathrm{G}}(y)\, dm_{\mathrm{K}}(t)$$
$$= \iint_{\mathrm{G}\times\mathrm{G}} g(y)f(x)\, dm_{\mathrm{G}}(x)\, dm_{\mathrm{G}}(y) \int_{\mathrm{K}} \omega(yxts)\, dm_{\mathrm{K}}(t)$$
$$= \omega(s) \iint \omega(yx)g(y)f(x)\, dm_{\mathrm{G}}(x)\, dm_{\mathrm{G}}(y),$$

in view of (22.6.7.1).

We have therefore

$$(U_{\omega}(f)\cdot \pi(\varepsilon_s) \mid \pi(\bar{g})) = \omega(s)(U_{\omega}(f)\cdot x_0 \mid \pi(\bar{g})),$$

and since the vectors $\pi(\bar{g})$ form a total set in E_{ω}, this establishes (22.9.4.1) for $z = \pi(\varepsilon_s)$, because $(\pi(\varepsilon_s) \mid x_0) = \omega(s)$. Since the vectors $\pi(\varepsilon_s)$ form a total set in E_{ω}, (22.9.4.1) for any $z \in \mathrm{E}_{\omega}$ follows by continuity. We have then

$$\mathrm{Tr}(U_{\omega}(f)) = (U_{\omega}(f)\cdot x_0 \mid x_0) = (\pi(f) \mid x_0)$$

which proves the formula (22.9.4.2).

(ii) It is immediately verified, from (22.9.4.1) and the definition of the adjoint of an operator (11.5.1), that for all $f \in \mathscr{L}^1_{\mathrm{C}}(\mathrm{G}/\mathrm{K})$ and $z \in \mathrm{E}_{\omega}$ we have

(22.9.4.6) $$U_{\omega}(f)^* \cdot z = (z \mid U_{\omega}(f)\cdot x_0)x_0.$$

Consequently, for f and g in $\mathscr{L}^1_{\mathrm{C}}(\mathrm{G}/\mathrm{K})$, we have

$$U_{\omega}(f)U_{\omega}(g)^* \cdot z = (z \mid U_{\omega}(g)\cdot x_0)(U_{\omega}(f)\cdot x_0),$$

and therefore

$$\begin{aligned}\mathrm{Tr}(U_{\omega}(f)U_{\omega}(g)^*) &= (U_{\omega}(f)\cdot x_0 \mid U_{\omega}(g)\cdot x_0)\\ &= (U_{\omega}(g)^*U_{\omega}(f)\cdot x_0 \mid x_0)\\ &= (U_{\omega}(\check{\bar{g}} * f)\cdot x_0 \mid x_0)\\ &= \bar{\mathscr{F}}(\check{\bar{g}} * f),\end{aligned}$$

because $\check{\bar{g}} * f$ belongs to $\mathscr{L}^1_{\mathbf{C}}(G/K)$. In fact, because $\check{\bar{g}} \in \mathscr{L}^1_{\mathbf{C}}(K\backslash G)$, $\check{\bar{g}} * f$ belongs to $\mathscr{L}^1_{\mathbf{C}}(K\backslash G/K)$. If f and g belong also to $\mathscr{L}^2_{\mathbf{C}}(G)$, then $\check{\bar{g}} * f$ is continuous and bounded **(14.10.7)**, hence belongs also to $\mathscr{L}^2_{\mathbf{C}}(K\backslash G/K)$. On the other hand, $\tilde{f} * f$ is a continuous function of positive type **(22.1.4)**; by virtue of **(22.2.1.3)**, it follows that if f and g belong to $\mathscr{L}^1_{\mathbf{C}}(G/K) \cap \mathscr{L}^2_{\mathbf{C}}(G/K)$, we have $\check{\bar{g}} * f \in \mathscr{P}^1(K\backslash G/K)$. The fact that $\bar{\mathscr{F}}(\check{\bar{g}} * f)$ is integrable therefore follows from **(22.8.10)**, and the formula **(22.9.4.4)** from the formula **(22.8.10.2)** giving the value of $((\check{\bar{g}} * f)^{\vee})(e) = (\check{\bar{g}} * f)(e)$.

(iii) Let (g_n) be a sequence of nonnegative functions in $\mathscr{K}(G)$, with their supports contained in a fixed compact set, such that $\int g_n \, dm_G = 1$ for each n, and such that the sequence $(f * g_n)$ converges uniformly to f on each compact subset of G **(14.11.2)**. The hypothesis that f belongs to $\mathscr{C}_{\mathbf{C}}(G/K)$ implies that the sequence $(f * g_n^\natural)$ also converges to f, uniformly on each compact set L. For we have

$$(f * g_n^\natural)(x) = \iiint f(s) g_n(ts^{-1}xt') \, dm_G(s) \, dm_K(t) \, dm_K(t')$$

$$= \iiint f(st) g_n(s^{-1}xt') \, dm_G(s) \, dm_K(t) \, dm_K(t')$$

$$= \iint f(s) g_n(s^{-1}xt') \, dm_G(s) \, dm_K(t')$$

$$= \int dm_K(t') \int f(s) g_n(s^{-1}xt') \, dm_G(s)$$

because $f(st) = f(s)$ for each $t \in K$. But the integral

$$\int f(s) g_n(s^{-1}xt') \, dm_G(s)$$

converges to $f(xt') = f(x)$ *uniformly* for $x \in L$ and $t' \in K$, and this proves our assertion.

Next, we require the following result, which generalizes **(22.7.1.4)**:

(22.9.4.7) *Let u, v be two functions belonging to $\mathscr{L}^1_{\mathbf{C}}(G)$. If either $u \in \mathscr{L}^1_{\mathbf{C}}(G/K)$ or $v \in \mathscr{L}^1_{\mathbf{C}}(K\backslash G)$, then $\mathscr{F}(u * v) = (\mathscr{F}u)(\mathscr{F}v)$ and $\bar{\mathscr{F}}(u * v) = (\bar{\mathscr{F}}u)(\bar{\mathscr{F}}v)$.*

If $u \in \mathscr{L}^1_{\mathbf{C}}(G/K)$, then by virtue of **(14.8.1)** we have

$$\mathscr{F}(u * v)(\omega) = \iint \omega(x^{-1})u(s)v(s^{-1}x)\, dm_G(s)\, dm_G(x)$$

$$= \iint \omega(x^{-1}s^{-1})u(s)v(x)\, dm_G(s)\, dm_G(x),$$

and since for almost all $s \in G$ we have $u(st) = u(s)$ for all $t \in K$,

$$\mathscr{F}(u * v)(\omega) = \iint \omega(x^{-1}t^{-1}s^{-1})u(s)v(x)\, dm_G(x)\, dm_G(s)$$

and therefore

$$\mathscr{F}(u * v)(\omega) = \iint u(s)v(x)\, dm_G(s)\, dm_G(x) \int \omega(x^{-1}t^{-1}s^{-1})\, dm_K(t)$$

$$= (\mathscr{F}u(\omega))(\mathscr{F}v(\omega))$$

by virtue of the functional equation **(22.6.7.1)**. The proof of the other assertion is similar.

(It should be observed that in general $\mathscr{F}(u * v) \neq (\mathscr{F}u)(\mathscr{F}v)$ if $u \in \mathscr{L}^1_{\mathbf{C}}(K\backslash G)$ and $v \in \mathscr{L}^1_{\mathbf{C}}(G/K)$.)

Reverting to the notation introduced earlier, we have therefore

$$\bar{\mathscr{F}}(\check{g}_n^\natural * f) = (\mathscr{F}g_n^\natural)(\bar{\mathscr{F}}f) = (\mathscr{F}g_n)(\bar{\mathscr{F}}f)$$

because $g_n^\natural \in \mathscr{K}(K\backslash G/K)$. But regularization **(14.11.1)** shows that $(\mathscr{F}g_n)(\omega)$ converges to 1 for each $\omega \in \mathbf{Z}$; since also $|(\mathscr{F}g_n)(\omega)| \leqq 1$ for each $\omega \in \mathbf{Z}$ **(22.7.1.5)**, the dominated convergence theorem proves that if $\bar{\mathscr{F}}f$ is $m_{\mathbf{Z}}$-integrable, then $\int_{\mathbf{Z}} \bar{\mathscr{F}}(\check{g}_n^\natural * f)(\omega)\, dm_{\mathbf{Z}}(\omega)$ converges to $\int_{\mathbf{Z}} \bar{\mathscr{F}}f(\omega)\, dm_{\mathbf{Z}}(\omega)$. Since, on the other hand, $\int f(s)g_n^\natural(s)\, dm_G(s)$ converges to $f(e)$, this establishes **(22.9.4.5)** for $s = e$; and the general case is reduced to this by replacing f by $\varepsilon_s * f$.

(22.9.5) *Remark.* We may obtain functions f satisfying the conditions of **(22.9.4(iii))** (and in fact forming a *total* set in $L^1(G/K)$), by taking $f = u * v$, where both u and v belong to $\mathscr{K}(G/K)$; for then we have

$$\bar{\mathscr{F}}(\varepsilon_s * f) = (\bar{\mathscr{F}}(\varepsilon_s * u))(\bar{\mathscr{F}}v)$$

by **(22.9.4.7)**, and the two factors on the right-hand side belong to $\mathscr{L}^2_{\mathbf{C}}(m_{\mathbf{Z}})$, so that $\bar{\mathscr{F}}(\varepsilon_s * f)$ is $m_{\mathbf{Z}}$-integrable, and moreover $f \in \mathscr{K}(G/K)$.

PROBLEMS

1. The hypotheses and notation are those of (22.6.11). Show that if the "exponential" α is a function of positive type on the group S, then the spherical function ω defined by (22.6.11.4) is of positive type on the group G (use the formula (22.21.10.1)). Consider the case where G, S, K are the groups of (22.3.9).

2. Let N be a simply connected nilpotent Lie group, $\mathfrak{n}$ its Lie algebra. The mapping $\exp: \mathfrak{n} \to \mathrm{N}$ is a diffeomorphism of $\mathfrak{n}$ onto N, and the Campbell–Hausdorff formula (Section 19.16, Problem 5) has only a finite number of terms. Let σ be an automorphism of N such that the derived automorphism σ_* of $\mathfrak{n}$ has no eigenvalue equal to 1. Show that the mapping $\varphi: x \mapsto \sigma(x)x^{-1}$ is a diffeomorphism of N onto itself. (Reduce to proving that the equation $\sigma_*(\mathbf{u}) = \log(\exp(\mathbf{v})\exp(\mathbf{u}))$ has a unique solution $\mathbf{u}$, whose coordinates relative to a basis of $\mathfrak{n}$ are polynomials in the coordinates of $\mathbf{v}$. Write the equation in the form $\sigma_*(\mathbf{u}) - \mathbf{u} = \mathbf{v} + \mathrm{P}(\mathbf{u}, \mathbf{v})$, and solve successively the equations $\sigma_*(\mathbf{u}_1) - \mathbf{u}_1 = \mathbf{v}$, $\sigma_*(\mathbf{u}_2) - \mathbf{u}_2 = \mathrm{P}(\mathbf{u}_1, \mathbf{v})$, $\ldots$, $\sigma_*(\mathbf{u}_k) - \mathbf{u}_k = \mathrm{P}(\mathbf{u}_1 + \cdots + \mathbf{u}_{k-1}, \mathbf{v}) - \mathrm{P}(\mathbf{u}_1 + \cdots + \mathbf{u}_{k-2}, \mathbf{v})$, by using the definition of a nilpotent Lie algebra.)

   Let n be the dimension of N, and let Ω be a left-invariant differential n-form on N. Show that ${}^t\varphi(\Omega) = (\prod_j (\alpha_j - 1))\Omega$, where the α_j $(1 \leqq j \leqq n)$ are the eigenvalues of σ_*. (Observe that $\det(\mathrm{Ad}(x)) = 1$ for each $x \in \mathrm{N}$, and use (16.9.9) and (19.16.1.2).)

3. Let G be a connected semisimple Lie group with finite center, and let $\mathrm{G} = \mathrm{KAN}$ be an Iwasawa decomposition of G, so that K is a maximal compact subgroup of G containing the center of G, A is a closed commutative subgroup isomorphic to $\mathbf{R}^n$, and N is a nilpotent group diffeomorphic to some $\mathbf{R}^m$ (21.21.10). Let $m_{\mathrm{K}}, m_{\mathrm{A}}, m_{\mathrm{N}}$ denote Haar measures on the (unimodular) groups K, A, N respectively.

   (a) Show that the mapping $(x, y, z) \mapsto xzy$ of $\mathrm{K} \times \mathrm{A} \times \mathrm{N}$ into G is a diffeomorphism of $\mathrm{K} \times \mathrm{A} \times \mathrm{N}$ onto G, and that there exists a Haar measure m_{G} on G such that, for all functions $f \in \mathcal{K}(\mathrm{G})$,

$$\int_{\mathrm{G}} f(s)\, dm_{\mathrm{G}}(s) = \iiint_{\mathrm{K} \times \mathrm{A} \times \mathrm{N}} f(xzy)\, dm_{\mathrm{K}}(x)\, dm_{\mathrm{A}}(y)\, dm_{\mathrm{N}}(z).$$

   (b) Deduce that there exists on G/A an invariant measure $m_{\mathrm{G/A}}$ (22.3.7) such that, if $\theta: \mathrm{K} \times \mathrm{N} \to \mathrm{G/A}$ is the diffeomorphism defined by $(x, z) \mapsto xz\mathrm{A}$, then for all functions $g \in \mathcal{K}(\mathrm{G/A})$

$$\int_{\mathrm{G/A}} g(\dot{s})\, dm_{\mathrm{G/A}}(\dot{s}) = \iint_{\mathrm{K} \times \mathrm{N}} g(\theta(x, z))\, dm_{\mathrm{K}}(x)\, dm_{\mathrm{N}}(z).$$

   (Use Section 14.4, Problem 2(a).)

   (c) Let $\mathfrak{a}$ be the Lie algebra of A, and as in (21.21.3) choose a Cartan subalgebra $\mathfrak{t} + i\mathfrak{t}'$ of the complexification of the Lie algebra of G, such that $\mathfrak{a} \subset i\mathfrak{t}'$, thus determining a root system $\mathbf{S}$. As in (21.21.6) let $\mathbf{S}''$ denote the set of these roots whose restriction to $\mathfrak{a}$ is not identically zero, and let $\mathbf{S}''_+$ denote the set of roots $\alpha \in \mathbf{S}''$ such that $\alpha(z_0) > 0$ for some definite point $z_0 \in \mathfrak{a}$ satisfying $\alpha(z_0) \neq 0$ for each root $\alpha \in \mathbf{S}''$. Let $\delta'' = \frac{1}{2} \sum_{\alpha \in \mathbf{S}''_+} \alpha$, and for each element $a \in \mathrm{A}$ let

$$\mathrm{D}''(a) = \prod_{\alpha \in \mathbf{S}''_+} \sinh(\tfrac{1}{2}\alpha(\log a)).$$

Show that for each function $f \in \mathscr{K}(G)$ we have

$$\int_G f(s)\,dm_G(s) = \iiint_{K\times A\times N} f(xyz)e^{2\delta''(\log y)}\,dm_K(x)\,dm_A(y)\,dm_N(z).$$

(Calculate the modulus of the automorphism $z \mapsto yzy^{-1}$ of N (where $y \in A$) with the help of **(16.9.9)**.)

For each function $f \in \mathscr{K}(G)$ and each $a \in A$, the mapping $s \mapsto f(sas^{-1})$ is constant on the left cosets of A, and may therefore be considered as belonging to $\mathscr{K}(G/A)$. Show that if $D''(a) \neq 0$ we have

$$2^r|D''(a)|\int_{G/A} f(sas^{-1})\,dm_{G/A}(\dot{s}) = e^{\delta''(\log a)}\iint_{K\times N} f(xazx^{-1})\,dm_K(x)\,dm_N(z)$$

where r is the number of roots in $\mathbf{S}''_+$. (Use Problem 2 for the automorphism $\sigma: z \mapsto a^{-1}za$ of N.)

(d) If we identify canonically $\mathfrak{a}$ and A **(19.8.7.1)**, the Weyl group W of the compact symmetric space associated with G/K (Section **21.21**, Problem 1) acts on A: if $w \in W$ is such that $w \cdot \mathbf{u} = \mathrm{Ad}(s) \cdot \mathbf{u}$ for $\mathbf{u} \in \mathfrak{a}$ and some $s \in K$, then w acts on A as the automorphism $y \mapsto sys^{-1}$. The relation $sAs^{-1} = A$ implies that for each $t \in G$ the coset $sts^{-1}A$ depends only on tA, thus defining a diffeomorphism $\varphi: tA \mapsto sts^{-1}A$ of G/A onto itself. Show that $\varphi(m_{G/A}) = m_{G/A}$. (For this purpose, use Section **14.4**, Problem 2: each function in $\mathscr{K}(G/A)$ can be written as $f^\flat$, where $f \in \mathscr{K}(G)$ and $f^\flat(\dot{t}) = \int_A f(ty)\,dm_A(y)$; put $f_s(t) = f(sts^{-1})$, so that $\int_G f(t)\,dm_G(t) = \int_G f_s(t)\,dm_G(t)$; observe that

$$\int_G f(t)\,dm_G(t) = \int_{G/A} f^\flat(\dot{t})\,dm_{G/A}(\dot{t}) \quad\text{and}\quad \int_G f_s(t)\,dm_G(t) = \int_{G/A} f_s^\flat(\dot{t})\,dm_{G/A}(\dot{t});$$

finally, observe that the Haar measure m_A is invariant under the finite group W **(14.3.6)** and deduce that

$$f_s^\flat(\dot{t}) = \int_A f(sts^{-1}y)\,dm_A(y).)$$

(e) Let $f \in \mathscr{K}(G)$ be such that $f(xsx^{-1}) = f(s)$ for all $s \in G$ and $x \in K$. For each $y \in A$ let

$$F_f(y) = e^{\delta''(\log y)}\int_N f(yz)\,dm_N(z).$$

For each element $w \in W$ we have $F_f(w(y)) = F_f(y)$. (Prove this when $D''(y) \neq 0$, by remarking that $|D''(y)|^2 = \prod_{\alpha\in\mathbf{S}''} |\sinh(\frac{1}{2}\alpha(\log y))|$ is invariant under the Weyl group, using (c), and proving with the help of (d) that

$$\int_{G/A} f(tsys^{-1}t^{-1})\,dm_{G/A}(\dot{t}) = \int_{G/A} f(styt^{-1}s^{-1})\,dm_{G/A}(\dot{t})$$

where $s \in K$ is defined as in (d).)

4. The hypotheses and notation remain the same as in Problem 3. For each $s \in G$ let $H(s)$ denote the unique element of the Lie algebra $\mathfrak{a}$ such that $s = x\exp(H(s)) \cdot z$ with $x \in K$ and

$z \in \mathrm{N}$. For each linear mapping λ: $\mathfrak{a} \to \mathbf{C}$, the function $z \mapsto e^{\lambda(\mathrm{H}(z))}$ is an "exponential" for the solvable group $\mathrm{S} = \mathrm{AN}$, and therefore

$$\omega_\lambda : s \mapsto \int_{\mathrm{K}} e^{(i\lambda - \delta'')(\mathrm{H}(sx))}\, dm_{\mathrm{K}}(x)$$

is a spherical function on G relative to K (**22.6.11**). Assuming that all the spherical functions on G relative to K are of this form, the mapping $\lambda \mapsto \omega_\lambda$ of $\mathfrak{a}^*_{(\mathbf{C})}$ onto $\mathbf{S}(\mathrm{G}/\mathrm{K})$ is continuous.

(a) Show that for each function $f \in \mathscr{K}(\mathrm{K}\backslash\mathrm{G}/\mathrm{K})$, we have

$$\mathscr{F}f(\omega_\lambda) = \int_{\mathrm{A}} e^{i\lambda(\log y)}\mathrm{F}_f(y)\, dm_{\mathrm{A}}(y).$$

(Use Problem 3.)

(b) The Weyl group W acts on $\mathfrak{a}^*_{(\mathbf{C})}$; deduce from (a) that $\omega_{w \cdot \lambda} = \omega_\lambda$ for all $\lambda \in \mathfrak{a}^*_{(\mathbf{C})}$ and all $w \in \mathrm{W}$. (Use Problem **3**(e).)

5. Let (G, K) be a Gelfand pair with G compact. With the notation of (**22.5.6**), show that the spherical function ω_ρ is given by the formula

$$\omega_\rho(s) = n_\rho \int_{\mathrm{K}} \chi_\rho(st)\, dm_{\mathrm{K}}(t).$$

6. With the notation of (**22.9**), give an example of a Gelfand pair (G, K) and functions $u \in \mathscr{L}^1_{\mathbf{C}}(\mathrm{K}\backslash\mathrm{G})$, $v \in \mathscr{L}^1_{\mathbf{C}}(\mathrm{G}/\mathrm{K})$ such that $\mathscr{F}(u * v) \neq (\mathscr{F}u)(\mathscr{F}v)$. (Take G to be compact and noncommutative.)

10. COMMUTATIVE HARMONIC ANALYSIS AND PONTRJAGIN DUALITY

Throughout the rest of this chapter, G will denote a *commutative* locally compact (separable and metrizable) group. We shall apply the theory developed in the preceding sections to the *Gelfand pair* $(\mathrm{G}, \{e\})$, so that $\mathrm{K}\backslash\mathrm{G}/\mathrm{K}$ is replaced by G in the notation.

(22.10.1) The functional equation (**22.6.7.1**) of spherical functions becomes now

(22.10.1.1) $$\omega(xy) = \omega(x)\omega(y);$$

in other words, a spherical function ω on G (relative to the subgroup $\{e\}$) is a *continuous homomorphism* of G into $\mathbf{C}^*$, such that $\omega(e) = 1$ and $|\omega(x)| \leq 1$ for all $x \in \mathrm{G}$; since $\omega(x^{-1}) = \omega(x)^{-1}$, it follows that $|\omega(x)| = 1$ for all $x \in \mathrm{G}$ and therefore ω is a *continuous homomorphism of* G *into the group* $\mathbf{U}$ of complex numbers of *absolute value* 1. These homomorphisms are called the *characters* of the commutative locally compact group G. (If G is *compact* (and commutative), this terminology agrees with that introduced in **(21.3.8)**.) If ω is a character of G, the relations $\omega(x^{-1}) = \omega(x)^{-1}$ and $|\omega(x)| = 1$ imply

(22.10.1.2) $$\omega(x^{-1}) = \overline{\omega(x)}.$$

Moreover, every character ω of G is a continuous function *of positive type*: for by using the relation $\omega(y^{-1}z) = \overline{\omega(y)}\omega(z)$, it is immediately seen that for each $f \in \mathscr{K}(\mathrm{G})$ we have

$$\langle \omega, \check{\bar{f}} * f \rangle = \left| \int \omega(x) f(x) \, dm_{\mathrm{G}}(x) \right|^2 \geqq 0.$$

The subspaces $\mathbf{S}(\mathrm{G}/\{e\})$ and $\mathbf{Z}(\mathrm{G}/\{e\})$ of the unit sphere in $\mathrm{L}^{\infty}_{\mathbf{C}}(\mathrm{G})$ **(22.6.8** and **22.7.2)** are therefore equal and may be canonically identified with the space $\mathbf{X}_0(\mathrm{A})$ of characters of the commutative Banach algebra $\mathrm{A} = \mathbf{C}\varepsilon_e + \mathrm{L}^1_{\mathbf{C}}(\mathrm{G})$ which are not identically zero on $\mathrm{L}^1_{\mathbf{C}}(\mathrm{G})$. The topologies induced on this space by the topology of $\mathscr{C}_{\mathbf{C}}(\mathrm{G})$ **(12.14.6)** and the weak topology of the dual $\mathrm{L}^{\infty}_{\mathbf{C}}(\mathrm{G})$ of $\mathrm{L}^1_{\mathbf{C}}(\mathrm{G})$ are the same, and make $\mathbf{S}(\mathrm{G}/\{e\})$ a *separable, metrizable, locally compact* space **(22.6.9)**.

(22.10.1.3) If ω is a character of G, the space E_ω of the corresponding continuous unitary representation U_ω **(22.9.1)** is the quotient of the algebra $\mathrm{A} = \mathbf{C}\varepsilon_e + \mathrm{L}^1_{\mathbf{C}}(\mathrm{G})$ by the maximal ideal consisting of the measures μ such that $\overline{\mathscr{F}}\mu(\omega) = 0$; in conformity with **(22.9.2.2)**, the irreducible representation U_ω is one-dimensional, and may be identified with the homomorphism $s \mapsto \overline{\omega(s)}$ of G into $\mathbf{C}^*$.

It is clear that the *product* $\omega_1 \omega_2$ of two characters of G and the *inverse* $\omega^{-1} = \bar{\omega}$ of a character of G are again characters of G, so that the set $\mathbf{S}(\mathrm{G}/\{e\})$ of characters of G is a *commutative group* with respect to multiplication; the identity element of this group is the *trivial* character of G, namely the constant function 1. Furthermore:

(22.10.2) *The topology of the set* $\mathbf{S}(\mathrm{G}/\{e\})$ *of characters of* G *is compatible with its group structure* **(12.8)**.

For if $\omega_1, \omega_2, \omega_1', \omega_2'$ are four characters of G, then for each $x \in \mathrm{G}$ we have

$$\begin{aligned} &|\omega_1'(x)\omega_2'(x) - \omega_1(x)\omega_2(x)| \\ &\quad = |(\omega_1'(x) - \omega_1(x))\omega_2'(x) + \omega_1(x)(\omega_2'(x) - \omega_2(x))| \\ &\quad \leqq |\omega_1'(x) - \omega_1(x)| + |\omega_2'(x) - \omega_2(x)| \end{aligned}$$

because $|\omega_2'(x)| = |\omega_1(x)| = 1$. If on some compact subset L of G we have

$$|\omega_1'(x) - \omega_1(x)| \leqq \tfrac{1}{2}\varepsilon, \qquad |\omega_2'(x) - \omega_2(x)| \leqq \tfrac{1}{2}\varepsilon$$

for all $x \in L$, it follows that

$$|\omega_1'(x)\omega_2'(x) - \omega_1(x)\omega_2(x)| \leqq \varepsilon$$

for all $x \in L$, and this establishes the continuity of the product in $\mathbf{S}(G/\{e\})$. The continuity of the inverse is clear, because $\omega^{-1} = \overline{\omega}$.

(22.10.3) Henceforth we shall denote by $\hat{G}$ the *separable, metrizable, locally compact commutative group* so defined, whose elements are the characters of G; it is called the *dual group* of G. The elements of $\hat{G}$ will usually be denoted by symbols such as $\hat{x}$, $\hat{y}$, $\hat{z}$, etc. We shall write $\langle x, \hat{x}\rangle$ in place of $\hat{x}(x)$, for $x \in G$ and $\hat{x} \in \hat{G}$. The mapping $(x, \hat{x}) \mapsto \langle x, \hat{x}\rangle$ of $G \times \hat{G}$ into $\mathbf{U}$ is *continuous* **(22.6.9)**.

(22.10.4) Once a Haar measure m_G has been fixed, the *Plancherel transformation* $\mu \mapsto \mu^{\Delta}$ is defined on the space $P(G) \subset M_C(G)$ of linear combinations of measures of positive type on G, and is a linear mapping into the set $M_C(\hat{G})$ of (complex) measures on $\hat{G}$; moreover, since here $\mu^{\natural} = \mu$, this mapping is *injective* **(22.8.1)**. If we replace m_G by am_G (where a is a real number > 0), μ^{Δ} is replaced by $a^{-1}\mu^{\Delta}$ **(22.7.11)**. For μ^{Δ} to be a *positive* measure, it is necessary and sufficient that μ should be a measure *of positive type*.

(22.10.5) *The canonical measure $m_{\hat{G}} = \varepsilon_e^{\Delta}$ on the group $\hat{G}$ is a Haar measure on $\hat{G}$.*

It is enough to prove that $m_{\hat{G}}$ is *invariant*, in other words that $\gamma(\hat{s})m_{\hat{G}} = m_{\hat{G}}$ for all $\hat{s} \in \hat{G}$. In view of the definition of the translate of a measure **(14.1.2)**, and **(22.7.4)**, we require to prove that

$$\text{(22.10.5.1)} \qquad \int \overline{\mathscr{F}}f(\hat{s}\hat{x})\overline{\overline{\mathscr{F}}g(\hat{s}\hat{x})}\, dm_{\hat{G}}(\hat{x}) = \int \overline{\mathscr{F}}f(\hat{x})\overline{\overline{\mathscr{F}}g(\hat{x})}\, dm_{\hat{G}}(\hat{x})$$

for any two functions $f, g \in \mathscr{K}(G)$. Now we have

$$\begin{aligned}\overline{\mathscr{F}}f(\hat{s}\hat{x}) &= \int \langle x, \hat{s}\hat{x}\rangle f(x)\, dm_G(x) \\ &= \int \langle x, \hat{x}\rangle\langle x, \hat{s}\rangle f(x)\, dm_G(x) \\ &= \overline{\mathscr{F}}f_1(\hat{x})\end{aligned}$$

where $f_1(x) = \langle x, \hat{s}\rangle f(x)$. If we define likewise $g_1(x) = \langle x, \hat{s}\rangle g(x)$, the left-hand side of **(22.10.5.1)** may be written as

$$\int \bar{\mathscr{F}} f_1(\hat{x}) \overline{\bar{\mathscr{F}} g_1(\hat{x})}\, dm_{\hat{G}}(\hat{x}).$$

By virtue of **(22.7.4)**, this integral is equal to

$$\int f_1(x)\overline{g_1(x)}\, dm_G(x);$$

but since $\langle x, \hat{s}\rangle\overline{\langle x, \hat{s}\rangle} = 1$, we have $f_1(x)\overline{g_1(x)} = f(x)\overline{g(x)}$, and the relation **(22.10.5.1)** is therefore a consequence of **(22.7.4)**.

The Haar measures m_G and $m_{\hat{G}}$ on G and its dual are said to be *associated*. If m_G is replaced by am_G, where a is a positive real number, then $m_{\hat{G}}$ is replaced by $a^{-1}m_{\hat{G}}$. *For the remainder of this section*, we shall suppose that the associated measures on G and $\hat{G}$ have been fixed.

(22.10.6) Recall **(22.8.5)** that the set $M_{\mathbf{C}}^1(G)$ of *bounded* measures on G is contained in P(G), and that for each measure $\mu \in M_{\mathbf{C}}^1(G)$ we have

(22.10.6.1)
$$\mu^{\Delta} = (\mathscr{F}\mu) \cdot m_{\hat{G}}$$

where the *Fourier transform* $\mathscr{F}\mu$ of the bounded measure μ is the bounded continuous function on $\hat{G}$ defined by

(22.10.6.2)
$$\mathscr{F}\mu(\hat{x}) = \int_G \overline{\langle x, \hat{x}\rangle}\, d\mu(x).$$

Since the support of $m_{\hat{G}}$ is the whole of $\hat{G}$, $\mathscr{F}\mu$ is the *only* continuous function on $\hat{G}$ which is a density for μ^{Δ} with respect to $m_{\hat{G}}$ **(13.14.4)**; consequently the linear mapping $\mu \mapsto \mathscr{F}\mu$ of $M_{\mathbf{C}}^1(G)$ into $\mathscr{C}_{\mathbf{C}}^{\infty}(\hat{G})$ is *injective*. Recall **(22.8.5.3)** that for $f \in \mathscr{L}_{\mathbf{C}}^2(G)$ and $\mu \in M_{\mathbf{C}}^1(G)$, we have

(22.10.6.3)
$$\mathscr{F}(\mu * f) = (\mathscr{F}\mu)(\mathscr{F}f)$$

almost everywhere on $\hat{G}$. Furthermore, the analogous equality for functions $f \in \mathscr{L}_{\mathbf{C}}^1(G)$ **(22.8.5.3)** generalizes here as follows: for *any two bounded measures* μ, ν on G, we have

(22.10.6.4)
$$\mathscr{F}(\mu * \nu) = (\mathscr{F}\mu)(\mathscr{F}\nu).$$

For by definition we have, for each $\hat{x} \in \hat{G}$ (14.5)

$$\begin{aligned} \mathscr{F}(\mu * \nu)(\hat{x}) &= \int \overline{\langle x, \hat{x} \rangle}\, d(\mu * \nu)(x) \\ &= \iint \overline{\langle yz, \hat{x} \rangle}\, d\mu(y)\, d\nu(z) \\ &= \iint \overline{\langle y, \hat{x} \rangle}\, \overline{\langle z, \hat{x} \rangle}\, d\mu(y)\, d\nu(z) \end{aligned}$$

which proves (22.10.6.4).

Next, recall (22.8.6) that for $\mu \in M^1_{\mathbf{C}}(G)$, the *Fourier cotransform* $\bar{\mathscr{F}}\mu$ of μ is the bounded continuous function $\mathscr{F}\check{\mu}$, the Fourier transform of $\check{\mu}$; so that we have

(22.10.6.5) $$\bar{\mathscr{F}}\mu(\hat{x}) = \int_G \langle x, \hat{x} \rangle\, d\mu(x)$$

and the relations

(22.10.6.6) $$\bar{\mathscr{F}}\mu(\hat{x}) = \mathscr{F}\mu(\hat{x}^{-1}) = \mathscr{F}\check{\mu}(\hat{x}) = \overline{\mathscr{F}\bar{\mu}(\hat{x})}$$

which may also be written in the form

(22.10.6.7) $$\bar{\mathscr{F}}\mu = \mathscr{F}\check{\mu} = (\mathscr{F}\mu)^{\vee} = \overline{\mathscr{F}\bar{\mu}}.$$

Here μ may be replaced by a function $f \in \mathscr{L}^1_{\mathbf{C}}(G)$.

From (22.8.6.1) we have

(22.10.6.8) $$\mathscr{F}\varepsilon_x(\hat{x}) = \overline{\langle x, \hat{x} \rangle}, \qquad \bar{\mathscr{F}}\varepsilon_x(\hat{x}) = \langle x, \hat{x} \rangle$$

and for all bounded measures μ on G and all $x \in G$ we have

(22.10.6.9) $$\begin{aligned} \mathscr{F}(\gamma(x)\mu)(\hat{x}) &= \mathscr{F}(\varepsilon_x * \mu)(\hat{x}) = \overline{\langle x, \hat{x} \rangle}\,\mathscr{F}\mu(\hat{x}), \\ \bar{\mathscr{F}}(\gamma(x)\mu)(\hat{x}) &= \bar{\mathscr{F}}(\varepsilon_x * \mu)(\hat{x}) = \langle x, \hat{x} \rangle\,\bar{\mathscr{F}}\mu(\hat{x}). \end{aligned}$$

(22.10.7) It follows from the continuity of the function $(x, \hat{x}) \mapsto \langle x, \hat{x} \rangle$ that, for each $x \in G$, the mapping $\hat{x} \mapsto \langle x, \hat{x} \rangle$ is a *character* of $\hat{G}$, which we denote by $\eta(x)$. The mapping $x \mapsto \eta(x)$ is evidently a homomorphism of the group G into the *dual* group $\hat{\hat{G}}$ of $\hat{G}$, called the *canonical homomorphism.*

(22.10.8) (Pontrjagin's duality theorem). *The canonical homomorphism* $\eta: \mathrm{G} \to \hat{\hat{\mathrm{G}}}$ *is an isomorphism of topological groups.*

(i) *η is continuous.* In view of the definition of the topology of $\mathscr{C}_{\mathrm{C}}(\hat{\mathrm{G}})$ **(12.14.6)**, we have to show that for each $\varepsilon > 0$ and each compact subset L of $\hat{\mathrm{G}}$, there exists a neighborhood V of e in G such that the relations $x \in \mathrm{V}$, $\hat{x} \in \mathrm{L}$ imply $|1 - \langle \hat{x}, \eta(x) \rangle| \leqq \varepsilon$, that is to say $|1 - \langle x, \hat{x} \rangle| \leqq \varepsilon$; but this follows from the fact that, for each compact neighborhood V_0 of e in G, the mapping $(x, \hat{x}) \mapsto \langle x, \hat{x} \rangle$ of $\mathrm{V}_0 \times \mathrm{L}$ into **U** is uniformly continuous **(3.16.5)**.

(ii) *η is injective.* Let V be a neighborhood of e in G, and let W be a neighborhood of e such that $\mathrm{W}^2 \subset \mathrm{V}$. Let $g \in \mathscr{K}(\mathrm{G})$ be a function $\geqq 0$ with support contained in W and such that $\int g^2 \, dm_{\mathrm{G}} = 1$. Then $h = \check{g} * g$ belongs to $\mathscr{K}(\mathrm{G})$, is $\geqq 0$, and has support contained in V, and $h(e) = 1$. Moreover **(22.8.11)**, h belongs to $\mathscr{P}^1(\mathrm{G})$ and therefore **(22.8.10.2)**, for each $x \in \mathrm{G}$,

(22.10.8.1)
$$h(x) - h(e) = \int_{\hat{\mathrm{G}}} (\langle x, \hat{x} \rangle - 1) \mathscr{F}h(\hat{x}) \, dm_{\hat{\mathrm{G}}}(\hat{x}).$$

Suppose that for some $x \neq e$ in G, $\eta(x)$ is the identity element $\hat{\hat{e}}$ of $\hat{\hat{\mathrm{G}}}$; then $\langle x, \hat{x} \rangle = 1$ *for all* $\hat{x} \in \hat{\mathrm{G}}$, and consequently $h(x) = 1$ by **(22.10.8.1)**. But this is absurd, since V can be chosen not to contain x.

(iii) *η is a homeomorphism of* G *onto the subgroup* $\eta(\mathrm{G})$ *of* $\hat{\hat{\mathrm{G}}}$. With the notation of (ii), the set V′ of elements $x \in \mathrm{G}$ such that $|h(x) - h(e)| \leqq \frac{1}{2}$ is a neighborhood of e contained in V, and it is enough to show that $\eta(\mathrm{V}')$ is the intersection with $\eta(\mathrm{G})$ of a neighborhood of $\hat{\hat{e}}$ in $\hat{\hat{\mathrm{G}}}$. Now it follows from **(22.10.8.1)** that when $\hat{\hat{\mathrm{G}}}$ is canonically identified with a subspace of $\mathrm{L}_{\mathbf{C}}^{\infty}(\hat{\mathrm{G}})$, $\eta(\mathrm{V}')$ is the intersection of $\eta(\mathrm{G})$ with $\hat{\hat{e}} + \mathrm{T}$, where T is the set of $\tilde{u} \in \mathrm{L}_{\mathbf{C}}^{\infty}(\hat{\mathrm{G}})$ such that $|\langle u, \mathscr{F}h \rangle| \leqq \frac{1}{2}$. Since $\mathscr{F}h \in \mathscr{L}_{\mathrm{C}}^1(\hat{\mathrm{G}})$, it follows that T is a neighborhood of 0 in the weak topology of $\mathrm{L}_{\mathbf{C}}^{\infty}(\hat{\mathrm{G}})$, considered as the dual of $\mathrm{L}_{\mathrm{C}}^1(\hat{\mathrm{G}})$. The conclusion therefore follows from **(22.6.9)**.

(iv) *We have* $\eta(\mathrm{G}) = \hat{\hat{\mathrm{G}}}$. Since $\eta(\mathrm{G})$ is a subgroup of $\hat{\hat{\mathrm{G}}}$ isomorphic to G, it is locally compact and therefore *closed* in $\hat{\hat{\mathrm{G}}}$ **(12.9.6)**. It is therefore enough to show that there exists in $\hat{\hat{\mathrm{G}}}$ no nonempty open set U such that $\mathrm{U} \cap \eta(\mathrm{G}) = \varnothing$. We shall use the following lemma:

(22.10.8.2) *If f' and g' are two functions in* $\mathscr{K}(\hat{\mathrm{G}})$, *there exists a function* $u \in \mathscr{P}^1(\mathrm{G})$ *such that* $\mathscr{F}u = f' * g'$.

Put $h' = f' * g' \in \mathscr{K}(\hat{G})$. The measure $\mu' = h' \cdot m_{\hat{G}}$ on $\hat{G}$ is bounded, and therefore (22.8.1) there exists a function $u \in \mathscr{P}(G)$ such that $(u \cdot m_G)^\Delta = \mu'$ and

$$u(x) = \int_{\hat{G}} \langle x, \hat{x} \rangle \, d\mu'(\hat{x}) = \int_{\hat{G}} \langle x, \hat{x} \rangle h'(\hat{x}) \, dm_{\hat{G}}(\hat{x}),$$

which signifies also that $u = (\bar{\mathscr{F}} h') \circ \eta$. Hence if we put $f = (\bar{\mathscr{F}} f') \circ \eta$ and $g = (\bar{\mathscr{F}} g') \circ \eta$, we have $u = fg$. But because f' and g' are in $\mathscr{L}^1_{\mathbf{C}}(\hat{G}) \cap \mathscr{L}^2_{\mathbf{C}}(\hat{G})$, they are Fourier transforms of functions f_1, g_1 belonging to $\mathscr{P}^2(G)$ (22.8.10), and it follows from (22.8.10.1) that $f_1 = f$ and $g_1 = g$. Since $u = fg$, we have $u \in \mathscr{L}^1_{\mathbf{C}}(G) \cap \mathscr{P}(G) = \mathscr{P}^1(G)$, and since $h' \cdot m_{\hat{G}} = (u \cdot m_G)^\Delta$, this implies that $h' = \mathscr{F} u$ (22.10.6.1).

We now return to the proof of (iv). If the open set $U \subset \hat{\hat{G}}$ is nonempty and does not intersect $\eta(G)$, there exists a function $f'' \geqq 0$ in $\mathscr{K}(\hat{\hat{G}})$, with support contained in U and not identically zero; then, by virtue of regularization (14.11.2), there exists a second function $g'' \geqq 0$ in $\mathscr{K}(\hat{\hat{G}})$ such that $f'' * g''$ has support contained in U and is not identically zero. By virtue of the lemma (22.10.8.2), with G replaced by $\hat{G}$, there exists a function $u' \in \mathscr{P}^1(\hat{G})$ such that $f'' * g'' = \mathscr{F} u'$, and u' is not identically zero (22.10.6). But the function $f'' * g''$ is zero on $\eta(G)$, that is to say we have

$$\int_{\hat{G}} \overline{\langle \hat{x}, \eta(x) \rangle} \, u'(\hat{x}) \, dm_{\hat{G}}(\hat{x}) = 0$$

for all $x \in G$, or equivalently

$$\int_{\hat{G}} \overline{\langle x, \hat{x} \rangle} \, u'(\hat{x}) \, dm_{\hat{G}}(\hat{x}) = 0.$$

Now u' belongs to $\mathscr{L}^1_{\mathbf{C}}(\hat{G}) \cap \mathscr{L}^2_{\mathbf{C}}(\hat{G})$, hence $u' = \mathscr{F} u$ with $u \in \mathscr{P}^2(G)$ (22.8.10), and we have

$$u(x) = \int_{\hat{G}} \overline{\langle x, \hat{x} \rangle} \, u'(\hat{x}) \, dm_{\hat{G}}(\hat{x}) = 0$$

for all $x \in G$, from the above and from (22.8.10.1); but then u' would be identically zero, which is absurd. Q.E.D.

(22.10.9) We shall henceforth *identify* $\hat{\hat{G}}$ with G by means of the canonical isomorphism η. With this identification, the mapping $\bar{\mathscr{F}}'$ of $M^1_{\mathbf{C}}(\hat{G})$ onto $\mathscr{P}(G)$ is identified with the *Fourier cotransform* (22.10.6.5), and will therefore be denoted henceforth by $\bar{\mathscr{F}}$. Furthermore, the Haar measure m_G is identified with the Haar measure $m_{\hat{\hat{G}}}$ on $\hat{\hat{G}}$ *associated with* $m_{\hat{G}}$ (22.10.5). For if

f, g are two functions in $\mathscr{P}^2(G)$, we have by **(22.8.10.1)** $f = \bar{\mathscr{F}}\mathscr{F}f$ and $g = \bar{\mathscr{F}}\mathscr{F}g$, so that a double application of **(22.7.5.1)** gives

$$\int f(x)\overline{g(x)}\,dm_G(x) = \int (\mathscr{F}f(\hat{x}))\overline{(\mathscr{F}g(\hat{x}))}\,dm_{\hat{G}}(\hat{x})$$

$$= \int (\bar{\mathscr{F}}\mathscr{F}f(x))\overline{(\bar{\mathscr{F}}\mathscr{F}g(x))}\,dm_{\hat{\hat{G}}}(x)$$

$$= \int f(x)\overline{g(x)}\,dm_{\hat{\hat{G}}}(x).$$

This proves our assertion, because the linear combinations of functions $f\bar{g}$, with f and g in $\mathscr{P}^2(G) \cap \mathscr{K}(G)$, form a set of functions in $\mathscr{K}(G)$ with the property that every function in $\mathscr{K}(G)$ is the uniform limit of a sequence of functions of this set, with supports contained in a fixed compact set **(22.8.11)**.

(22.10.10) The fact that G can be identified with the dual of its dual $\hat{G}$ allows us to complete the results of Sections **22.7** and **22.8**. The Plancherel transformation, defined on the space $P(\hat{G})$, will again be denoted by $\mu' \mapsto \mu'^{\Delta}$, and its image by $P'(G) \subset M_c(G)$. The following diagram recapitulates the various subspaces of $P(G) \cap P'(G)$ which have been considered:

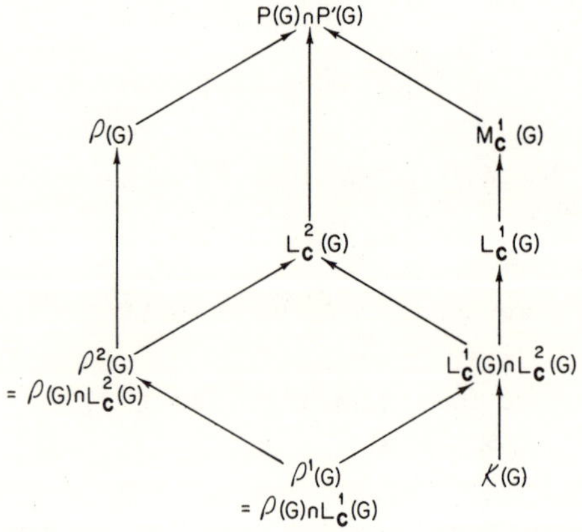

The arrows are the canonical injections. There are no other inclusion relations between these spaces which are valid for all locally compact commutative groups.

In the following table, the second column gives the images under the Plancherel transformation (defined on P(G)) of the subspaces in the first column:

(22.10.10.2)

μ	μ^{Δ}
$\mathscr{P}(G)$	$M_{\mathbf{C}}^1(\hat{G})$
$L_{\mathbf{C}}^2(G)$	$L_{\mathbf{C}}^2(\hat{G})$
$M_{\mathbf{C}}^1(G)$	$\mathscr{P}(\hat{G})$
$\mathscr{P}^2(G)$	$L_{\mathbf{C}}^1(\hat{G}) \cap L_{\mathbf{C}}^2(\hat{G})$
$L_{\mathbf{C}}^1(G) \cap L_{\mathbf{C}}^2(G)$	$\mathscr{P}^2(\hat{G})$
$\mathscr{P}^1(G)$	$\mathscr{P}^1(\hat{G})$

In the three subspaces $\mathscr{P}(G)$, $L_{\mathbf{C}}^2(G)$, and $M_{\mathbf{C}}^1(G)$, the bijection inverse to the Plancherel transformation $\mu \mapsto \mu^{\Delta}$ is the mapping

(22.10.10.3) $$\mu' \mapsto ((\mu')^{\vee})^{\Delta}.$$

In other words, for a measure μ on G belonging to one of these subspaces we have

(22.10.10.4) $$\mu = ((\mu^{\Delta})^{\vee})^{\Delta}.$$

Moreover, in the subspaces $L_{\mathbf{C}}^2(G)$ and $M_{\mathbf{C}}^1(G)$, the Plancherel transformation is related to the Fourier transformation by the formulas

(22.10.10.5) $$(f \cdot m_G)^{\Delta} = \mathscr{F}f \cdot m_{\hat{G}}$$

for $\tilde{f} \in L_{\mathbf{C}}^2(G)$, and

(22.10.10.6) $$\mu^{\Delta} = \mathscr{F}\mu \cdot m_{\hat{G}}$$

for $\mu \in M_{\mathbf{C}}^1(G)$. Furthermore, we have the *duality relation*

(22.10.10.7) $$\langle \mu, \nu'^{\Delta} \rangle = \langle \mu^{\Delta}, \nu' \rangle$$

in the following three cases:

(I) $\mu \in M_{\mathbf{C}}^1(G)$, $\nu' \in M_{\mathbf{C}}^1(\hat{G})$, μ^{Δ} and ν'^{Δ} being identified with the functions $\mathscr{F}\mu$ and $\mathscr{F}\nu'$ by (22.10.10.6);

(II) $\mu \in L^2_{\mathbf{C}}(G)$, $\nu' \in L^2_{\mathbf{C}}(\hat{G})$, μ and ν' being identified with their densities relative to m_G and $m_{\hat{G}}$, respectively, and μ^Δ, ν'^Δ with the Fourier transforms of these densities;

(III) $\mu \in \mathscr{P}(G)$, $\nu' \in \mathscr{P}(\hat{G})$, identified with their densities relative to m_G and $m_{\hat{G}}$, respectively.

(22.10.11) Practically all the results summarized in **(22.10.10)** have already been proved. The fact that the mapping $\tilde{f} \mapsto \overline{\mathscr{F}}\mathscr{F}\tilde{f}$ of $L^2_{\mathbf{C}}(G)$ into itself is the identity follows from the fact that it is an automorphism of this space **(22.7.5)** and reduces to the identity on the dense subspace $\mathscr{P}^2(G)$ **(22.8.10)**. The fact that $\mathscr{P}(G) \subset P'(G)$ follows from **(22.7.10)** applied to the group $\hat{G}$, and the identification of G with the dual of $\hat{G}$; the formula **(22.10.10.6)** in $\mathscr{P}(G)$ and in $M^1_{\mathbf{C}}(G)$ also follows from **(22.7.10)**, again bearing in mind this identification. Case (II) of **(22.10.10.7)** is another form of **(22.7.5.1)**, because we may write $\mu = f \cdot m_G$ and $\nu' = (\overline{\mathscr{F}}g) \cdot m_{\hat{G}}$, and use the fact that $\mathscr{F}(\overline{\mathscr{F}}g) = \bar{g}$ almost everywhere. Cases (I) and (III) express the same property in two different forms, in view of **(22.10.10.4)**; case (I) is a simple application of the Lebesgue–Fubini theorem, since both $\int \mathscr{F}\mu(\hat{x})\, d\nu'(\hat{x})$ and $\int \mathscr{F}\nu'(x)\, d\mu(x)$ are equal to $\iint \overline{\langle x, \hat{x}\rangle}\, d\mu(x)\, d\nu'(\hat{x})$, the measure $\mu \otimes \nu'$ being bounded and the function $(x, \hat{x}) \mapsto \langle x, \hat{x}\rangle$ continuous and bounded.

(22.10.12) The properties of the Plancherel and Fourier transformations in regard to convolution may be recapitulated in the formula

(22.10.12.1)
$$(\mu * \nu)^\Delta = \mu^\Delta \cdot \nu^\Delta$$

where the measure ν belongs to one of the three subspaces $\mathscr{P}(G)$, $L^2_{\mathbf{C}}(G)$ and $M^1_{\mathbf{C}}(G)$, and the measure μ to either of the last two of these three spaces, so that μ^Δ is of the form $f' \cdot m_{\hat{G}}$ and the right-hand side of **(22.10.12.1)** is to be interpreted as denoting the measure $f' \cdot \nu^\Delta$ of density f' relative to ν^Δ. This formula is valid in the following cases:

(I) μ and ν both belong to $M^1_{\mathbf{C}}(G)$, in which case the same is true of $\mu * \nu$;

(II) $\mu \in M^1_{\mathbf{C}}(G)$, $\nu = f \cdot m_G$ with $f \in \mathscr{L}^2_{\mathbf{C}}(G)$, in which case $\mu * f \in \mathscr{L}^2_{\mathbf{C}}(G)$;

(III) $\mu \in M^1_{\mathbf{C}}(G)$, $\nu = f \cdot m_G$ with $f \in \mathscr{P}(G)$, in which case $\mu * f \in \mathscr{P}(G)$;

(IV) $\mu = f \cdot m_G$, $\nu = g \cdot m_G$ with $f, g \in \mathscr{L}^2_{\mathbf{C}}(G)$, in which case $f * g \in \mathscr{P}(G)$.

Case (I) has been dealt with in (22.10.6.4), case (II) in (22.10.6.3), and case (IV) in (22.8.8). There remains case (III) to be considered. Clearly we may assume that μ and f are of positive type. Then for each function $g \in \mathscr{K}(\mathrm{G})$ we have

$$\begin{aligned}\langle \mu * f, g * \check{g} \rangle &= \iiint f(s^{-1}x)g(y)\check{g}(y^{-1}x)\, dm_{\mathrm{G}}(x)\, dm_{\mathrm{G}}(y)\, d\mu(s)\\ &= \iiint f(x)g(y)\check{g}(y^{-1}sx)\, dm_{\mathrm{G}}(x)\, dm_{\mathrm{G}}(y)\, d\mu(s)\\ &= \iiint f(x)g(sy)\check{g}(y^{-1}x)\, dm_{\mathrm{G}}(x)\, dm_{\mathrm{G}}(y)\, d\mu(s)\\ &= \int d\mu(s) \int f(x)((\gamma(s^{-1})g * \check{g})(x)\, dm_{\mathrm{G}}(x)\end{aligned}$$

by virtue of the invariance of m_{G}. By (22.7.1.6) we have

$$\overline{\mathscr{F}}((\gamma(s^{-1})g) * \check{g})(\hat{x}) = \overline{\langle s, \hat{x} \rangle}\, |\overline{\mathscr{F}}g(\hat{x})|^2;$$

if we put $\nu' = (f \cdot m_{\mathrm{G}})^{\Delta}$, which is by hypothesis a positive bounded measure on $\hat{\mathrm{G}}$, it follows from (22.7.10.2) that

$$\int f(x)((\gamma(s^{-1})g) * \check{g})(x)\, dm_{\mathrm{G}}(x) = \int_{\hat{\mathrm{G}}} \overline{\langle s, \hat{x} \rangle}\, |\overline{\mathscr{F}}g(\hat{x})|^2\, d\nu'(\hat{x})$$

and consequently

$$\langle \mu * f, g * \check{g} \rangle = \int_{\hat{\mathrm{G}}} |\overline{\mathscr{F}}g(\hat{x})|^2 \mathscr{F}\mu(\hat{x})\, d\nu'(\hat{x}) \geqq 0$$

because by hypothesis $\mathscr{F}\mu \geqq 0$ on $\hat{\mathrm{G}}$. Hence we see that $\mu * f$ is a continuous function of positive type, having regard to (14.9.2). The formula above moreover shows that

(22.10.12.2) $$((\mu * f) \cdot m_{\mathrm{G}})^{\Delta} = (\mathscr{F}\mu) \cdot (f \cdot m_{\mathrm{G}})^{\Delta},$$

bearing in mind (22.7.4).

The formula (22.10.10.3) allows us to express these results in the following equivalent form: the formula

(22.10.12.3) $$(f \cdot \mu)^{\Delta} = (f \cdot m_{\mathrm{G}})^{\Delta} * \mu^{\Delta}$$

is valid in each of the following cases:

(I) $f \in \mathscr{P}(G)$, $\mu = g \cdot m_G$ with $g \in \mathscr{P}(G)$, in which case $fg \in \mathscr{P}(G)$;
(II) $f \in \mathscr{P}(G)$, $\mu = g \cdot m_G$ with $g \in \mathscr{L}_\mathbf{C}^2(G)$, in which case $fg \in \mathscr{L}_\mathbf{C}^2(G)$;
(III) $f \in \mathscr{P}(G)$, $\mu \in M_\mathbf{C}^1(G)$, in which case $f \cdot \mu \in M_\mathbf{C}^1(G)$;
(IV) $f \in \mathscr{L}_\mathbf{C}^2(G)$, $\mu = g \cdot m_G$ with $g \in \mathscr{L}_\mathbf{C}^2(G)$, in which case $fg \in \mathscr{L}_\mathbf{C}^1(G)$.

Of the subspaces of $M_\mathbf{C}(G)$ in the diagram (21.10.10.1),

$$M_\mathbf{C}^1(G), L_\mathbf{C}^1(G), L_\mathbf{C}^1(G) \cap L_\mathbf{C}^2(G), \mathscr{P}^1(G), \mathscr{K}(G)$$

are *algebras with respect to convolution*; and

$$\mathscr{P}(G), \mathscr{P}^2(G), \mathscr{P}^1(G), \mathscr{K}(G)$$

are *algebras with respect to the usual product* of functions.

Remarks

(22.10.13) The fact that the Fourier transform of a function $f \in \mathscr{L}_\mathbf{C}^1(G)$ tends to 0 at infinity is known as the *Riemann–Lebesgue theorem*. The fact that the continuous functions of positive type on G are precisely the functions of the form

$$x \mapsto \int_{\hat{G}} \langle x, \hat{x} \rangle \, d\mu'(\hat{x})$$

where μ' is a bounded positive measure on $\hat{G}$ (22.7.10) is known as *Bochner's theorem*. The fact that for two functions f, $g \in \mathscr{L}_\mathbf{C}^2(G)$, the functions $\mathscr{F}f$ and $\mathscr{F}g$ belong to $\mathscr{L}_\mathbf{C}^2(\hat{G})$ and satisfy

$$\int f(x)\overline{g(x)} \, dm_G(x) = \int \mathscr{F}f(\hat{x})\overline{\mathscr{F}g(\hat{x})} \, dm_{\hat{G}}(\hat{x})$$

is known as *Plancherel's theorem*. The formula $f = \bar{\mathscr{F}}\mathscr{F}f$, or equivalently

$$f(x) = \int \langle x, \hat{x} \rangle \, dm_{\hat{G}}(\hat{x}) \int \overline{\langle y, \hat{x} \rangle} f(y) \, dm_G(y),$$

valid for all functions $f \in \mathscr{P}^1(G)$, is called *Fourier's reciprocity formula*.

(22.10.14) From the lemma (22.10.8.2) it follows that if L' is a compact subset of $\hat{G}$ and if U' is an open neighborhood of L', then there exists a function $u \in \mathscr{P}^1(G)$ such that $\mathscr{F}u$ is a continuous function on $\hat{G}$ with values in [0, 1], equal to 1 on L' and to 0 on the complement of U'. For if V' is a compact symmetric neighborhood of $\hat{e}$ such that $L'V'^3 \subset U'$ (3.17.11 and 12.9.1) we may take in (22.10.8.2) f' with values in [0, 1], equal to 1 on L'V' and to 0 outside $L'V'^2$, and g' positive, with support contained in V' and such that $\int g'\, dm_{\hat{G}} = 1$.

(22.10.15) It should be observed that there does not necessarily exist any nonnegligible function $u \in \mathscr{L}_{\mathbf{C}}^1(G)$ such that *both u and its Fourier transform $\mathscr{F}u$ have compact support* **(22.18.1)**.

(22.10.16) Recall **(22.8.11)** that the spaces $\mathscr{P}(G)$, $\mathscr{P}^1(G)$ and $\mathscr{P}^2(G)$ are in general *not closed* in $\mathscr{C}_{\mathbf{C}}^\infty(G)$ (in other words, they are not complete with respect to the norm $\|f\|$ of the space $\mathscr{B}_{\mathbf{C}}(G)$) (Problem 14). The same is true of the image of $L_{\mathbf{C}}^1(\hat{G})$ under the Fourier transformation (Problem 13). On the other hand, the set $\mathscr{P}_+(G)$ of continuous functions of positive type on G is closed in $\mathscr{C}_{\mathbf{C}}^\infty(G)$ **(22.1.10)**.

PROBLEMS

1. (a) With the notation of **(22.10.9)**, let U' and V' be two compact symmetric neighborhoods of $\hat{e}$ in $\hat{G}$. Let $f = F\varphi_{U'}$, $g = F\varphi_{U'V'}$, $h = (m_{\hat{G}}(U'))^{-1}fg$. Show that

$$N_1(h) \leqq (m_{\hat{G}}(U'V')/m_{\hat{G}}(U'))^{1/2}$$

and that for all $y \in G$

$$N_1(\gamma(y)h - h) \leqq 2(m_{\hat{G}}(U'V')/m_{\hat{G}}(U'))^{1/2} \sup_{\hat{x} \in U'V'} |\langle y, \hat{x}\rangle - 1|.$$

(Write $\gamma(y)h - h$ in the form

$$(\gamma(y)f - f)\gamma(y)g + f(\gamma(y)g - g),$$

and use Plancherel's theorem and the Cauchy–Schwarz inequality.) We have $0 \leqq \mathscr{F}h(\hat{x}) \leqq 1$ for all $\hat{x} \in \hat{G}$, $\mathscr{F}h(\hat{x}) = 1$ for $\hat{x} \in V'$, and $\operatorname{Supp}(\mathscr{F}h) \subset U'^2V'$.

(b) Deduce from (a) that for each $\alpha > 1$, each $\varepsilon > 0$ and each compact subset K of G, there exists a continuous function $h \in L_{\mathbf{C}}^1(G)$ such that $N_1(h) \leqq \alpha$, $N_1(\gamma(y)h - h) \leqq \varepsilon$ for all $y \in K$, and such that $\mathscr{F}h(\hat{x}) = 1$ at all points of a neighborhood of $\hat{e}$ in $\hat{G}$.

(c) For each function $f \in \mathscr{L}_{\mathbf{C}}^1(G)$, each $\alpha > 1$ and each $\varepsilon > 0$, there exists a continuous function $h \in L_{\mathbf{C}}^1(G)$ such that $N_1(h) \leqq \alpha$, $\mathscr{F}h(\hat{x}) = 1$ at all points of some neighborhood of $\hat{e}$ in $\hat{G}$, and such that

$$N_1\left(f * h - \left(\int f(x)\, dm_G(x)\right)h\right) \leqq \varepsilon.$$

2. (a) With the notation of **(22.10.9)**, let $f \in \mathscr{L}_{\mathbf{C}}^1(G)$, let $\hat{x}_0 \in \hat{G}$, and let Φ be a holomorphic complex function defined on a neighborhood of the point $\mathscr{F}f(\hat{x}_0)$. Show that there exists a function $g_{\hat{x}_0} \in \mathscr{L}_{\mathbf{C}}^1(G)$ such that $\mathscr{F}g_{\hat{x}_0}(\hat{x}) = \Phi(\mathscr{F}f(\hat{x}))$ in a neighborhood of $\hat{x}_0$. (Use the Taylor expansion of Φ, Problem 1(c) and **(22.7.1.5)**.)

(b) Let $f \in \mathscr{L}^1_{\mathbf{C}}(G)$, let K′ be a compact subset of $\hat{G}$, and let Φ be a holomorphic complex function defined on a neighborhood of the compact subset $\mathscr{F}f(K')$ of **C**. Show that there exists a function $g \in L^1_{\mathbf{C}}(G)$ such that $\mathscr{F}g(\hat{x}) = \Phi(\mathscr{F}f(\hat{x}))$ for all $\hat{x} \in K'$ (*Wiener–Lévy theorem*). (Use (22.10.14) and a suitably chosen partition of unity.)
(c) Let $f \in \mathscr{L}^1_{\mathbf{C}}(G)$ be such that its Fourier transform $\mathscr{F}f$ does not vanish at any point of $\hat{G}$. Show that for each function $g \in \mathscr{L}^1_{\mathbf{C}}(G)$ such that $\mathscr{F}g \in \mathscr{K}(\hat{G})$, there exists a function $h \in \mathscr{L}^1_{\mathbf{C}}(G)$ such that $g = h * f$. (Use (b).)

3. With the notation of (22.10.9), show that for each function $f \in \mathscr{L}^1_{\mathbf{C}}(G)$ and each $\varepsilon > 0$ there exists a function $u \in \mathscr{L}^1_{\mathbf{C}}(G)$ such that $\mathscr{F}u$ has compact support in $\hat{G}$ and such that $N_1(u * f - f) \leq \varepsilon$. (Use the fact that $\mathscr{K}(G) \cap \mathscr{P}^1(G)$ is dense in $L^1_{\mathbf{C}}(G)$ (22.8.11).)

4. (a) For each function $f \in \mathscr{L}^1_{\mathbf{C}}(G)$, show that the closed ideal of $L^1_{\mathbf{C}}(G)$ generated by $\tilde{f}$ is equal to the closed vector subspace generated by the classes of the translates $\gamma(s)f$ for $s \in G$. (Reduce to proving that the distance from $(h * f)^{\sim}$ to this subspace is zero when $h \in \mathscr{L}^1_{\mathbf{C}}(G)$ is compactly supported; then use the continuity of the function $x \mapsto \gamma(x)f$ (14.10.6.3).)
(b) In order that the set of classes of the functions $\gamma(s)f$, where $s \in G$, should be *total* in $L^1_{\mathbf{C}}(G)$, it is necessary and sufficient that the continuous function $\mathscr{F}f$ should be nonzero at each point of $\hat{G}$ (*Wiener's approximation theorem*). (Use (a) and Problem 2(c).)
(c) Suppose that $\mathscr{F}f$ is $\neq 0$ at each point of $\hat{G}$ and that, for some function $u \in \mathscr{L}^1_{\mathbf{C}}(G)$, there exists a constant C such that the continuous function $f * u - C \int f(x)\, dm_G(x)$ tends to 0 at infinity (13.20.6). Show that for *each* function $g \in \mathscr{L}^1_{\mathbf{C}}(G)$, the continuous function

$$g * u - C\left(\int g(x)\, dm_G(x)\right)$$

tends to 0 at infinity (*Wiener's Tauberian theorem*). (Use (b) to reduce to the case where g is a linear combination of a finite number of translates $\gamma(x_i)f$.)
(d) Show that there exist functions $f \in \mathscr{L}^1_{\mathbf{C}}(G)$ such that $\mathscr{F}f(\hat{x}) > 0$ for all $\hat{x} \in \hat{G}$. (Use (3.18.3) and (22.10.4) to obtain a function $f \in \mathscr{L}^2_{\mathbf{C}}(G)$ such that $\mathscr{F}f(\hat{x}) > 0$ for all $\hat{x} \in \hat{G}$, and then consider f^2.) (Cf. Section 22.17, Problem 5.)

5. With the notation of (22.10.9), let $f \in \mathscr{L}^2_{\mathbf{C}}(G)$. In order that the set of classes of functions $\gamma(s)f$, where $s \in G$, should be *total* in $L^2(G)$, it is necessary and sufficient that the set of $\hat{x} \in \hat{G}$ such that $\mathscr{F}f(\hat{x}) = 0$ should be of measure zero relative to $m_{\hat{G}}$. (Observe that if g is orthogonal to all the classes of the functions $\gamma(s)f$, the function $\mathscr{F}f \cdot \mathscr{F}g$ is $m_{\hat{G}}$-negligible.) Deduce that the left regular representation of G is topologically cyclic (cf. Problem 4(d)).

6. With the notation of (22.10.9), show that the characters of G form a *total* set in the *Fréchet space* $\mathscr{C}_{\mathbf{C}}(G)$ (cf. Problem 7). (Use (22.8.11) and Fourier's reciprocity formula for functions in $\mathscr{P}^1(G)$.)

7. With the notation of (22.10.9), the *closed* subspace of the Banach space $\mathscr{C}^{\infty}_{\mathbf{C}}(G)$ of bounded continuous functions on G, generated by the characters of G, is called the space of *almost periodic functions* on G and is denoted by $\mathscr{P}p(G)$; it is contained in the closure $\bar{\mathscr{P}}(G)$ of $\mathscr{P}(G)$ in $\mathscr{C}^{\infty}_{\mathbf{C}}(G)$ (and therefore consists of uniformly continuous functions on G), but is *not*

in general contained in $\mathscr{P}(G)$ (Problem 12). The set $\mathscr{P}p(G)$ is a star subalgebra of $\mathscr{C}_{\mathbf{C}}^{\infty}(G)$. If G is compact, $\mathscr{P}p(G) = \mathscr{C}_{\mathbf{C}}^{\infty}(G) = \mathscr{C}_{\mathbf{C}}(G)$.

(a) Let f be an almost periodic function on G, hence the uniform limit of a sequence (g_n), where each of the functions g_n is a linear combination of a finite sequence $(\hat{x}_{mn})_{1 \leq m \leq r_n}$ of characters of G. Let Γ be the at most denumerable subgroup of $\hat{G}$ generated by the $\hat{x}_{mn}$. Show that if $\hat{\Gamma}$ denotes the metrizable compact group dual to the *discrete* group Γ (22.14.1), there exists a continuous homomorphism φ of G into $\hat{\Gamma}$ and a continuous function u on $\hat{\Gamma}$ such that $f(x) = u(\varphi(x))$ for all $x \in G$.

(b) Conversely, for every continuous homomorphism φ of G into a metrizable compact commutative group H, and every continuous complex function u on H, the function $x \mapsto u(\varphi(x))$ is almost periodic on G (apply (21.3.4) to the group H).

(c) If f is an almost periodic function on G, the set of translates $\gamma(s)f$, where $s \in G$, is relatively compact in the Banach space $\mathscr{C}_{\mathbf{C}}^{\infty}(G)$. (Use (a) to reduce to the case where G is compact.)

(d) Deduce from (a) and (b) that, for each continuous mapping ψ of **C** into itself and each almost periodic function f, the function $\psi \circ f$ is almost periodic.

(e) If f is almost periodic, then so also is $f * \mu$ for any bounded measure $\mu \in M_{\mathbf{C}}^{1}(G)$ (use (14.9.2)).

8. (a) Let X be a metrizable compact space, and endow the space $\mathscr{C}_X(X)$ with the metrizable topology defined in (16.27.4.1). Show that if $\mathscr{H}(X)$ is the group of homeomorphisms of X onto itself, the topology induced on $\mathscr{H}(X)$ by that of $\mathscr{C}_X(X)$ is compatible with the group structure of $\mathscr{H}(X)$. (Observe that if $u_0 \in \mathscr{H}(X)$, then u_0^{-1} is uniformly continuous on X.) Moreover, the mapping $(u, x) \mapsto u(x)$ of $\mathscr{C}_X(X) \times X$ into X is continuous.

(b) Let G be a topological group acting continuously on a Banach space F (12.10), so that the set of mappings $x \mapsto s \cdot x$ of F into itself, for $s \in G$, is equicontinuous (7.5). A vector $x_0 \in F$ is said to be *almost periodic* for the action of G if the closure X of the orbit of x_0 in F is *compact*. Then X is stable under G. For each $s \in G$, let $\varphi(s)$ denote the homeomorphism $x \mapsto s \cdot x$ of X onto itself, and let Γ be the subgroup of $\mathscr{H}(X)$ consisting of the $\varphi(s)$, $s \in G$. Show that the closure $\bar{\Gamma}$ of Γ in $\mathscr{H}(X)$ is a metrizable compact group, and that φ is a continuous homomorphism of G into $\bar{\Gamma}$ (use Ascoli's theorem (7.5.7)).

(c) Suppose that the mappings $x \mapsto s \cdot x$ of F into itself (for all $s \in G$) are isometries. In order that x_0 should be almost periodic for the action of G it is necessary and sufficient that, for each $\varepsilon > 0$, there should exist a finite number of elements $s_j \in G$ $(1 \leq j \leq m)$ such that for each $t \in G$ there exists an index j for which $\|x_0 - (s_j^{-1}t) \cdot x_0\| \leq \varepsilon$.

9. (a) Deduce from Problems 7(c) and 8 that if G is a locally compact commutative group and f is a bounded continuous function on G, the following conditions are equivalent:

(α) f is almost periodic;

(β) the set $\{\gamma(s)f \colon s \in G\}$ is relatively compact in $\mathscr{C}_{\mathbf{C}}^{\infty}(G)$;

(γ) for each $\varepsilon > 0$, there exists a finite number of elements $s_j \in G$ such that for each $t \in G$ there exists an index j for which $|f(x) - f(s_j^{-1}tx)| \leq \varepsilon$ for all $x \in G$.

(To show that (β) implies (α), let X denote the closure of the set of $\gamma(s)f$ in $\mathscr{C}_{\mathbf{C}}^{\infty}(G)$, $\varphi(s)$ the mapping $g \mapsto \gamma(s)g$ of X onto itself, and Γ the image $\varphi(G)$; observe that if we put $u(\sigma) = (\sigma f)(e)$ for $\sigma \in \bar{\Gamma}$, then $f(x) = u(\varphi(x))$ for all $x \in G$.)

(b) Let φ be a continuous homomorphism of G into a compact metrizable space H, such that $\varphi(G)$ is dense in H, and let μ be the Haar measure on H with total mass 1. For an almost periodic function $f = u \circ \varphi$, where $u \in \mathscr{C}_{\mathbf{C}}^{\infty}(H)$, let $M(f) = \int_H u \, d\mu$. Show that if

$W(f)$ is the closed convex hull of the $\gamma(s)f$ in $\mathscr{C}_{\mathbf{C}}^{\infty}(G)$, where s runs through G (Section **12.14**, Problem 13), there exists a unique constant function in $W(f)$, whose value is $M(f)$. (Use the uniform continuity of u and a suitable covering of H by sets of the form $\gamma(x_j)V$, where V is a neighborhood of e in H, to show (with the aid of a partition of unity) that for each $\varepsilon > 0$ there exist a finite number of points $s_j \in G$ and numbers $c_j \geqq 0$ such that $\sum_j c_j = 1$ and $|M(f) - \sum_j c_j u(\varphi(s_j)x)| \leqq \varepsilon$ for all $x \in H$. The uniqueness of the constant $M(f)$ results from the facts that $M(\gamma(s)f) = M(f)$ for all $s \in G$ and $|M(f)| \leqq \|f\|$: a relation of the form $|A - \sum_j c_j' f(s_j' s)| \leqq \varepsilon$, valid for all $s \in G$, with $c_j' \geqq 0$ and $\sum_j c_j' = 1$, implies (since $M(A) = A$) the relation $|A - \sum_j c_j' M(f)| = |A - M(f)| \leqq \varepsilon$.)

The number $M(f)$, defined independently of the expression $f = u \circ \varphi$ of the almost periodic function f, is called the *mean* of f over G. The mapping $f \mapsto M(f)$ is a continuous linear form on the Banach space $\mathscr{P}p(G)$.

(c) If $\hat{x}, \hat{y}$ are two distinct characters of G, show that $M(\hat{x}\bar{\hat{y}}) = 0$ (cf. **21.3.2**). Deduce that if $\hat{y}$ is a character of G distinct from the characters $\hat{x}_{mn}$ in Problem 7, we have $M(\bar{\hat{y}}f) = 0$. For every almost periodic function f on G, the set Ξ_f of characters $\hat{x}$ of G such that $M(\bar{\hat{x}}f) \neq 0$ is therefore at most denumerable, and we have $M(|f|^2) = \sum_{\hat{x} \in \Xi_f} |M(\bar{\hat{x}}f)|^2$. The relation $M(|f|^2) = 0$ is equivalent to $f = 0$ (**21.3.2**).

(d) For each at most denumerable subset Ξ of $\hat{G}$, the set $\mathscr{P}p_{\Xi}(G)$ of almost periodic functions f on G such that $\Xi_f \subset \Xi$ is a closed vector subspace of $\mathscr{P}p(G)$. Show that for each $\hat{x} \in \Xi$ there exists a number $c_{n,\hat{x}}$ such that for each function $f \in \mathscr{P}p_{\Xi}(G)$, f is the uniform limit in G of the continuous functions

$$L_n(f) = \sum_{\hat{x} \in \Xi} c_{n,\hat{x}} M(\bar{\hat{x}}f)\hat{x},$$

the family on the right-hand side being summable in the topology of uniform convergence (use (**21.3.3**) and (**14.11.1**)).

(e) Show that if f is almost periodic and $\mu \in M_{\mathbf{C}}^1(G)$, we have $M(f * \mu) = M(f)\mu(G)$.

(f) If f and g are two almost periodic functions on G, show that for each character $\hat{x} \in \hat{G}$ we have

$$M(\bar{\hat{x}}fg) = \sum_{\hat{y}\hat{z} = \hat{x}} M(\bar{\hat{y}}f)M(\bar{\hat{z}}g),$$

the family on the right-hand side being absolutely summable.

10. With the notation of (**22.10.9**), let (μ_n) be a sequence of bounded complex measures on G, such that the sequence $(\|\mu_n\|)$ is *bounded*. To say that the sequence $(\mathscr{F}\mu_n)$ converges pointwise in $\hat{G}$ signifies that the sequences $(\mathscr{R}\mu_n)$ and $(\mathscr{I}\mu_n)$ converge in $M_{\mathbf{R}}^1(G)$ with respect to the weak topology corresponding to the vector space of real-valued almost periodic functions on G (Problem 7). If we denote by $E_{9/4}$ the vector space of the real parts of the functions in $\mathscr{P}(G)$, and by $\mathscr{T}_{9/4}$ the corresponding weak topology on $M_{\mathbf{R}}^1(G)$, then the preceding condition is implied by the convergence of the sequence (μ_n) in the topology $\mathscr{T}_{9/4}$, and *a fortiori* by its convergence in the topology $\mathscr{T}_{5/2}$ (Section **14.9**, Problem 5) and by its convergence in the topology $\mathscr{T}_3$ (Section **13.20**, Problem 1; we denote by $\mathscr{T}_j$ the topology on $M_{\mathbf{C}}^1(G) = M_{\mathbf{R}}^1(G) \oplus iM_{\mathbf{R}}^1(G)$ which is the product of the topologies $\mathscr{T}_j$ on the two factors).

(a) Show that (under the hypothesis that the sequence $(\|\mu_n\|)$ is bounded) if the sequence $(\mathscr{F}\mu_n)$ converges pointwise in $\hat{G}$, then the sequence (μ_n) converges in the topology $\mathscr{T}_{9/4}$ and *a fortiori* in the topology $\mathscr{T}_2$. (Use Bochner's theorem (**22.10.13**).)

(b) Take $G = \hat{G} = \mathbf{R}$ (22.14.3). The sequence of measures ε_n converges to 0 with respect to $\mathcal{T}_2$, but does not converge with respect to $\mathcal{T}_{9/4}$. If we take

$$\mu_n = n^{-1}(\varepsilon_1 + \varepsilon_2 + \cdots + \varepsilon_n),$$

the sequence (μ_n) converges to 0 with respect to $\mathcal{T}_{9/4}$, but the sequence $(\mathcal{F}\mu_n)$ does not converge pointwise to 0; its limit is a discontinuous function.
(c) Take $G = \hat{G} = \mathbf{R}$ (22.14.3), and consider the sequence (μ_n) of measures on $\mathbf{R}$ such that μ_n has as density with respect to Lebesgue measure the function g_n defined by $g_n(x) = x/n^2$ for $-n \leqq x \leqq n$, and $g_n(x) = 0$ otherwise. Show that (μ_n) converges to 0 with respect to the topology $\mathcal{T}_{9/4}$, but not with respect to $\mathcal{T}_{5/2}$ (and hence that there are uniformly continuous bounded functions on $\mathbf{R}$ which do not belong to $\overline{\mathcal{P}(\mathbf{R})}$). The sequence $(\mathcal{F}\mu_n)$ converges pointwise to 0 in $\mathbf{R}$, but the convergence is not uniform in any neighborhood of 0.
(d) Again take $G = \hat{G} = \mathbf{R}$ (22.14.3) and consider the sequence of measures $\mu_n = \varepsilon_n - \varepsilon_{n+(1/n)}$. The sequence (μ_n) converges to 0 in the topology $\mathcal{T}_{5/2}$ but not in $\mathcal{T}_3$; the sequence $(\mathcal{F}\mu_n)$ converges to 0 uniformly on each compact subset of $\mathbf{R}$, but not on the whole of $\mathbf{R}$.
(e) If the sequence (μ_n) converges vaguely to μ, then we always have

$$\|\mu\| \leqq \liminf_{n\to\infty} \|\mu_n\|.$$

Show that if $\|\mu_n\| \to \|\mu\|$ as $n \to \infty$, then $\mu_n \to \mu$ in the topology $\mathcal{T}_3$ (but there exist sequences which converge to 0 in the topology $\mathcal{T}_3$ without the sequence $(\|\mu_n\|)$ converging to 0). (We may restrict our attention to the case where the sequence $(|\mu_n|)$ is vaguely convergent (13.4.2). Argue by contradiction to show that the condition of Section 13.20, Problem 1(b) is satisfied, by showing that otherwise there would exist a number $\alpha > 0$ such that $|\mu|(K) \leqq \|\mu\| - \alpha$ for all compact subsets K of G.)
(f) Show that if the sequence (μ_n) converges to μ with respect to the topology $\mathcal{T}_3$, then the sequence $(\mathcal{F}\mu_n)$ converges uniformly to $\mathcal{F}\mu$ on each compact subset of $\hat{G}$ (observe that this sequence is equicontinuous).
(g) Suppose that the μ_n are positive and such that $\mathcal{F}\mu_n$ converges pointwise in $\hat{G}$ to a function φ which is continuous at the identity element $\hat{e}$. Show that under these conditions the sequence (μ_n) converges with respect to the topology $\mathcal{T}_3$ to a positive measure ν such that $\varphi = \mathcal{F}\nu$. (Observe first that the sequence $(\|\mu_n\|) = (\mu_n(1))$ is bounded; let V′ be a neighborhood of $\hat{e}$ in which $|\varphi(\hat{x}) - \varphi(\hat{e})| \leqq \varepsilon$; then the set H of elements $x \in G$ such that $\mathcal{R}(1 - \langle x, \hat{x}\rangle) \leqq \frac{1}{2}$ for all $\hat{x} \in V'$ is a neighborhood of e in G; show that $\limsup_{n\to\infty} \mu_n(\complement H) \leqq 2\varepsilon$.)
(h) Suppose that the sequence $(\|\mu_n\|)$ is bounded and that the sequence $(\mathcal{F}\mu_n)$ converges pointwise in $\hat{G}$ to a function φ which is continuous at all points of $\hat{G}$. Show that (μ_n) converges to a bounded measure μ with respect to the topology $\mathcal{T}_{9/4}$ (but not necessarily with respect to $\mathcal{T}_{5/2}$, as (c) above shows) and that $\mathcal{F}\mu = \varphi$. (Choose a vaguely convergent subsequence of (μ_n) (13.4.2); if μ is its limit, show that we have $\langle \mathcal{F}\mu, g'\rangle = \langle \varphi, g'\rangle$ for all functions g' of the form $u' * v'$, where u' and v' are in $\mathcal{K}(\hat{G})$, by using (22.10.10.7).)

11. Take $G = \hat{G} = \mathbf{R}$ (22.14.3), and consider the sequence of measures μ_n, where μ_n is the measure whose density with respect to Lebesgue measure is the function g_n defined by $g_n(x) = x/n^{3/2}$ for $|x| \leqq n$, and $g_n(x) = 0$ otherwise. The sequence $(\|\mu_n\|)$ tends to $+\infty$; the sequence $(\mathcal{F}\mu_n)$ converges pointwise to 0 in $\mathbf{R}$, but is not uniformly bounded.

12. Let G be a (separable, metrizable) locally compact commutative group. Show that if G is not compact, the irreducible representations U_ω corresponding to the characters of G (22.9.2) are not equivalent to any subrepresentations of the regular representation of G (21.1.9). (By regularization, it comes to the same thing to say that there exists no continuous function $f \in L^2_{\mathbf{C}}(G)$ which is a nonzero eigenvector for the regular representation; use the Fourier transformation, observing that the support of $\mathscr{F}f$ cannot be a discrete set unless $\hat{G}$ is discrete.)

13. With the notation of (22.10.9), we propose to show that unless G is *finite*, the image of $L^1_{\mathbf{C}}(G)$ under the Fourier transformation $\mathscr{F}$ cannot be the whole of $\mathscr{C}^0_{\mathbf{C}}(\hat{G})$. (*I. Segal's theorem*). The proof goes as follows:
(a) If $\mathscr{F}(L^1_{\mathbf{C}}(G)) = \mathscr{C}^0_{\mathbf{C}}(\hat{G})$, there would exist $c > 0$ such that $\|\mathscr{F}f\| \geqq c \cdot N_1(f)$ by Banach's theorem (12.16.8). By duality ((13.17.1) and (22.10.10.7)), the continuous linear mapping $\mu' \mapsto (\mathscr{F}\mu')^{\sim}$ would be a bijection of $M^1_{\mathbf{C}}(\hat{G})$ onto $L^\infty_{\mathbf{C}}(G)$.
(b) If G is not discrete, there are functions in $\mathscr{L}^\infty_{\mathbf{C}}(G)$ which are not equivalent to a continuous function.
(c) By Banach's theorem (12.16.8), there would exist $c' > 0$ such that $N_\infty(\mathscr{F}\mu') \geqq c'\|\mu'\|$, and *a fortiori* $\|\mathscr{F}f'\| \geqq c' \cdot N_1(f')$ for all functions $f' \in \mathscr{L}^1_{\mathbf{C}}(\hat{G})$. By (b) it follows that $\hat{G}$ must also be discrete.

14. Deduce from Problem **13** that if G is not finite, the subspace $\mathscr{P}(G)$ is not closed in $\mathscr{C}^\infty_{\mathbf{C}}(G)$. (Observe that by Banach's theorem we should again have $\|\mathscr{F}\mu'\| \geqq c'\|\mu'\|$ for all $\mu' \in M^1_{\mathbf{C}}(\hat{G})$.) Likewise, $\mathscr{P}^1(G)$ and $\mathscr{P}^2(G)$ are not closed in $\mathscr{C}^\infty_{\mathbf{C}}(G)$ unless G is finite. (Observe that their closure is $\mathscr{C}^0_{\mathbf{C}}(G)$ (22.8.11).)

15. Let G be a unimodular locally compact group containing a commutative normal subgroup A (such that the relation $s^2 = e$ in A implies $s = e$) and a compact subgroup K such that the mapping $(t, x) \mapsto tx$ of $K \times A$ into G is a homeomorphism (22.6.3).

Let $\mathscr{K}^0(A)$ denote the subspace of $\mathscr{K}(A)$ consisting of functions f such that $f(txt^{-1}) = f(x)$ for all $t \in K$ and $x \in A$.
(a) For each function $f \in \mathscr{K}(A)$, put

$$f^0(x) = \int_K f(txt^{-1})\, dm_K(t)$$

where m_K is the normalized Haar measure on K. The mapping $f \mapsto f^0$ is a projection of $\mathscr{K}(A)$ onto $\mathscr{K}^0(A)$, which extends to a continuous projection, given by the same formula, of $L^1_{\mathbf{C}}(A)$ onto the subspace $L^{1,0}_{\mathbf{C}}(A)$ consisting of the classes of functions $f \in \mathscr{L}^1_{\mathbf{C}}(A)$ such that for all $t \in K$ we have $f(txt^{-1}) = f(x)$ for almost all x. Show that if $\tilde{f} \in L^{1,0}_{\mathbf{C}}(A)$ and $\tilde{g} \in L^1_{\mathbf{C}}(A)$, then $(f * g)^0 = f * g^0$ almost everywhere. Now extend the above projection to a projection $\mu \mapsto \mu^0$ of $\mathbf{C}\varepsilon_e + L^1_{\mathbf{C}}(A) = B$ onto the subalgebra $B^0 = \mathbf{C}\varepsilon_e + L^{1,0}_{\mathbf{C}}(A)$ by defining $\varepsilon^0_e = \varepsilon_e$, and then we have $(\mu * \nu)^0 = \mu * \nu^0$ for $\mu \in B^0$ and $\nu \in B$.
(b) Let φ be a continuous mapping of B into C such that $\varphi(\varepsilon_e) = 1$ and $\varphi(\mu^0 * \nu) = \varphi(\mu)\varphi(\nu)$ for $\mu, \nu \in B$, so that the restriction of φ to B^0 is a character of B^0. Show that there exists a character of B which agrees with φ on B^0. (Let $\mathfrak{n}$ be the kernel of $\varphi \mid B_0$, which is a maximal ideal of B^0; show that there exists a maximal ideal $\mathfrak{m}$ of B such that $\mathfrak{m} \cap B^0 = \mathfrak{n}$, by considering the ideal in B consisting of the μ such that $(\mu * \nu)^0 \in \mathfrak{n}$ for all $\nu \in B$.)
(c) Let ω be a spherical function on G relative to K. Show that we have

$$\int_K \omega(stxt^{-1})\, dm_K(t) = \omega(s)\omega(x)$$

for $x \in A$ and $s \in G$. (Start with the formula $\omega(s) = (U_\omega(s) \cdot x_0 | x_0)$ and observe that for each $x \in A$ the operator $\int_K U_\omega(txt^{-1})\, dm_K(t)$ commutes with the $U_\omega(y)$ for $y \in A$ and the $U_\omega(t)$ for $t \in K$, and is therefore a scalar.) Deduce that if we put $\varphi(\mu) = \int \omega(x)\, d\mu(x)$ for each measure μ on A, we have $\varphi(\mu^0 * \nu) = \varphi(\mu)\varphi(\nu)$ for $\mu, \nu \in B$. Hence show that there exists a character χ of A such that

$$\varphi(f) = \int f^0(x)\chi(x)\, dm_A(x).$$

(Use (b) and the fact that $\varphi(f) = \varphi(f^0)$ for $f \in L^1_{\mathbf{C}}(A)$.) Conclude that

$$\omega(x) = \int_K \chi(txt^{-1})\, dm_K(t)$$

for all $x \in A$.

16. Let G be a (metrizable, separable) locally compact commutative group. Show that for $f \in \mathscr{L}^2_{\mathbf{C}}(G)$ to be tame (Section **22.2**, Problem 2) it is necessary and sufficient that its Fourier transform $\mathscr{F}f$ should be essentially bounded.

17. Let G be a locally compact commutative group. For each number p such that $1 < p < 2$ and each function $f \in \mathscr{L}^1_{\mathbf{C}}(G) \cap \mathscr{L}^2_{\mathbf{C}}(G)$, f belongs to $\mathscr{L}^p_{\mathbf{C}}(G)$. Show that $\mathscr{F}f$ belongs to $\mathscr{L}^q_{\mathbf{C}}(\hat{G})$, where $p^{-1} + q^{-1} = 1$, and that $N_q(\mathscr{F}f) \leq N_p(f)$ (*Hausdorff–Young theorem*). (Apply the Riesz–Thorin theorem, cf. Section **13.17**, Problem 7.)

Deduce that $\mathscr{F}$ extends to a continuous linear mapping with norm ≤ 1 of $L^p_{\mathbf{C}}(G)$ into $L^q_{\mathbf{C}}(\hat{G})$. Show that $L^p_{\mathbf{C}}(G) \subset P(G)$ and that for each $f \in \mathscr{L}^p_{\mathbf{C}}(G)$ we have $(f \cdot m_G)^\wedge = \mathscr{F}f \cdot m_{\hat{G}}$. Show that if G is discrete, the mapping $\mathscr{F}$ is not surjective from $L^p_{\mathbf{C}}(G)$ onto $L^q_{\mathbf{C}}(\hat{G})$ (use Section **13.21**, Problem 14). Show that if G is compact, there exist functions $f \in \mathscr{C}_{\mathbf{C}}(G) \subset \mathscr{L}^q_{\mathbf{C}}(G)$ whose Fourier transform $\mathscr{F}f$ does not belong to $\mathscr{L}^p_{\mathbf{C}}(\hat{G})$ (same reference).

11. DUAL OF A SUBGROUP AND OF A QUOTIENT GROUP

(22.11.1) Let G, G′ be two (separable, metrizable) locally compact commutative groups, and let $u: G \to G'$ be a *continuous homomorphism*. For each character $\hat{y}' \in \hat{G}'$, the function $\hat{y}' \circ u$ is a character of G, i.e. an element $\hat{x}$ of $\hat{G}$, which we denote by ${}^t u(\hat{y}')$. The mapping ${}^t u: \hat{G}' \to \hat{G}$ so defined is called the *transpose* of u and is characterized by the formula

(22.11.1.1)
$$\langle x, {}^t u(\hat{y}')\rangle = \langle u(x), \hat{y}'\rangle$$

for all $x \in G$ and $\hat{y}' \in \hat{G}'$. Since we have $(\hat{y}'\hat{z}') \circ u = (\hat{y}' \circ u)(\hat{z}' \circ u)$ for any two characters $\hat{y}'$, $\hat{z}'$ of G′, it follows that ${}^t u$ is a *homomorphism* of $\hat{G}'$ into $\hat{G}$.

Moreover, this homomorphism is *continuous*. For let L be a compact subset of G, and let ε be a positive real number; then $u(\mathrm{L})$ is a compact subset of G′ **(3.17.9)**, and the set of characters $\hat{y}' \in \hat{\mathrm{G}}'$ such that $|1 - \langle x', \hat{y}' \rangle| \leqq \varepsilon$ for all $x' \in u(\mathrm{L})$ is therefore a neighborhood of the identity element in $\hat{\mathrm{G}}'$; for each point $\hat{y}'$ in this neighborhood we have $|1 - \langle x, {}^t u(\hat{y}') \rangle| \leqq \varepsilon$ for each $x \in \mathrm{L}$, by virtue of **(22.11.1.1)**; this establishes the continuity of ${}^t u$ at the identity element of $\hat{\mathrm{G}}'$, and hence everywhere **(12.8.4)**.

If $v: \mathrm{G}' \to \mathrm{G}''$ is a continuous homomorphism of G′ into a locally compact commutative group G″, it follows immediately from **(22.11.1.1)** that

(22.11.1.2) $${}^t(v \circ u) = {}^t u \circ {}^t v.$$

Finally, since the duals of $\hat{\mathrm{G}}$ and $\hat{\mathrm{G}}'$ are canonically identified with G and G′, it follows from **(22.11.1.1)** that

(22.11.1.3) $${}^t({}^t u) = u.$$

(22.11.2) Let G be a locally compact commutative group, $\hat{\mathrm{G}}$ the dual of G; an element $x \in \mathrm{G}$ and an element $\hat{x} \in \hat{\mathrm{G}}$ are said to be *orthogonal* if $\langle x, \hat{x} \rangle = 1$. For each subset A of G, the set of elements $\hat{x} \in \hat{\mathrm{G}}$ orthogonal to *all* the elements of A is evidently a *closed subgroup* of $\hat{\mathrm{G}}$ **(3.15.1)**, called the *annihilator* of A and denoted by $\mathrm{A}^{\perp}$. If H is the subgroup of G generated by A, it is clear that $\bar{\mathrm{H}}^{\perp} = \mathrm{H}^{\perp} = \bar{\mathrm{A}}^{\perp} = \mathrm{A}^{\perp}$. The annihilator of G is $\{\hat{e}\}$ and the annihilator of $\{e\}$ is $\hat{\mathrm{G}}$.

(22.11.3) *Let* H *be a closed subgroup of the locally compact commutative group* G, *and let* $j: \mathrm{H} \to \mathrm{G}$ *and* $p: \mathrm{G} \to \mathrm{G/H}$ *be the corresponding canonical injection and surjection. Then the transpose* ${}^t p$ *of* p *is an isomorphism of* $(\mathrm{G/H})\hat{}$ *onto the closed subgroup* $\mathrm{H}^{\perp}$ *of* $\hat{\mathrm{G}}$, *and the transpose* ${}^t j$ *of* j *is a strict morphism* **(12.12.7)** *of* $\hat{\mathrm{G}}$ *onto* $\hat{\mathrm{H}}$, *with kernel* $\mathrm{H}^{\perp}$.

(i) A character $\hat{x}$ of G is of the form $\hat{z}'' \circ p$, where $\hat{z}''$ is a character of G/H, if and only if $\hat{x}$ is trivial on H **(12.10.6)**, so that ${}^t p((\mathrm{G/H})\hat{}) = \mathrm{H}^{\perp}$. Moreover, if ${}^t p(\hat{z}'')$ is the identity element $\hat{e}$ of $\hat{\mathrm{G}}$, then for each $x \in \mathrm{G}$ we have $1 = \langle x, {}^t p(\hat{z}'') \rangle = \langle p(x), \hat{z}'' \rangle$, so that $\hat{z}''$ is the identity element of $(\mathrm{G/H})\hat{}$. Hence ${}^t p$ is an *injective* homomorphism of $(\mathrm{G/H})\hat{}$ into $\hat{\mathrm{G}}$. Let us show that ${}^t p$ is a homeomorphism of $(\mathrm{G/H})\hat{}$ onto its image $\mathrm{H}^{\perp}$. We may obtain a fundamental system of neighborhoods of the identity element of $(\mathrm{G/H})\hat{}$ by forming, for each $\varepsilon > 0$ and each compact subset L″ of G/H, the set $\mathrm{U}(\varepsilon, \mathrm{L}'')$ of $\hat{z}'' \in (\mathrm{G/H})\hat{}$ such that $|\langle z'', \hat{z}'' \rangle - 1| \leqq \varepsilon$ for all $z'' \in \mathrm{L}''$. There exists a compact subset L of G such that $p(\mathrm{L}) = \mathrm{L}''$ **(12.10.9)**. Let V be the neighborhood of $\hat{e}$ in $\hat{\mathrm{G}}$ consisting of the characters $\hat{x}$ such that $|\langle x, \hat{x} \rangle - 1| \leqq \varepsilon$ for all $x \in \mathrm{L}$. Then if ${}^t p(\hat{z}'') \in \mathrm{V}$, we have $|\langle p(x), \hat{z}''\rangle - 1| \leqq \varepsilon$ for all $x \in \mathrm{L}$, and consequently $\hat{z}'' \in \mathrm{U}(\varepsilon, \mathrm{L}'')$; in other words, $\mathrm{U}(\varepsilon, \mathrm{L}'')$ contains the neighborhood ${}^t p^{-1}(\mathrm{V})$ of the identity element of $(\mathrm{G}/\mathrm{H})\hat{}$, which proves our assertion.

(ii) A character $\hat{x}$ of G satisfies $\hat{x} \circ j = 1$ if and only if $\hat{x}$ is orthogonal to H, so that the kernel of ${}^t j$ is $\mathrm{H}^\perp$. Put $\mathrm{G}' = (\hat{\mathrm{G}}/\mathrm{H}^\perp)\hat{}$, so that $\hat{\mathrm{G}}/\mathrm{H}^\perp = \hat{\mathrm{G}}'$. Since every continuous homomorphism of $\hat{\mathrm{G}}$ into $\hat{\mathrm{G}}'$ is the transpose of a continuous homomorphism of G′ into G **(22.11.1.3)**, the canonical surjection of $\hat{\mathrm{G}}$ onto $\hat{\mathrm{G}}'$ may be written in the form ${}^t\psi$, where $\psi: \mathrm{G}' \to \mathrm{G}$ is a continuous homomorphism. On the other hand, we have ${}^t j = u \circ {}^t\psi$, where $u: \hat{\mathrm{G}}' \to \hat{\mathrm{H}}$ is a continuous homomorphism **(12.10.6)**, and we may write $u = {}^t\eta$, where $\eta: \mathrm{H} \to \mathrm{G}'$ is a continuous homomorphism. We have then **(22.11.1.2)** $j = \psi \circ \eta$, hence $\psi(\mathrm{G}') \supset \mathrm{H}$. On the other hand, since ${}^t p((\mathrm{G}/\mathrm{H})\hat{}) = \mathrm{H}^\perp$, we have ${}^t(p \circ \psi)((\mathrm{G}/\mathrm{H})\hat{}) = \{\hat{e}\}$; this implies that for each $y' \in \mathrm{G}'$ we have $\langle (p \circ \psi)(y'), \hat{z}''\rangle = 1$ for all $\hat{z}'' \in (\mathrm{G}/\mathrm{H})\hat{}$, hence $(p \circ \psi)(\mathrm{G}')$ consists only of the identity element of G/H, which implies that $\psi(\mathrm{G}') \subset \mathrm{H}$. We have therefore shown that $\psi(\mathrm{G}') = \mathrm{H}$. Part (i) of the proof, with G and H replaced by $\hat{\mathrm{G}}$ and $\mathrm{H}^\perp$, now shows that ψ is an *isomorphism* of G′ onto the closed subgroup H. Since $j = \psi \circ \eta$, η is an isomorphism of H onto G′, ${}^t\eta$ is an isomorphism of $\hat{\mathrm{G}}'$ onto $\hat{\mathrm{H}}$, and hence ${}^t j$ is a strict morphism of $\hat{\mathrm{G}}$ onto $\hat{\mathrm{H}}$, with kernel $\mathrm{H}^\perp$.

Henceforth we shall *identify* the dual of H with $\hat{\mathrm{G}}/\mathrm{H}^\perp$ by means of ${}^t\eta$, and the dual of G/H with $\mathrm{H}^\perp$ by means of ${}^t p$.

(22.11.4) (i) *For each subgroup* H *of* G, $(\mathrm{H}^\perp)^\perp = \bar{\mathrm{H}}$.

(ii) *For each family* (H_α) *of closed subgroups of* G, *the annihilator of the subgroup of* G *generated by the* H_α *is* $\bigcap_\alpha \mathrm{H}_\alpha^\perp$; *the annihilator of* $\bigcap_\alpha \mathrm{H}_\alpha$ *is the closure of the subgroup of* $\hat{\mathrm{G}}$ *generated by the* $\mathrm{H}_\alpha^\perp$.

(i) Suppose first that H is closed. Then, with the notation of **(22.11.3)**, we have seen that ${}^t j$ is a surjective strict morphism of $\hat{\mathrm{G}}$ onto $\hat{\mathrm{H}}$, with kernel $\mathrm{H}^\perp$; hence $j = {}^t({}^t j)$ is an isomorphism of H onto $(\mathrm{H}^\perp)^\perp$, by reason of **(22.11.3)**, and therefore we have $\mathrm{H} = (\mathrm{H}^\perp)^\perp$ in this case. In general we have $\bar{\mathrm{H}} \subset (\mathrm{H}^\perp)^\perp = (\bar{\mathrm{H}}^\perp)^\perp = \bar{\mathrm{H}}$.

(ii) The first assertion is obvious, and the second follows from the first by interchanging G and $\hat{\mathrm{G}}$, and using (i).

(22.11.5) *Let* $u: \mathrm{G} \to \mathrm{G}'$ *be a continuous homomorphism of locally compact commutative groups.*

(i) *The annihilator of* $u(\mathrm{G})$ *is the kernel of* ${}^t u$. *For* ${}^t u$ *to be injective it is necessary and sufficient that* $u(\mathrm{G})$ *should be dense in* G′.

(ii) *In order that u should be a strict morphism* (resp. *a surjective strict morphism*, resp. *an injective strict morphism*) *it is necessary and sufficient that* ${}^t u$ *should be a strict morphism* (resp. *an injective strict morphism*, resp. *a surjective strict morphism*).

(i) In order that $\hat{x}' \in \hat{G}'$ should satisfy the condition ${}^t u(\hat{x}') = \hat{e}$, it is necessary and sufficient that $\langle u(x), \hat{x}' \rangle = 1$ for each $x \in G$, by **(22.11.1.1)**, which means that $\hat{x}'$ is orthogonal to $u(G)$. Hence it follows from **(22.11.4)** that $\overline{u(G)}$ is the annihilator of the kernel of ${}^t u$, which proves the second assertion.

(ii) If u is a surjective strict morphism, ${}^t u$ is an injective strict morphism by **(22.11.3)**. If u is an injective strict morphism, u is an isomorphism of G onto the subgroup $u(G)$ of G′; consequently $u(G)$ is locally compact and hence closed in G′ **(12.9.6)**. It now follows from **(22.11.3)** that ${}^t u$ is a surjective strict morphism. Finally, since every strict morphism u is the composition of an injective strict morphism, an isomorphism, and a surjective strict morphism **(12.12.7)**, it follows that ${}^t u$ is also strict, by reason of **(22.11.1.2)**.

PROBLEMS

1. (a) With the notation of **(22.11.3)**, show that every character of H is the restriction of some character of G.

(b) If μ is a positive measure on G/H, there exists a unique positive measure μ^* on G such that $\mu^*(f) = \mu(f^\flat)$ for all functions $f \in \mathcal{K}(G)$, in the notation of Section **14.4**, Problem 2. If h is the function defined in Section **22.3**, Problem 6, and if π: G → G/H is the canonical mapping, then in order that a function g defined on G/H should be μ-integrable it is necessary and sufficient that the function $h \cdot (g \circ \pi)$ should be μ^*-integrable, and we have

$$\int_{G/H} g(\dot{x})\, d\mu(\dot{x}) = \int_{G} h(x) g(\pi(x))\, d\mu^*(x).$$

(c) Show that every continuous function of positive type on H extends to a continuous function of positive type on G. (Use Bochner's theorem, and (b) applied to $\hat{G}$ and $\hat{G}/H^\perp$.)

2. Let G be a (separable, metrizable) locally compact group, H a closed normal commutative subgroup of G.

(a) Let p be a continuous function of positive type on G, and let U be a continuous unitary representation of G on a separable Hilbert space E such that $p(s) = (U(s) \cdot x_0 \mid x_0)$, where x_0 is a totalizing vector for U **(22.1.3)**. If the restriction of p to H is a character χ of H, we have $U(t) \cdot x_0 = \chi(t^{-1}) x_0$ for all $t \in H$; for all $t \in H$, the vectors $U(s) \cdot x_0$ $(s \in G)$ are eigenvectors of the unitary operator $U(t)$, for the eigenvalues $\chi(st^{-1}s^{-1})$.

(b) Deduce from (a) that if χ is a character of H such that, for each neighborhood V of e in G, there exist $s \in V$ and $t \in H$ such that $\chi(st^{-1}s^{-1}) \neq \chi(t^{-1})$, then χ cannot be the restriction to H of a continuous function of positive type on G. (Use the fact that two eigenvectors of $U(t)$ corresponding to distinct eigenvalues are orthogonal in E.)

(c) Show that for the group G of matrices

$$\begin{pmatrix} x & y \\ 0 & 1 \end{pmatrix},$$

where $x \in \mathbf{R}^*$ and $y \in \mathbf{R}$, and the subgroup H of matrices

$$\begin{pmatrix} 1 & y \\ 0 & 1 \end{pmatrix},$$

there exists a character χ of H satisfying the condition in (b).

3. (a) With the notation of (**22.11.3**), a closed subgroup H of G is compact if and only if the subgroup $H^{\perp}$ of $\hat{G}$ is open in $\hat{G}$.
(b) Let (H_n) be a decreasing sequence of compact subgroups of G, such that $\bigcap_n H_n = \{e\}$. Show that $\hat{G}$ is the union of the open subgroups $H_n^{\perp}$. Moreover, if $\pi_n : G \to G/H_n$ is the canonical homomorphism, the mapping $x \mapsto (\pi_n(x))$ is an isomorphism of G onto a closed subgroup of the group $\prod_n (G/H_n)$.

4. (a) With the notation of (**22.11.3**), suppose that the groups G and $\hat{G}$ are written additively. Then the mapping $x \mapsto nx$, for each positive integer n, is a continuous homomorphism of G into itself, whose transpose is the homomorphism $\hat{x} \mapsto n\hat{x}$. Let $G^{(n)}$ and $G_{(n)}$ denote the image and kernel, respectively, of the homomorphism $x \mapsto nx$. Then $G_{(n)}$ is a closed subgroup of G, whereas the group $G^{(n)}$ is not necessarily closed in G (Section **12.2**, Problem 3). We have $(G^{(n)})^{\perp} = \hat{G}_{(n)}$ and $(G_{(n)})^{\perp} = \overline{\hat{G}^{(n)}}$.
(b) The intersection $G^{(\infty)}$ of the subgroups $G^{(n)}$ is a subgroup of G. The elements of $G^{(\infty)}$ are said to be *of infinite height* in G. The union $G_{(\infty)}$ of the $G_{(n)}$ is the subgroup of G consisting of the elements of finite order, also called the *torsion* elements of G. The group G is said to be a *torsion* group (resp. a *torsion-free* group) if $G_{(\infty)} = G$ (resp. $G_{(\infty)} = \{e\}$). We have $(G^{(\infty)})^{\perp} = \overline{\hat{G}_{(\infty)}}$.
(c) Show that if G is compact, then $G^{(\infty)}$ is the identity component of G. (Observe that in this case the groups $G^{(n)}$ are closed in G. Observe also that if H is the identity component of G, then G/H is totally disconnected (**12.10.12**), and hence that H is the intersection of the open subgroups of G which contain H (Section **12.9**, Problem 3), and that $H^{\perp} = \hat{G}_{(\infty)}$ by Problem 3(a).)
(d) The locally compact group $\mathbf{Q}_p$ (Section **14.3**, Problem 6) is totally disconnected, but all its elements are of infinite height.
(e) Let G be the subgroup of the product group $\mathbf{Q}_p^{\mathbf{N}}$ consisting of the sequences $x = (x_n)_{n \in \mathbf{N}}$ such that $x_n \in \mathbf{Z}_p$ for all but finitely many values of n. If $\mathfrak{B}$ is the set of neighborhoods of 0 in the compact metrizable group $\mathbf{Z}_p^{\mathbf{N}}$, $\mathfrak{B}$ is a fundamental system of neighborhoods of 0 for a topology on G compatible with its group structure, with respect to which G is metrizable, separable, and locally compact, and $H = \mathbf{Z}_p^{\mathbf{N}}$ is an open compact subgroup of G. Show that the subgroup $G^{(\infty)}$ of elements of infinite height in G is a proper dense subgroup.
(f) Let G be the compact group which is the product of the groups $\mathbf{Z}/n\mathbf{Z}$, $n \geqq 2$. Show that the group $G_{(\infty)}$ of torsion elements of G is a proper dense subgroup of G.

5. A (separable, metrizable) locally compact commutative group G is said to be *divisible* if $G = G^{(\infty)}$. If G is compact, an equivalent condition is that G is connected (Problem 4), or that the discrete group $\hat{G}$ is torsion-free.

(a) Let G be a locally compact commutative group, H an open subgroup of G; also let G′ be a divisible group, f a continuous homomorphism of H into G′, and x an element of $\complement$H. Show that there exists a continuous homomorphism g of the (open) subgroup H′ of G generated by H and x, into the group G′, which extends f. (Distinguish two cases, according as there exists or does not exist an integer $n > 0$ such that $nx \in$ H. In the first case, let k be the least such integer, and observe that there exists $y \in$ H such that $f(kx) = kf(y)$.)
(b) Deduce from (a) that there exists a continuous homomorphism $\bar{f}$ of G into G′ which extends f. (Observe that the discrete group G/H is generated by an at most denumerable family of elements, and proceed by induction.)
(c) Show that if G is connected and locally isomorphic to $\mathbf{R}^n$ (**16.9.9.4**), then G is isomorphic to a group of the form $\mathbf{R}^{n-p} \times \mathbf{T}^p$ for some integer $p \leq n$. (Argue as in (**16.30.7**), using the fact that $\mathbf{R}^n$ is simply connected; then apply (**19.7.9.1**).)
(d) Show that if G is locally isomorphic to $\mathbf{R}^n$, it is isomorphic to a group of the form $\mathbf{R}^m \times \mathbf{T}^p \times$ D, where D is a discrete group. (Observe that the connected component H of G is an open subgroup isomorphic to $\mathbf{R}^m \times \mathbf{T}^p$, hence is divisible, and apply (b) to the identity mapping of H into itself.)

6. (a) Let G, G′ be two (separable, metrizable) locally compact commutative groups, u a continuous homomorphism of G into G′. If μ is a bounded measure on G, $u(\mu)$ its image under u (Section **13.9**, Problem 24), we have $\mathscr{F}(u(\mu)) = (\mathscr{F}\mu) \circ {}^t u$.
(b) If σ is an automorphism of G, show that for each function $f \in \mathscr{L}^1_{\mathbf{C}}(\mathrm{G})$ we have $(\mathscr{F}(f \circ \sigma))(\hat{x}) = (\mathrm{mod}\ \sigma)^{-1}((\mathscr{F}f)({}^t\sigma^{-1}(\hat{x})))$ (cf. (**14.3.6.1**)).

7. With the notation of (**22.11.3**), let $\mu \in \mathrm{M}^1_{\mathbf{C}}(\mathrm{G})$, and suppose that $\mu \neq 0$.
(a) For a character $\hat{x} \in \hat{\mathrm{G}}$, the following conditions are equivalent: (1) $\hat{x} \cdot \mu = \mu$; (2) $\gamma(\hat{x})\mathscr{F}\mu = \mathscr{F}\mu$; (3) the support of μ is contained in the kernel $\hat{x}^{-1}(1)$ of $\hat{x}$.
(b) Let H be the closed subgroup of G generated by Supp(μ). Show that $\mathrm{H}^\perp$ is the subgroup of all $\hat{x} \in \hat{\mathrm{G}}$ such that $\gamma(\hat{x})\mathscr{F}\mu = \mathscr{F}\mu$.
(c) The measures $\hat{x} \cdot \mu$, as $\hat{x}$ runs through $\hat{\mathrm{G}}$, are pairwise distinct. Show that if there exists $c > 0$ such that $\|\hat{x} \cdot \mu - \hat{y} \cdot \mu\| \geqq c$ for $\hat{x} \neq \hat{y}$, then the group H is compact. (There exists a compact subset K of H such that $|\mu|(\mathrm{H} - \mathrm{K}) \leqq c/4$; also the set V of $\hat{x} \in \hat{\mathrm{H}}$ such that $|1 - \langle x, \hat{x}\rangle| \leqq c/(3\|\mu\|)$ for all $x \in$ K is a neighborhood of $\hat{e}$ in H; show that the hypothesis implies that V = $\{\hat{e}\}$.)

8. Let G be a metrizable infinite compact commutative group and let μ be a measure on G; μ has a unique expression as $\mu' + \mu''$, where μ' is a measure with base the normalized Haar measure m_{G}, and μ'' is disjoint from m_{G} (**13.18.4**).
(a) As $\hat{x}$ runs through the denumerable group $\hat{\mathrm{G}}$, the set of measures $\hat{x} \cdot \mu$ on G is relatively compact in $\mathrm{M}_{\mathbf{C}}(\mathrm{G})$ with respect to the vague topology. Show that if $(\hat{x}_n)$ is a sequence of distinct points in $\hat{\mathrm{G}}$ such that the sequence $(\hat{x}_n \cdot \mu)$ converges in the vague topology with limit ν, then it converges to ν also in the topology $\mathscr{T}_6$ (in the notation of Section **13.20**, Problem 2).
(b) Show that ν is a measure with base $|\mu''|$, hence disjoint from m_{G}. (Use the Riemann–Lebesgue theorem to show that the sequence $(\hat{x}_n \cdot \mu')$ converges to 0 in the topology $\mathscr{T}_6$.)
(c) For each closed subgroup H of G, show that either there are only finitely many distinct measures $\varphi_{\mathrm{H}}\hat{x}_n \cdot \mu$, or else $\varphi_{\mathrm{H}} \cdot \nu$ is disjoint from Haar measure m_{H}. (Apply (b) to the measures $\varphi_{\mathrm{H}}\hat{x}_n \cdot \mu$, considered as measures on H.)

9. (a) Prove the following lemma: let λ be a bounded positive measure on a separable, metrizable, locally compact space X. Let $f = g + ih$ be a complex λ-measurable function such that $|f| \leqq 1$; put $\int g\,d\lambda = k \cdot \|\lambda\|$, and assume that $0 < k < 1$. Then we have

$$\int |1 - f|\,d\lambda \leqq \|\lambda\|(1 - k + \sqrt{1 - k^2}).$$

(Observe that $\|\lambda\| \geqq \left|\int g\,d\lambda + i\int h\,d\lambda\right|$.)

(b) Let G be a metrizable compact commutative group, μ a nonzero complex measure on G; suppose that there exists $c > 0$ such that $\|\hat{x} \cdot \mu - \hat{y} \cdot \mu\| \geqq c$ for all $\hat{x} \neq \hat{y}$ in $\hat{G}$. Show that if there exists a complex μ-measurable function f such that $|f| \leqq 1$, $\mathscr{R}\left(\int f\hat{x}_1\,d\mu\right) \geqq k\|\mu\|$ and $\mathscr{R}\left(\int f\hat{x}_2\,d\mu\right) \geqq k\|\mu\|$ for two distinct points $\hat{x}_1$, $\hat{x}_2$ in $\hat{G}$, with $0 < k < 1$, then we must have

$$1 - k + \sqrt{1 - k^2} \geqq c/2\|\mu\|.$$

(If $\mu = h|\mu|$, where $|h(x)| = 1$ for all $x \in G$, apply (a) to the measure $|\mu|$ and the functions $\hat{x}_1 fh$ and $\hat{x}_2 fh$, by using the inequality

$$c \leqq \|\hat{x}_1 \cdot \mu - \hat{x}_2 \cdot \mu\| = \int |\hat{x}_1 - \hat{x}_2|\,d|\mu|$$

and the fact that $|\hat{x}_1(x)| = |\hat{x}_2(x)| = 1$ for all $x \in G$.)

10. Let G be a (separable, metrizable) locally compact commutative group. Let $E(G) \subset M_C^1(G)$ denote the set of bounded measures μ on G such that $\mathscr{F}\mu$ takes only *integer* values on $\hat{G}$.
(a) Show that E(G) is a *ring* with respect to convolution and contains all *idempotent* measures (i.e., such that $\mu * \mu = \mu$). For each measure $\mu \in E(G)$ and each character $\hat{x} \in \hat{G}$, we have $\hat{x} \cdot \mu \in E(G)$. Each nonzero $\mu \in E(G)$ satisfies $\|\mu\| \geqq 1$.
(b) Show that for every measure $\mu \in E(G)$, the closed subgroup of G generated by $\mathrm{Supp}(\mu)$ is compact. (Use Problem 7.)
(c) Henceforth suppose that G is *compact*. Let $\mu \in E(G)$, $\mu \neq 0$, and let $(\hat{x}_n)$ be a sequence of distinct elements of $\hat{G}$, such that the sequence $(\hat{x}_n \cdot \mu)$ converges vaguely to a measure ν. Then we have $\|\nu\| \leqq \|\mu\| - (12\|\mu\|)^{-1}$. (We have $\|\nu\| \leqq \|\mu\|$. For each k such that $0 < k < \|\nu\|/\|\mu\|$, show that there exists a continuous complex-valued function f on G such that $\|f\| \leqq 1$ and $\mathscr{R}\left(\int f\,d\nu\right) > k\|\mu\|$, and that there exist two distinct indices m, n such that $\mathscr{R}\left(\int f\hat{x}_m\,d\mu\right) > k\|\mu\|$ and $\mathscr{R}\left(\int f\hat{x}_n\,d\mu\right) > k\|\mu\|$. Then use Problem 9(b) to show that $k \leqq 1 - (12\|\mu\|^2)^{-1}$.)
(d) Let $\mu \in E(G)$, $\mu \neq 0$, and let $A \subset M_C(G)$ be the set of measures $\hat{x} \cdot \mu$ for which $\int \hat{x}\,d\mu \neq 0$. If $\bar{A}$ is the closure of A in the vague topology, show that $\bar{A} \subset E(G)$, $0 \notin \bar{A}$, and that if $\delta = \inf_{\lambda \in \bar{A}} \|\lambda\|$, there exists $\nu \in \bar{A}$ such that $\|\nu\| = \delta$. Show that the set B of measures $\hat{x} \cdot \nu$ such that $\int \hat{x} \cdot d\nu \neq 0$ is contained in $\bar{A}$ and is *finite*. (Argue by contradiction, using (c).)

(e) If H is the closed subgroup of G generated by Supp(ν), show that we have $\nu = \sum_{i=1}^{q} n_i \hat{x}_i \cdot m_{\mathrm{H}}$ for a finite number of characters $\hat{x}_i$, with coefficients $n_i \in \mathbf{Z}$. (Use (d) to show that the restrictions to H of the characters $\hat{x}$ such that $\int \hat{x}\, d\nu \neq 0$ are finite in number.)

(f) Deduce that there exist finitely many closed subgroups H_i $(1 \leqq i \leqq p)$ of G, and for each i a finite number of characters $\hat{x}_{ij}$ $(1 \leqq j \leqq q_i)$ such that

$$\mu = \sum_{i=1}^{p} \left(\sum_{j=1}^{q} n_{ij} \hat{x}_{ij} \right) m_{\mathrm{H}_i}$$

with coefficients $n_{ij} \in \mathbf{Z}$ (*P. Cohen's theorem*). (Distinguish the cases $\nu \in \mathrm{A}$ and $\nu \notin \mathrm{A}$. In the latter case, deduce from Problem 8(c) that we may write $\nu = \varphi_{\mathrm{H}} \hat{x}_1 \cdot \mu$ for some $\hat{x}_1 \in \hat{\mathrm{G}}$; if $\mu_1 = \bar{\hat{x}}_1 \cdot \mu$, show that μ_1 and $\mu - \mu_1$ are disjoint and hence that

$$\|\mu - \mu_1\| \leqq \|\mu\| - 1.$$

Then argue by induction on the least integer $> \|\mu\|$.)

(g) Deduce from (f) that the measure $\sum_{n=0}^{\infty} \varepsilon_n$ on $\mathbf{R}$ is not the Plancherel transform of a measure of positive type (cf. Section **22.17**, Problem 5).

12. POISSON'S FORMULA

(22.12.1) Let G be a locally compact commutative group, H a closed subgroup of G, and m_{G}, m_{H} Haar measures on G, H, respectively. Then there exists a unique Haar measure $m_{\mathrm{G/H}}$ on G/H such that

$$\text{(22.12.1.1)} \qquad \int_{\mathrm{G}} f(x)\, dm_{\mathrm{G}}(x) = \int_{\mathrm{G/H}} dm_{\mathrm{G/H}}(\dot{x}) \int_{\mathrm{H}} f(xt)\, dm_{\mathrm{H}}(t)$$

for all functions $f \in \mathscr{K}(\mathrm{G})$, where $\dot{x}$ is the canonical image of x in G/H (**14.4.2**); we shall denote this measure by $m_{\mathrm{G}}/m_{\mathrm{H}}$.

(22.12.2) *Let* G *be a (separable, metrizable) locally compact commutative group,* H *a closed subgroup of* G*;* m_{G}, m_{H} *Haar measures on* G, H, *respectively, and* $m_{\mathrm{G/H}}$ *the Haar measure* $m_{\mathrm{G}}/m_{\mathrm{H}}$ *on* G/H. *Identify* $(\mathrm{G/H})^{\wedge}$ *canonically with* $\mathrm{H}^{\perp}$, *and let* $m_{\mathrm{H}^{\perp}}$ *denote the Haar measure on* $\mathrm{H}^{\perp}$ *associated with* $m_{\mathrm{G/H}}$ (**22.10.5**). *Let* $f \in \mathscr{L}^1_{\mathbf{C}}(\mathrm{G})$, *and suppose that the restriction to* $\mathrm{H}^{\perp}$ *of the bounded continuous function* $\mathscr{F}f$ *is* $m_{\mathrm{H}^{\perp}}$*-integrable. Then, for* almost all $x \in \mathrm{G}$, *the function* $t \mapsto f(xt)$, *defined on* H, *is* m_{H}*-integrable, and we have*

$$\text{(22.12.2.1)} \qquad \int_{\mathrm{H}} f(xt)\, dm_{\mathrm{H}}(t) = \int_{\mathrm{H}^{\perp}} \langle x, \hat{s} \rangle \mathscr{F}f(\hat{s})\, dm_{\mathrm{H}^{\perp}}(\hat{s}).$$

The first assertion follows from (14.4.5), which also shows that the function $\dot{x} \mapsto g(\dot{x}) = \int_{H} f(xt)\, dm_H(t)$, defined almost everywhere on G/H, is $m_{G/H}$*-integrable*. The Fourier transform $\mathscr{F}g$ of g, considered as a function on $H^{\perp}$, is given by

$$\mathscr{F}g(\hat{s}) = \int_{G/H} \overline{\langle \dot{x}, \hat{s} \rangle}\, dm_{G/H}(\dot{x}) \int_{H} f(xt)\, dm_H(t)$$

for $\hat{s} \in H^{\perp}$. But $\langle \dot{x}, \hat{s} \rangle = \langle xt, \hat{s} \rangle$ for $x \in \dot{x}$ and $t \in H$, and hence we may write

$$\begin{aligned}\mathscr{F}g(\hat{s}) &= \int_{G/H} dm_{G/H}(\dot{x}) \int_{H} \overline{\langle xt, \hat{s} \rangle} f(xt)\, dm_H(t)\\ &= \int_{G} \overline{\langle x, \hat{s} \rangle} f(x)\, dm_G(x) = \mathscr{F}f(\hat{s})\end{aligned}$$

by virtue of (14.4.5), since the function $x \mapsto \overline{\langle x, \hat{s} \rangle} f(x)$ is m_G-integrable. By hypothesis, the function $\mathscr{F}g$, continuous and bounded on $H^{\perp}$, is therefore $m_{H^{\perp}}$-integrable, and consequently belongs to $\mathscr{L}^2_{\mathbf{C}}(m_{H^{\perp}}) \cap \mathscr{L}^1_{\mathbf{C}}(m_{H^{\perp}})$. It follows (22.10.10) that g is almost everywhere equal to a function in $\mathscr{P}^2(G/H) \cap \mathscr{L}^1_{\mathbf{C}}(G/H) = \mathscr{P}^1(G/H)$, and by Fourier's reciprocity formula we have

$$g(\dot{x}) = \int_{H^{\perp}} \langle x, \hat{s} \rangle \mathscr{F}g(\hat{s})\, dm_{H^{\perp}}(\hat{s})$$

almost everywhere with respect to $m_{G/H}$; in view of the relation $\mathscr{F}g(\hat{s}) = \mathscr{F}f(\hat{s})$ for $\hat{s} \in H^{\perp}$, the proof is complete.

(22.12.3) *With the notation of* (22.12.2), *suppose that*

(1) *f is m_G-integrable;*
(2) *the restriction of $\mathscr{F}f$ to $H^{\perp}$ is $m_{H^{\perp}}$-integrable;*
(3) *for* all $x \in G$, *the function $t \mapsto f(xt)$ on H is m_H-integrable;*
(4) *the function $x \mapsto \int_{H} f(xt)\, dm_H(t)$ is continuous on G.*

Then we have

(22.12.3.1)
$$\int_{H} f(t)\, dm_H(t) = \int_{H^{\perp}} \mathscr{F}f(\hat{s})\, dm_{H^{\perp}}(\hat{s})$$

(*Poisson's formula*).

With the notation of the proof of (22.12.2), the right-hand side of (22.12.2.1), which is equal to $\bar{\mathscr{F}}\mathscr{F}g(\dot{x})$, is always a continuous function on G/H; the hypotheses imply that the same is true of the left-hand side of (22.12.2.1), hence the equality (22.12.2.1) holds *everywhere* in G. If we now set $x = e$ in this equality, we have (22.12.3.1).

(22.12.4) *With the notation of* (22.12.2):

(i) *If* m_{H} (resp $m_{\mathrm{H}^\perp}$) *is canonically identified with a measure on* G (resp $\hat{\mathrm{G}}$) (13.1.7), $m_{\mathrm{H}^\perp}$ *is the Plancherel transform of the measure of positive type* m_{H} (22.2.3).

(ii) *If* $\hat{\mathrm{H}}$ *and* $\hat{\mathrm{G}}/\mathrm{H}^\perp$ *are canonically identified, the Haar measure* $m_{\hat{\mathrm{H}}}$ *associated with* m_{H} *is equal to* $m_{\hat{\mathrm{G}}}/m_{\mathrm{H}^\perp}$.

(i) Let g and h be two functions in $\mathscr{K}(\mathrm{G})$. Then $\check{\tilde{h}} * g = f$ satisfies the four conditions of (22.12.3). For we have $f \in \mathscr{K}(\mathrm{G})$, which shows immediately that conditions (1) and (3) are satisfied; condition (4) is satisfied by virtue of (14.1.5.5); finally, as to condition (2), if we put $f_1(\dot{x}) = \int_{\mathrm{H}} f(xt)\, dm_{\mathrm{H}}(t)$, then we have seen in the course of the proof of (22.12.2) that $\mathscr{F}f_1(\hat{s}) = \mathscr{F}f(\hat{s}) = \mathscr{F}g(\hat{s})\overline{\mathscr{F}h(\hat{s})}$ for $\hat{s} \in \mathrm{H}^\perp$; if also we put $g_1(\dot{x}) = \int_{\mathrm{H}} g(xt)\, dm_{\mathrm{H}}(t)$, $h_1(\dot{x}) = \int_{\mathrm{H}} h(xt)\, dm_{\mathrm{H}}(t)$, then likewise we have $\mathscr{F}g_1(\hat{s}) = \mathscr{F}g(\hat{s})$ and $\mathscr{F}h_1(\hat{s}) = \mathscr{F}h(\hat{s})$ for $\hat{s} \in \mathrm{H}^\perp$; since $\mathscr{F}g_1$ and $\mathscr{F}h_1$ belong to $\mathscr{L}^2_{\mathbf{C}}(m_{\mathrm{H}^\perp})$, it follows that $\mathscr{F}f_1$ belongs to $\mathscr{L}^1_{\mathbf{C}}(m_{\mathrm{H}^\perp})$ and therefore condition (2) is satisfied. Poisson's formula now gives

$$\int_{\mathrm{G}} (\check{\tilde{h}} * g)(x)\, dm_{\mathrm{H}}(x) = \int_{\hat{\mathrm{G}}} \mathscr{F}g(\hat{s})\overline{\mathscr{F}h(\hat{s})}\, dm_{\mathrm{H}^\perp}(\hat{s})$$

and this, together with the fact that $\check{m}_{\mathrm{H}^\perp} = m_{\mathrm{H}^\perp}$, shows that $m_{\mathrm{H}^\perp} = m_{\mathrm{H}}^{\wedge}$ by virtue of (22.7.4).

(ii) The measure $m_{\hat{\mathrm{H}}}$ is in any case proportional to $m_{\hat{\mathrm{G}}}/m_{\mathrm{H}^\perp}$, and it is therefore enough to show that for *one* non-negligible function $f' \in \mathscr{L}^1_{\mathbf{C}}(m_{\hat{\mathrm{G}}})$ we have

(22.12.4.1) $$\int_{\hat{\mathrm{G}}} f'(\hat{y})\, dm_{\hat{\mathrm{G}}}(\hat{y}) = \int_{\hat{\mathrm{G}}/\mathrm{H}^\perp} dm_{\hat{\mathrm{H}}}(\dot{\hat{y}}) \int_{\mathrm{H}^\perp} f'(\hat{y}\hat{s})\, dm_{\mathrm{H}^\perp}(\hat{s}).$$

Let $f \in \mathscr{K}(\mathrm{G})$. For $x \in \mathrm{G}$ and $\hat{y} \in \hat{\mathrm{G}}$, put

$$\psi(x, \hat{y}) = \int_{\mathrm{H}} f(xt)\langle t, \hat{y}\rangle\, dm_{\mathrm{H}}(t).$$

This function ψ is clearly continuous on $\mathrm{G} \times \hat{\mathrm{G}}$ (14.1.5.5). For fixed x in G, we have $\psi(x, \hat{y}\hat{s}) = \psi(x, \hat{y})$ for all $\hat{s} \in H^\perp$, so that $\psi(x, \hat{y})$ depends only on the class $\dot{\hat{y}}$ of $\hat{y}$ in $\hat{G}/H^\perp$; the mapping $\dot{\hat{y}} \mapsto \psi(x, \hat{y})$ is the Fourier cotransform of the function $t \mapsto f(xt)$ on H (which belongs to $\mathscr{K}(H)$); hence Plancherel's theorem gives

$$\text{(22.12.4.2)} \qquad \int_{\hat{G}/H^\perp} |\psi(x, \hat{y})|^2 \, dm_{\hat{H}}(\dot{\hat{y}}) = \int_H |f(xt)|^2 \, dm_H(t).$$

On the other hand, for fixed $\hat{y}$ we have $\langle xt', \hat{y}\rangle\psi(xt', \hat{y}) = \langle x, \hat{y}\rangle\psi(x, \hat{y})$ for all $t' \in H$ by the invariance of m_H, therefore $\langle x, \hat{y}\rangle\psi(x, \hat{y})$ depends only on the class $\dot{x}$ of x in G/H, and the same is true of $|\langle x, \hat{y}\rangle\psi(x, \hat{y})| = |\psi(x, \hat{y})|$. The Fourier cotransform of the function $\dot{x} \mapsto \langle x, \hat{y}\rangle\psi(x, \hat{y})$ on G/H (which belongs to $\mathscr{K}(G/H)$) is the following function on $H^\perp$:

$$\begin{aligned} \hat{s} \mapsto & \int_{G/H} \langle \dot{x}, \hat{s}\rangle\langle x, \hat{y}\rangle\psi(x, \hat{y}) \, dm_{G/H}(\dot{x}) \\ &= \int_{G/H} dm_{G/H}(\dot{x}) \int_H \langle \hat{y}\hat{s}, xt\rangle f(xt) \, dm_H(t) \\ &= \int_G \langle \hat{y}\hat{s}, x\rangle f(x) \, dm_G(x) \\ &= \overline{\mathscr{F}}f(\hat{y}\hat{s}) \end{aligned}$$

in view of **(22.12.1.1)**. Plancherel's theorem therefore gives

$$\text{(22.12.4.3)} \qquad \int_{G/H} |\psi(x, \hat{y})|^2 \, dm_{G/H}(\dot{x}) = \int_{H^\perp} |\overline{\mathscr{F}}f(\hat{y}\hat{s})|^2 \, dm_{H^\perp}(\hat{s}).$$

If we now make use of the Plancherel theorem for G and **(22.12.1.1)**, we deduce from **(22.12.4.2)** and **(22.12.4.3)** that

$$\begin{aligned} \int_{\hat{G}} |\overline{\mathscr{F}}f(\hat{y})|^2 \, dm_{\hat{G}}(\hat{y}) &= \int_G |f(x)|^2 \, dm_G(x) \\ &= \int_{G/H} dm_{G/H}(\dot{x}) \int_H |f(xt)|^2 \, dm_H(t) \\ &= \int_{G/H} dm_{G/H}(\dot{x}) \int_{\hat{G}/H^\perp} |\psi(x, \hat{y})|^2 \, dm_{\hat{H}}(\dot{\hat{y}}) \\ &= \int_{\hat{G}/H^\perp} dm_{\hat{H}}(\dot{\hat{y}}) \int_{G/H} |\psi(x, \hat{y})|^2 \, dm_{G/H}(\dot{x}) \\ &= \int_{G/H^\perp} dm_{\hat{H}}(\dot{\hat{y}}) \int_{H^\perp} |\overline{\mathscr{F}}f(\hat{y}\hat{s})|^2 \, dm_{H^\perp}(\hat{s}) \end{aligned}$$

which is the formula **(22.12.4.1)** for $f' = |\mathscr{F}f|^2$, a function which is $m_{\hat{G}}$-integrable and nonnegligible, provided that f is not identically zero.

Q.E.D.

PROBLEMS

1. With the notation of **(22.12.2)**, take $G = \hat{G} = \mathbf{R}$ and $H = H^{\perp} = \mathbf{Z}$ (cf. **(22.14.4.5)**). Also take

$$f(x) = \begin{cases} (e^{-\pi x} \sin \pi x)/\pi x & \text{for } x \geqq 1 \\ 0 & \text{for } x \leqq 1. \end{cases}$$

The function f is integrable with respect to Lebesgue measure; the sum $\sum_{n \in \mathbf{Z}} |f(x+n)|$ converges; and the mapping $x \mapsto \sum_{n \in \mathbf{Z}} f(x+n)$ is continuous on $\mathbf{R}$. Nevertheless, $\mathscr{F}f$ is not integrable, and we have $\sum_{n \in \mathbf{Z}} |\mathscr{F}f(n)| = +\infty$. (Calculate the derivative of $\mathscr{F}f$.)

2. (a) Let $a > 0$, $b > 0$ and let f be the continuous real-valued function defined on $\mathbf{R}$ which is equal to b when $x = 0$, is zero for $|x| \geqq a$, and is linear in each of the intervals $[-a, 0]$ and $[0, a]$. We have

$$(\mathscr{F}f)(t) = \frac{b \sin^2 \pi a t}{\pi^2 a t^2}.$$

Deduce that if $g(x) = \sum_{n=-N}^{N} f(x+n)$, then

$$(\mathscr{F}g)(t) = \frac{b \sin^2 \pi a t}{\pi^2 a t^2} \cdot \frac{\sin(2N+1)\pi t}{\sin \pi t}.$$

(b) Let g_k denote the function g constructed in (a), when $a = 1/k$, $b = 1/k^3$ and $N = k^2$. Then the function $h = \sum_{k=1}^{\infty} (-1)^k g_k$ is continuous and integrable, and also $\mathscr{F}h$ is integrable, so that $h \in \mathscr{P}^1(\mathbf{R})$; we have $\sum_{n \in \mathbf{Z}} |h(x+n)| < +\infty$ when $x \notin \mathbf{Z}$, the mapping $x \mapsto \sum_{n \in \mathbf{Z}} h(x+n)$ is continuous for $x \notin \mathbf{Z}$, and tends to a finite limit as x approaches an integer. Nevertheless $\mathscr{F}h(n)$ does not tend to 0 as n tends to $\pm\infty$. (Use the estimation of Lebesgue's constant given in Section **11.6**, Problem 2(b).)

3. (a) With the notation of **(22.12.2)** and of Section **14.4**, Problem 2, show that if $u, v \in \mathscr{K}(G)$ and $w = u * v$, then $w^{\flat} = u^{\flat} * v^{\flat}$.

(b) For each function $g \in \mathscr{K}(G)$, show that $\mathscr{F}(g^{\flat})$ is the restriction to $H^{\perp}$ (identified with the dual of G/H) of the Fourier transform $\mathscr{F}g$.

(c) Deduce from (a) and (b) that the restriction of $\mathscr{F}w$ to $H^{\perp}$ is $m_{H^{\perp}}$-integrable and that we have the Poisson formula

$$\int_H w(t)\, dm_H(t) = \int_{H^{\perp}} \mathscr{F}w(\hat{s})\, dm_{H^{\perp}}(\hat{s}).$$

(Compare with Problem 2.)

4. Let $K \subset \mathbf{R}^n$ be a symmetric compact convex set having 0 as an interior point. Let V denote the Lebesgue measure of $2K = K + K$. Show that if 0 is the only point of $\mathbf{Z}^n$ which belongs to 2K, then

$$2^n = V + \frac{1}{2^{2n}V} \sum_{\substack{m \in \mathbf{Z}^n \\ m \neq 0}} \left| \int_K e^{-2\pi i(x|m)}\, dx \right|^2.$$

(Show that Poisson's formula (with $G = \mathbf{R}^n$ and $H = \mathbf{Z}^n$) may be applied to the function $f = \varphi_K * \varphi_K$.) Hence give another proof of Minkowski's theorem (Section **14.2**, Problem 2).

5. (a) For each $x = (x_1, \ldots, x_n) \in \mathbf{R}^n$, put $r(x) = (x_1^2 + \cdots + x_n^2)^{1/2}$. If f_0 is a complex-valued function defined on $]0, +\infty[$ and measurable with respect to Lebesgue measure, the function $f\colon x \mapsto f_0(r(x))$ is measurable with respect to Lebesgue measure on $\mathbf{R}^n$; for it to be integrable, it is necessary and sufficient that the function $t \mapsto t^{n-1}f_0(t)$ should be integrable on $]0, +\infty[$. When this is so, show that the Fourier transform $\mathscr{F}f$ is of the form $\xi \mapsto F_0(r(\xi))$, where F_0 is the continuous function on $]0, +\infty[$ given by the formula

$$F_0(t) = 2\pi t^{-(n-2)/2} \int_0^{+\infty} f_0(s) s^{n/2} J_{(n-2)/2}(2\pi ts)\, ds,$$

where $J_{(n-2)/2}$ is the Bessel function given by the formula of Section **17.11**, Problem 2(c). (To calculate $\mathscr{F}f$, first use the formula (**16.24.9.1**), then evaluate the integral on S_{n-1}; we may limit ourselves to calculating $\mathscr{F}f(\xi)$ when $\xi = (0, \ldots, 0, t)$ and apply (**16.24.9.2**).)
(b) For $\lambda > 0$ and $x > 0$ we have

$$J_\lambda(x) = \left(\frac{x}{2}\right)^\lambda \sum_{m=0}^{\infty} \frac{(-x^2)^m}{2^{2m} m!\, \Gamma(m + \lambda + 1)}$$

and therefore

$$\frac{J_{\lambda+1}(x)}{x^{\lambda+1}} = -\frac{1}{x}\frac{d}{dx}\left(\frac{J_\lambda(x)}{x^\lambda}\right).$$

Also, as $x \to +\infty$, we have $J_\lambda(x) = O(x^{-1/2})$.† Deduce that if in (a) we take $f = \varphi_B$, where B is the ball with center 0 and radius 1 in $\mathbf{R}^n$, there exists a constant C_n such that

$$|\mathscr{F}\varphi_B(\xi)| \leq C_n (r(\xi))^{-(n+1)/2}$$

(integrate by parts). (When $\lambda = 0$, observe that $J_0' = -J_1$.)
(c) For each number $\rho > 0$, let B_ρ be the open ball with center 0 and radius ρ in $\mathbf{R}^n$. For each $x \in \mathbf{R}^n$, let $U_\rho(x)$ denote the number of points of $\mathbf{Z}^n$ in the ball $-x + B_\rho$, with center $-x$ and radius ρ; we have $U_\rho(x) = \sum_{m \in \mathbf{Z}^n} \varphi_{B_\rho}(x + m)$. If V_ρ is the Lebesgue measure of B_ρ, put $R_\rho(x) = U_\rho(x) - V_\rho$. Let f be a nonnegative function in $\mathscr{D}(\mathbf{R}^n)$, with support contained in a ball with center 0 and radius $h < \frac{1}{2}$, such that $\int f(x)\, dx = 1$. Show that

$$(f * R_\rho)(x) = \sum_{\substack{m \in \mathbf{Z}^n \\ m \neq 0}} \mathscr{F}f(m) \mathscr{F}\varphi_{B_\rho}(m) \exp(2\pi i(x|m))$$

† See my book, *Infinitesimal Calculus*, Boston (Houghton–Mifflin), 1971.

by using Poisson's formula. Deduce that for each number $k > 1$ we have

$$\text{(1)} \qquad |(f * \mathrm{R}_\rho)(x)| \leqq \sum_{1 \leq r(m) \leq k} |\mathscr{F}\varphi_{\mathrm{B}_\rho}(m)| + \mathrm{N}_2(f)\Big(\sum_{r(m) > k} |\mathscr{F}\varphi_{\mathrm{B}_\rho}(m)|^2\Big)^{1/2}.$$

Deduce that there exists a constant C'_n, depending only on n, such that

$$\text{(2)} \qquad |(f * \mathrm{R}_\rho)(x)| \leqq \mathrm{C}'_n(\rho^{(n-1)/2}k^{(n-1)/2} + \mathrm{N}_2(f)\rho^{(n-1)/2}k^{-1/2}).$$

(Use the inequality in (b), and majorize the sums by integrals.)

(d) We have

$$(f * \mathrm{U}_{\rho-h})(x) \leqq \mathrm{U}_\rho(x) \leqq (f * \mathrm{U}_{\rho+h})(x)$$

and therefore

$$(f * \mathrm{R}_{\rho-h})(x) - (\mathrm{V}_\rho - \mathrm{V}_{\rho-h}) \leqq \mathrm{R}_\rho(x) \leqq (f * \mathrm{R}_{\rho+h})(x) + (\mathrm{V}_{\rho+h} - \mathrm{V}_\rho).$$

Take $h = \rho^{-\alpha}$ and $k = \rho^\beta$, where $\alpha > 0$ and $\beta > 0$. Show that there exists a constant B_n depending only on n, α and β, such that

$$|\mathrm{R}_\rho(x)| \leqq \mathrm{B}_n(\rho^{n-1-\alpha} + \rho^{(n-1)(1+\beta)/2} + \rho^{(n\alpha+n-1-\beta)/2}).$$

(Use the estimate (2) and observe that for a suitable choice of f as a function of h we may assume that $\mathrm{N}_2(f) \leqq ch^{-n/2}$.)

By choosing α and β suitably, deduce that there exists a constant A_n such that

$$|\mathrm{R}_\rho(x)| \leqq \mathrm{A}_n\rho^{n(n-1)/(n+1)}$$

(*Sierpinski's theorem*).

13. DUAL OF A PRODUCT

(22.13.1) *Let $\mathrm{G}_1, \ldots, \mathrm{G}_n$ be locally compact commutative groups, and let j_k be the canonical injection of G_k into $\mathrm{G} = \prod_{1 \leq i \leq n} \mathrm{G}_i$. Then the mapping $({}^t j_k)_{1 \leq k \leq n}$ of $\hat{\mathrm{G}}$ into $\prod_{1 \leq i \leq n} \hat{\mathrm{G}}_i$ is an isomorphism.*

By induction on n we reduce to the case $n = 2$. Since ${}^t j_1$ and ${}^t j_2$ are surjective strict morphisms of $\hat{\mathrm{G}}$ onto $\hat{\mathrm{G}}_1$ and $\hat{\mathrm{G}}_2$ respectively, $({}^t j_1, {}^t j_2)$ is a surjective strict morphism of $\hat{\mathrm{G}}$ onto $\hat{\mathrm{G}}_1 \times \hat{\mathrm{G}}_2$, by virtue of the criterion of **(12.12.7)**. If $\mathrm{G}'_1 = j_1(\mathrm{G}_1)$, $\mathrm{G}'_2 = j_2(\mathrm{G}_2)$, then the kernel of ${}^t j_1$ is $\mathrm{G}'^{\perp}_1$ and that of ${}^t j_2$ is $\mathrm{G}'^{\perp}_2$, hence the kernel of $({}^t j_1, {}^t j_2)$ is the intersection $\mathrm{G}'^{\perp}_1 \cap \mathrm{G}'^{\perp}_2$; but since $\mathrm{G}'_1 \cup \mathrm{G}'_2$ generates G, we have $\mathrm{G}'^{\perp}_1 \cap \mathrm{G}'^{\perp}_2 = \{\hat{e}\}$ by **(22.11.4)**, and the proof is complete.

We shall identify $\hat{\mathrm{G}}$ with the product $\prod_{1 \leq i \leq n} \hat{\mathrm{G}}_i$ by means of the isomorphism $({}^t j_k)$; a character of G is therefore identified with an element $(\hat{x}_1, \ldots, \hat{x}_n)$

of this product, and the duality between $G = \prod_{1 \leq i \leq n} G_i$ and $\hat{G} = \prod_{1 \leq i \leq n} \hat{G}_i$ is expressed by the formula

(22.13.1.1) $$\langle (x_1, \ldots, x_n), (\hat{x}_1, \ldots, \hat{x}_n) \rangle = \prod_{1 \leq i \leq n} \langle x_i, \hat{x}_i \rangle.$$

(22.13.2) With the same notation, let m_{G_i} be a Haar measure on G_i $(1 \leqq i \leqq n)$, and let m_G be the product Haar measure on G: $m_G = \bigotimes_{i=1}^{n} m_{G_i}$ **(14.2.7)**. Then the associated Haar measure $m_{\hat{G}}$ on $\hat{G}$ is identified with the product $\bigotimes_{i=1}^{n} m_{\hat{G}_i}$ of the Haar measures $m_{\hat{G}_i}$ associated with m_{G_i}. To see this, consider for each $i = 1, \ldots, n$ a function $f_i \in \mathscr{K}(G_i)$, not identically zero, and let f be the function $(x_1, \ldots, x_n) \mapsto f_1(x_1) \cdots f_n(x_n)$, which belongs to $\mathscr{K}(G)$; it follows from **(22.13.1.1)** that the Fourier transform of f is the function $(\hat{x}_1, \ldots, \hat{x}_n) \mapsto \mathscr{F}f_1(\hat{x}_1) \cdots \mathscr{F}f_n(\hat{x}_n)$. Plancherel's theorem applied to G and to each of the G_i then gives

$$\begin{aligned} \int |f(x)|^2 \, dm_G(x) &= \prod_{i=1}^{n} \int |f_i(x_i)|^2 \, dm_{G_i}(x_i) \\ &= \prod_{i=1}^{n} \int |\mathscr{F}f_i(\hat{x}_i)|^2 \, dm_{\hat{G}_i}(\hat{x}_i) \\ &= \int |\mathscr{F}f(\hat{x})|^2 \, dm_{\hat{G}}(\hat{x}) \end{aligned}$$

which proves our assertion.

PROBLEMS

1. Let G be a (separable, metrizable) locally compact commutative group generated by a symmetric compact neighborhood V of e.
   (a) There exist a finite number of points $x_1, \ldots, x_m$ in G such that V^2 is contained in the union of the $x_i V$ $(1 \leqq i \leqq m)$. If D_0 is the subgroup of G generated by the x_i $(1 \leqq i \leqq m)$, show that $G = D_0 V$.
   (b) Let J be a subset of $\{1, 2, \ldots, m\}$ which is maximal with respect to the property that the subgroup D of D_0 generated by the x_i, $i \in J$, is topologically isomorphic to $\mathbf{Z}^{\text{Card}(J)}$, and in particular is closed in G **(12.9.6)**. Let $p: G \to G/D$ be the canonical homomorphism, and let $i \notin J$. Show that the subgroup H_i of G/D generated by $p(x_i)$ cannot be topologically isomorphic to **Z**. Deduce (Section **12.9**, Problem 10) that $\bar{H}_i$ is compact, and conclude that G/D is compact.

(c) Show that G is isomorphic to a group of the form $\mathbf{R}^p \times \mathbf{Z}^q \times K$, where K is compact. (Observe that $\hat{G}$ is locally isomorphic to $\hat{D} \cong \mathbf{T}^n$, and use (22.13.1) and Section 22.11, Problem 5.)

2. Show that every (separable, metrizable) locally compact commutative group G is isomorphic to a product $\mathbf{R}^p \times G_1$, where G_1 contains a compact open subgroup. (Let V be a compact neighborhood of e in G, and let H be the open subgroup of G generated by V, which is isomorphic to $\mathbf{R}^p \times \mathbf{Z}^q \times K$, where K is a compact group; hence there exists a continuous surjective homomorphism $q: H \to \mathbf{R}$; now use Section 22.11, Problem 5(b).)
(b) Let G be a locally compact commutative group which has an open compact subgroup. Its identity component C is the intersection of the open subgroups of G, and is compact. The union F of the compact subgroups of G is an open subgroup of G, namely the set of all $x \in G$ such that the subgroup generated by x is relatively compact. The group $F^\perp$ is the identity component of $\hat{G}$, and the subgroup $C^\perp$ is the union of the compact subgroups of $\hat{G}$. The group G/F is discrete, and all its elements other than the identity element have infinite order. The group F/C is totally disconnected, and so is its dual $C^\perp/F^\perp$.

3. (a) Let (G_n) be a sequence of (separable, metrizable) locally compact commutative groups, and for each n let H_n be an open compact subgroup of G_n. In the product group $\prod_n G_n$, consider the subgroup G of all $x = (x_n)$ such that $x_n \in H_n$ for all but finitely many values of n; G therefore contains the subgroup $H = \prod_n H_n$. Consider on H the product topology, with respect to which H is a compact metrizable group. If $\mathfrak{B}$ is the set of all neighborhoods of the identity element in H for this topology, then $\mathfrak{B}$ is a fundamental system of neighborhoods of the identity element of G for a topology on G compatible with its group structure. With this topology, G is separable, metrizable and locally compact; H is an open compact subgroup of G; and G/H is a discrete group isomorphic to the direct sum of the groups G_n/H_n. G is said to be the *local direct sum* of the groups G_n, relative to the compact open subgroups H_n.
(b) Show that the dual $\hat{G}$ is isomorphic to the local direct sum of the groups $\hat{G}_n$, relative to the open compact subgroups $H_n^\perp$.

14. EXAMPLES OF DUALITY

(22.14.1) *Let* G *be a locally compact commutative group. Then* G *is compact if and only if* $\hat{G}$ *is discrete. If* G *is infinite, the normalized Haar measures on* G *and* $\hat{G}$ (14.3.5) *are then associated.*

We have already remarked (22.6.10) that when G is compact, $\hat{G}$ (which is then a *Hilbert basis* of $L^2_{\mathbf{C}}(G)$ (21.3.2)) is discrete. On the other hand, if G is discrete, then $L^1_{\mathbf{C}}(G)$ contains the identity element ε_e of $M^1_{\mathbf{C}}(G)$, and its spectrum is therefore *compact*, and may be identified with $\hat{G}$ (22.6.8). The first assertion follows from this and from Pontrjagin's duality theorem. If G is a discrete group and if we take as Haar measure m_G the measure with mass 1 at each point, the Fourier transform of the characteristic function $\varphi_{\{e\}}$ is the constant function 1; hence Plancherel's theorem shows that the associated Haar measure $m_{\hat{G}}$ is the Haar measure with total mass 1, and the proof is complete.

(22.14.2) For a *compact* commutative group G, the diagrams **(22.10.10.1)** for G and its *discrete* dual group $\hat{G}$ simplify, as follows:

$$\begin{array}{ccc} & \mathrm{M_C(G) = M_C^1(G) = P(G) = P'(G)} & \mathrm{P(\hat{G}) = P'(\hat{G})} = \mathscr{P}(\hat{G}) \\ & \uparrow & \uparrow \\ & \mathrm{L_C^1(G)} & \mathrm{L_C^2(\hat{G})} = \mathscr{P}^2(\hat{G}) \\ & \uparrow & \uparrow \\ \text{(22.14.2.1)} & \mathrm{L_C^2(G)} & \mathrm{M_C^1(\hat{G}) = L_C^1(\hat{G})} = \mathscr{P}^1(\hat{G}) \\ & \uparrow & \uparrow \\ & \mathscr{K}(\mathrm{G}) = \mathscr{C}(\mathrm{G}) & \mathscr{K}(\hat{\mathrm{G}}) \\ & \uparrow & \\ & \mathscr{P}(\mathrm{G}) = \mathscr{P}^1(\mathrm{G}) = \mathscr{P}^2(\mathrm{G}) & \end{array}$$

(22.14.3) (i) *For each $y \in \mathbf{R}$, let ω_y be the character $x \mapsto \exp(2\pi i x y)$ of the additive group* **R**. *The mapping $y \mapsto \omega_y$ is an isomorphism of* **R** *onto its dual* $\hat{\mathbf{R}}$, *and if* **R** *is identified with its dual group by means of this isomorphism, Lebesgue measure is associated to itself.*

(ii) *For each $y \in \mathbf{R}$, let $\dot{y}$ be the coset of y in $\mathbf{T} = \mathbf{R}/\mathbf{Z}$, and let $\varpi_{\dot{y}}$ be the character $n \mapsto \exp(2\pi i n y)$ of* **Z** *(which depends only on $\dot{y}$). Then the mapping $\dot{y} \mapsto \varpi_{\dot{y}}$ is an isomorphism of* **T** *onto the dual* $\hat{\mathbf{Z}}$ *of the discrete group* **Z**.

It is clear that ω_y is a character of **R** and that $\theta\colon y \mapsto \omega_y$ is a homomorphism of **R** into $\hat{\mathbf{R}}$, which is evidently injective because $\exp(2\pi i x y) = 1$ for *all* $x \in \mathbf{R}$ necessarily implies $y = 0$ **(9.5.5)**. Next, for each compact interval $|x| \leqq \mathrm{M}$ in **R** and each $\varepsilon > 0$, the continuity of the product xy implies that there exists $\delta > 0$ such that the relation $|y| \leqq \delta$ implies

$$|\exp(2\pi i x y) - 1| \leqq \varepsilon$$

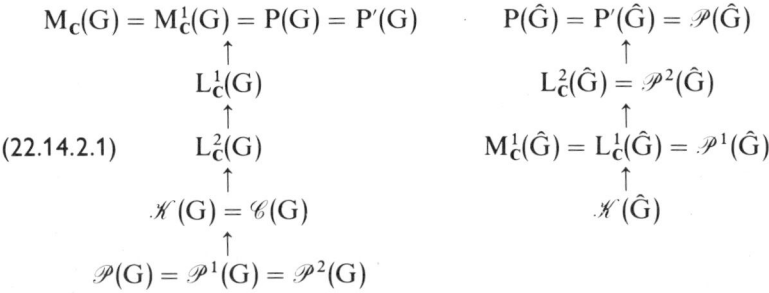

for $|x| \leqq \mathrm{M}$. This shows that the homomorphism θ is continuous. Since $\langle x, \theta(y)\rangle = \langle y, \theta(x)\rangle$ for all $x, y \in \mathbf{R}$, it follows that θ is equal to its transpose, hence **(22.11.5)** $\theta(\mathbf{R})$ is *dense* in $\hat{\mathbf{R}}$. Finally, we shall show that θ is an isomorphism of **R** onto the subgroup $\theta(\mathbf{R})$ of $\hat{\mathbf{R}}$; this will imply that $\theta(\mathbf{R})$ is closed in $\hat{\mathbf{R}}$ **(12.9.6)** and consequently that θ is an isomorphism of **R** onto $\hat{\mathbf{R}}$. We may obtain a fundamental system of neighborhoods of the identity element of $\hat{\mathbf{R}}$ by forming, for each $\varepsilon > 0$ and each $\mathrm{M} > 0$, the set $\mathrm{U}(\varepsilon, \mathrm{M})$ of characters $\hat{x}$ such that $|\langle x, \hat{x}\rangle - 1| \leqq \varepsilon$ whenever $|x| \leqq \mathrm{M}$. Now let $x_0 \in \mathbf{R}$ be such that $\exp(2\pi i x_0) \neq 1$, and put $\varepsilon = |\exp(2\pi i x_0) - 1|$. For each $\mathrm{M} > 0$, the set of $y \in \mathbf{R}$ which satisfy $|\omega_y(x) - 1| < \varepsilon$ whenever $|x| \leqq \mathrm{M}$ satisfy *a* *fortiori* the relation $|y^{-1}x_0| > \mathrm{M}$, so that $|y| < \mathrm{M}^{-1}|x_0|$. This means that the set $\theta^{-1}(\mathrm{U}(\varepsilon, \mathrm{M}))$ is contained in the neighborhood $|y| < \mathrm{M}^{-1}|x_0|$ of 0 in **R**, and hence establishes that θ is an isomorphism of **R** onto $\hat{\mathbf{R}}$.

If we identify $\hat{\mathbf{R}}$ with **R** by means of θ, the annihilator of the subgroup **Z** of **R** in $\hat{\mathbf{R}}$ is **Z** itself, and assertion (ii) therefore follows from (22.11.3). Observe now that if $m_{\mathbf{Z}}$ and $m_{\mathbf{T}}$ are the normalized Haar measures on **Z** and **T**, and if $m_{\mathbf{R}}$ is Lebesgue measure on **R**, it is immediate that $m_{\mathbf{T}} = m_{\mathbf{R}}/m_{\mathbf{Z}}$ (22.12.1). By (22.14.1), the Haar measure on $\mathbf{T} = \hat{\mathbf{Z}}$ associated to $m_{\mathbf{Z}}$ is $m_{\mathbf{T}}$, and the Haar measure on $\mathbf{Z} = \hat{\mathbf{T}}$ associated to $m_{\mathbf{T}}$ is $m_{\mathbf{Z}}$. The fact that the Haar measure on $\hat{\mathbf{R}} = \mathbf{R}$ associated to $m_{\mathbf{R}}$ is $m_{\mathbf{R}}$ itself now follows from (22.12.4).

(22.14.4) From now on we shall almost always identify **R** with its dual group by means of the isomorphism $y \mapsto \omega_y$. The dual of the group $\mathbf{R}^n$ is then canonically identified with $\mathbf{R}^n$, and Lebesgue measure on $\mathbf{R}^n$ is self-associated (22.13). For $x, y \in \mathbf{R}^n$ we have, with this identification,

(22.14.4.1)
$$\langle x, y \rangle = \exp(2\pi i(x \mid y))$$

where $(x|y) = \sum_{j=1}^{n} x_j y_j$ if $x = (x_j)$ and $y = (y_j)$ (the usual Euclidean scalar product). The Fourier transform of a bounded measure μ on $\mathbf{R}^n$ is therefore given by the formula

(22.14.4.2)
$$\mathscr{F}\mu(y) = \int_{\mathbf{R}^n} \exp(-2\pi i(x \mid y))\, d\mu(x),$$

and if we write $\int_{\mathbf{R}^n} f(x)\, dx$ for the integral with respect to Lebesgue measure in $\mathbf{R}^n$, the Fourier transform of a function $f \in \mathscr{L}^1_{\mathbf{C}}(\mathbf{R}^n)$ is given by

(22.14.4.3)
$$(\mathscr{F}f)(y) = \int_{\mathbf{R}^n} \exp(-2\pi i(x \mid y)) f(x)\, dx,$$

and Fourier's reciprocity formula, for $f \in \mathscr{P}^2(\mathbf{R}^n)$, is given by

(22.14.4.4)
$$f(x) = \int_{\mathbf{R}^n} \exp(2\pi i(x \mid y)) \mathscr{F}f(y)\, dy.$$

(Of course, $\mathscr{F}f$ is given by the formula (22.14.4.3) only if in addition $f \in \mathscr{P}^1(\mathbf{R}^n)$.)

We shall also spell out Poisson's formula (22.12.3.1) for the case where $G = \mathbf{R}$ and $H = \mathbf{Z}$: it becomes

(22.14.4.5) $$\sum_{n \in \mathbf{Z}} \mathscr{F}f(n) = \sum_{n \in \mathbf{Z}} f(n)$$

and is valid under the following conditions:

(1) f is Lebesgue-integrable on $\mathbf{R}$;
(2) $\sum_{n \in \mathbf{Z}} |f(x+n)| < +\infty$ for all $x \in \mathbf{R}$;
(3) the function $x \mapsto \sum_{n \in \mathbf{Z}} f(x+n)$ is continuous on $\mathbf{R}$;
(4) $\sum_{n \in \mathbf{Z}} |\mathscr{F}f(n)| < +\infty$.

Remarks

(22.14.5) The dual of the cyclic group $\mathbf{Z}/n\mathbf{Z}$ of order n is isomorphic to the annihilator in $\mathbf{T}$ of the subgroup $n\mathbf{Z}$ of $\mathbf{Z}$; this subgroup of $\mathbf{T}$ is the image of the subgroup of $\mathbf{R}$ consisting of the numbers y such that $\exp(2\pi i n y) = 1$, i.e., the group $n^{-1}\mathbf{Z}$. It is immediate that the quotient of $\mathbf{T}$ by this subgroup is a cyclic group of order n. Since every *finite* commutative group G is a product of cyclic groups (A.26.4), it follows that the dual group $\hat{G}$ is a group *isomorphic* to G (but not canonically). The argument of (22.14.1) shows that if we take on G the Haar measure with mass 1 at each point, then the associated Haar measure on $\hat{G}$ has total mass 1.

(22.14.6) A *lattice* in $\mathbf{R}^n$ is a discrete subgroup M which is a free $\mathbf{Z}$-module generated by a basis of $\mathbf{R}^n$. Every basis of the $\mathbf{Z}$-module M is then a basis of $\mathbf{R}^n$. An equivalent definition (19.7.9.1) is that M is a closed subgroup of $\mathbf{R}^n$ such that $\mathbf{R}^n/M$ is compact (hence isomorphic to $\mathbf{T}^n$). It is then preferable to consider an n-dimensional vector space E over $\mathbf{R}$ in place of $\mathbf{R}^n$, and not to identify E with its dual $\hat{E}$; $\hat{E}$ may be identified with the vector space dual E' of E (A.9.1), an element $x' \in E'$ being identified with the character $x \mapsto \exp(2\pi i\langle x, x'\rangle)$. It is prudent here to avoid the use of the notation $\langle x, \hat{x}\rangle$ introduced in (22.10.3), and to reserve for the symbol $\langle x, x'\rangle$ the meaning attached to it in (A.9.1). If M is a lattice in E, its *annihilator* in the sense of (21.11.2) is the set of $x' \in E'$ such that $\langle x, x'\rangle$ is an *integer for each* $x \in M$. It is immediate that if $(a_j)_{1 \le j \le n}$ is a basis of the $\mathbf{Z}$-module M, and therefore also a basis of the vector space E, then the $x' \in E'$ which satisfy the above condition are the vectors of the form $\sum_{j=1}^{n} z_j a_j^*$, where $(a_j^*)_{1 \leq j \leq n}$ is the basis of E′ *dual* to the basis (a_j) of E (A.9.2) and the z_j are integers. We may therefore identify the annihilator of M in the above sense with the *dual* of the **Z**-module M (A.9.1); it is therefore denoted by M* and is called the *dual lattice* (in E′) of the lattice $M \subset E$. We have $(M^*)^* = M$ (22.11.4).

Now let $N \subset M$ be a second lattice in E. Then there exists a basis $(a_j)_{1 \leq j \leq n}$ of M and positive integers α_j $(1 \leqq j \leqq n)$ such that α_j divides α_{j+1} for $1 \leqq j \leqq n-1$, and such that the $\alpha_j a_j$ form a basis of N (A.26.3). The $\alpha_j^{-1} a_j^*$ therefore form a basis of N*; the finite commutative group N*/M* is isomorphic to M/N, and may be canonically identified with the *dual* of M/N. It is enough to observe that the group dual to E/N (in the sense of (22.10.3)) is identified with N*, and M* with the annihilator of M/N in N*, so that the dual of M/N is canonically identified with N*/M* (22.11.3).

PROBLEMS

1. Let p be a prime number. For each $x \in \mathbf{Q}_p$ (Section 14.3, Problem 6) there exists a unique number of the form q/p^ν in **Q**, with ν an integer $\geqq 0$ and $0 \leqq q < p^\nu$, such that $x - (q/p^\nu) \in \mathbf{Z}_p$. Put $\lambda(x) = q/p^\nu$; then for any two p-adic numbers x, x' we have $\lambda(x + x') \equiv \lambda(x) + \lambda(x') \pmod{\mathbf{Z}}$. The function $\chi_0 \colon x \mapsto \exp(2\pi i \lambda(x))$ is a character of the locally compact additive group $\mathbf{Q}_p$, and the kernel of χ_0 is $\mathbf{Z}_p$. For each $y \in \mathbf{Q}_p$, let ω_y be the character $x \mapsto \chi_0(xy)$ of the additive group $\mathbf{Q}_p$. Show that the mapping $y \mapsto \omega_y$ is an isomorphism of $\mathbf{Q}_p$ onto its dual group $\hat{\mathbf{Q}}_p$. If we identify $\mathbf{Q}_p$ with its dual by means of this isomorphism, the Haar measure β on $\mathbf{Q}_p$ such that $\beta(\mathbf{Z}_p) = 1$ is self-associated. (This is the *normalized* Haar measure on $\mathbf{Q}_p$.)

   For each $y \in \mathbf{Q}_p$, let $\dot{y}$ be the image of y in $\mathbf{Q}_p/\mathbf{Z}_p$, and let $\varpi_{\dot{y}}$ be the character $z \mapsto \chi_0(zy)$ of $\mathbf{Z}_p$ (which depends only on $\dot{y}$); then the mapping $\dot{y} \mapsto \varpi_{\dot{y}}$ is an isomorphism of the discrete group $\mathbf{Q}_p/\mathbf{Z}_p$ onto the dual $\hat{\mathbf{Z}}_p$ of $\mathbf{Z}_p$. (Argue as in (22.14.3).)

2. With the notation of Problem 1, the topology induced on $\mathbf{Q}_p^* = \mathbf{Q}_p - \{0\}$ by that of $\mathbf{Q}_p$ is compatible with the structure of the multiplicative group $\mathbf{Q}_p^*$, which is therefore locally compact and totally disconnected. The group $\mathbf{Q}_p^*$ is isomorphic to the product of the discrete additive group **Z** and the compact group U of units of $\mathbf{Z}_p$.

   Let $\mathfrak{p} = p\mathbf{Z}_p$ be the maximal ideal of $\mathbf{Z}_p$. For each $a \in U$, the sequence $(a^{p^n})_{n \geq 0}$ converges in U to a limit α such that $\alpha \equiv a \pmod{\mathfrak{p}}$ and $\alpha^p = \alpha$. (Show by induction that $a^{p^n} \equiv a^{p^{n-1}} \pmod{\mathfrak{p}^n}$.) All the roots of the polynomial $X^{p-1} - 1$ (in an algebraic closure of $\mathbf{Q}_p$) lie in $\mathbf{Q}_p$ and are pairwise incongruent mod $\mathfrak{p}$. (Take for a the roots of the congruence $x^{p-1} - 1 \equiv 0 \pmod{p}$.) Deduce that if $p > 2$ the group U is the direct product of the subgroup $V = 1 + \mathfrak{p}$ and the finite subgroup consisting of the $(p-1)$st roots of unity. If $p = 2$, U is the direct product of the subgroup $1 + \mathfrak{p}^2$ and a group of order 2.

   If $p > 2$ and if $a \in 1 + \mathfrak{p}$, $a \neq 1$, the mapping $n \mapsto a^n$ is continuous on **Z** with respect to the p-adic topology, and extends to a continuous homomorphism, written $x \mapsto a^x$, of $\mathbf{Z}_p$

into U; this homomorphism is injective, and its image is $1 + \mathfrak{p}$ if $a \notin 1 + \mathfrak{p}^2$. For $p = 2$, the mapping $n \mapsto 5^n$ is continuous on $\mathbf{Z}$ with respect to the 2-adic topology, and extends to an isomorphism $x \mapsto 5^x$ of $\mathbf{Z}_2$ onto $1 + \mathfrak{p}^2$.

Deduce from these results that $\mathbf{Q}_p^*$ is isomorphic to the product

$$\mathbf{Z} \times \mathbf{Z}_p \times (\mathbf{Z}/(p-1)\mathbf{Z})$$

if $p > 2$, and to the product

$$\mathbf{Z} \times \mathbf{Z}_2 \times (\mathbf{Z}/2\mathbf{Z})$$

if $p = 2$.

If β is a Haar measure on $\mathbf{Q}_p$, show that the linear form $\beta^*: f \mapsto \int f(x)|x|_p^{-1}\, d\beta(x)$ (Section 14.3, Problem 6) is a Haar measure on the group $\mathbf{Q}_p^*$. The *normalized* Haar measure on $\mathbf{Q}_p^*$ is the measure β^* such that $\beta(\mathrm{U}) = 1$, i.e., the restriction of $(p/(p-1))\beta$, where β is the normalized measure on $\mathbf{Q}_p$.

3. (a) Let $\mathbf{a} = (a_0, a_1, a_2, \ldots)$ be an infinite sequence of integers > 1. Consider the sequence of finite groups $\mathrm{G}_r = \mathbf{Z}/(a_0 a_1 \cdots a_r)\mathbf{Z}$ and the product G of the G_r, which is metrizable, compact and totally disconnected. For each r, let $\varphi_r: \mathrm{G}_r \to \mathrm{G}_{r-1}$ be the canonical homomorphism. The set $\mathbf{Z}_\mathbf{a}$ of all $z = (z_r) \in \mathrm{G}$ such that $z_{r-1} = \varphi_r(z_r)$ for all $r \geqq 1$ is a compact totally disconnected subgroup of G. If $a_i = p$ (p a prime number) for all i, this group $\mathbf{Z}_\mathbf{a}$ is the group $\mathbf{Z}_p$ of p-adic integers (Section 12.9, Problem 4).
(b) Show that the homomorphism $\chi \mapsto \chi(1)$ is an isomorphism of the dual group $\hat{\mathbf{Z}}_\mathbf{a}$ onto the subgroup of $\mathbf{U}$ consisting of the numbers $\exp(2\pi i q/(a_0 a_1 \cdots a_r))$ for all $r \in \mathbf{N}$ and all $q \in \mathbf{Z}$, endowed with the discrete topology.
(c) Consider the product group $\mathbf{R} \times \mathbf{Z}_\mathbf{a}$ and the discrete subgroup B of all pairs (n, n) with $n \in \mathbf{Z}$ (the group $\mathbf{Z}$ being identified with the dense subgroup of $\mathbf{Z}_\mathbf{a}$ consisting of the elements $z = (z_r)$ in which the z_r are the canonical images of some element of $\mathbf{Z}$ in the G_r). Put $\mathbf{T}_\mathbf{a} = (\mathbf{R} \times \mathbf{Z}_\mathbf{a})/\mathrm{B}$. Show that $\mathbf{T}_\mathbf{a}$ is a metrizable compact connected group. If $a_i = p$ (prime) for all i, the group $\mathbf{T}_\mathbf{a}$ is isomorphic to the p-adic solenoid $\mathbf{T}_p$ defined in Section 12.9, Problem 5. Give an analogous definition of $\mathbf{T}_\mathbf{a}$ as a closed subgroup of $\mathbf{T}^\mathbf{N}$, and generalize to $\mathbf{T}_\mathbf{a}$ the properties of $\mathbf{T}_p$.
(d) Show that the dual group $\hat{\mathbf{T}}_\mathbf{a}$ is canonically isomorphic to the subgroup of the additive group $\mathbf{Q}$ consisting of the numbers $q/(a_0 a_1 \cdots a_r)$ for all $r \in \mathbf{N}$ and $q \in \mathbf{Z}$, endowed with the discrete topology.
(e) Deduce that the dual $\hat{\mathbf{Q}}$ of the additive group $\mathbf{Q}$ is isomorphic to $\mathbf{T}_\mathbf{a}$ for $\mathbf{a} = (2, 3, 4, 5, 6, \ldots)$.

4. A Hausdorff topological group G is said to *monothetic* (resp. *solenoidal*) if there exists a continuous homomorphism of $\mathbf{Z}$ (resp. $\mathbf{R}$) into G with dense image (which implies that G is commutative).
(a) A totally disconnected infinite compact metrizable group G is monothetic if and only if it is isomorphic to one of the groups $\mathbf{Z}_\mathbf{a}$ (Problem 3). An equivalent condition is that $\hat{\mathrm{G}}$ should be isomorphic to a torsion subgroup of $\mathbf{T}$, with the discrete topology.
(b) A compact metrizable group G is monothetic if and only if $\hat{\mathrm{G}}$ is isomorphic to an at most denumerable subgroup of $\mathbf{T}$, with the discrete topology.
(c) For a compact metrizable group G, the following conditions are equivalent:

(α) G is solenoidal;

(β) $\hat{G}$ is isomorphic to an at most denumerable subgroup of $\mathbf{R}$, with the discrete topology;

(γ) $\hat{G}$ is at most denumerable and torsion-free;

(δ) G is a quotient of $\hat{\mathbf{Q}}^{\mathbf{N}}$.

(To show that (γ) implies (β), embed $\hat{G}$ in a $\mathbf{Q}$-vector space with an at most denumerable basis.)

5. Let G be a (separable, metrizable) locally compact commutative group, in additive notation. For each prime number p, the *p-primary component* of G is the set G_p of elements $x \in G$ for which the mapping $n \mapsto nx$ of $\mathbf{Z}$ into G extends to a continuous homomorphism of $\mathbf{Z}_p$ into G. If G is discrete, G_p is the set of $x \in G$ of finite order equal to a power of p.
(a) Show that if G is totally disconnected, then G_p is a closed subgroup of G. (Observe that G has a fundamental system of neighborhoods of 0 consisting of open subgroups (Section 12.9, Problem 3).)
(b) If G is compact and totally disconnected, G is isomorphic to the product of its primary components. (Observe that a discrete torsion group is the direct sum of its primary components, by using Bézout's identity.)
(c) Suppose that both G and $\hat{G}$ are locally compact and totally disconnected, and let H be a compact open subgroup of G (Section 22.13, Problem 2). Show that G is isomorphic to the local direct sum of its primary components G_p, relative to the subgroups $H \cap G_p$ (Section 22.13, Problem 3). (Observe that G/H is a torsion group, and deduce, with the aid of (b), that for every compact open subgroup H′ containing H and x, the component x_p of x in the p-primary component of H′ is independent of H′ and belongs to G_p; then define a continuous homomorphism of G′, the local direct sum of the G_p relative to the subgroups $H \cap G_p$, into G, and show that this homomorphism is bijective.)

6. A locally compact, totally disconnected commutative group (written additively) is said to be *p-primary* (for a prime number p) if it is equal to its p-primary component. A discrete commutative group is said to be a *p-group* if all its elements have finite order equal to a power of p. In such a group G, an element x is said to be *of height* p^r if there exists an element $y \in G$ such that $x = p^r y$ but no element $z \in G$ such that $x = p^{r+1}z$. A subgroup H of G is said to be *pure* if, for each $x \in H$ and $y \in G$ such that $x = p^r y$ for some integer r, there exists $z \in H$ such that $x = p^r z$.
(a) Let G be a discrete commutative p-group and let H be a pure subgroup of G such that the elements of G/H have finite height. If $H \neq G$, there exists in G/H an element $\dot{x}$ of height 1 and order $p^r > 1$, such that $\dot{y} = p^{r-1}\dot{x}$ is of order p and height p^{r-1}. Show that the subgroup generated by $\dot{x}$ in G/H is pure. Show next that there exists $x \in G$ of order p^r belonging to $\dot{x}$, and that if H_x is the subgroup generated by x, the sum $H + H_x$ is direct and is a pure subgroup of G.
(b) Let G be a compact metrizable p-primary torsion group. Show that G is the product of a (finite or infinite) sequence of cyclic groups $\mathbf{Z}/p^{r_i}\mathbf{Z}$, where the exponents r_i are bounded. (Show that $\hat{G}$ is the direct sum of a sequence of cyclic groups, by arguing by induction and using (a) above.)

7. (a) Show that a (separable, metrizable) locally compact commutative torsion group is the product of a finite number of its nondiscrete primary components G_{p_i} $(1 \leq i \leq m)$ and a discrete group, equal to the direct sum of the other primary components of G. (By considering a compact open subgroup H of G, show first that $H^{\perp}$ is a torsion group, and deduce that both G and $\hat{G}$ are totally disconnected. Then show that $H \cap G_p = \{0\}$ for all

but a finite set of prime numbers, by using Problem 6, and finally apply Problem 5(c).)
(b) Consider the product group $G_0 = (\mathbf{Z}/p^2\mathbf{Z})^{\mathbf{N}}$, in which the topology is not the product topology, but is defined as follows: consider the set $\mathfrak{B}$ of neighborhoods of 0 in the subgroup $H = (p\mathbf{Z}/p^2\mathbf{Z})^{\mathbf{N}}$, endowed with the product of the discrete topologies on the factors $p\mathbf{Z}/p^2\mathbf{Z}$; then $\mathfrak{B}$ is a fundamental system of neighborhoods of 0 for a topology $\mathcal{T}$ on G_0, which is compatible with the group structure and metrizable, but not separable. If G is the subgroup of G_0 generated by H and an at most denumerable set of elements of $\complement H$, then G is separable, metrizable and locally compact, and all its elements are of order $\leq p^2$. Show that G can be chosen so that it is not isomorphic to the local direct sum (Problem 5) of a sequence (G_n) of groups isomorphic to $\mathbf{Z}/p^2\mathbf{Z}$, relative to the subgroups $H_n = pG_n$. (Observe that in such a local direct sum, if e_n is a generator of G_n, the pe_n form a basis of pG considered as a vector space over the field $\mathbf{Z}/p\mathbf{Z}$, and this basis is topologically free in the sense that none of the pe_n belongs to the closed subgroup of G generated by the pe_m, $m \neq n$.)

8. Let G be a compact metrizable commutative group and let σ be an automorphism of G. A Haar measure on G is then invariant under σ (**14.3**).
(a) Show that if σ is ergodic (Section **13.9**, Problem 13), the orbits $\neq \{\hat{e}\}$ of the discrete group $\hat{G}$ under the action of the group consisting of the ${}^t\sigma^n$ ($n \in \mathbf{Z}$) are necessarily infinite. (If ${}^t\sigma^n(\hat{x}) = \hat{x}$ for some $\hat{x} \neq \hat{e}$, the sum of the ${}^t\sigma^k(\hat{x})$ for $1 \leq k \leq n$ is invariant under ${}^t\sigma$ and nonconstant on G.)
(b) Conversely, if the orbits in $\hat{G}$ other than $\{\hat{e}\}$ under the action of the group generated by ${}^t\sigma$ are infinite, there exists an at most denumerable set I and a bijection $(i, n) \mapsto \hat{x}_{i,n}$ of $I \times \mathbf{Z}$ onto $\hat{G} - \{\hat{e}\}$ such that ${}^t\sigma(\hat{x}_{i,n}) = \hat{x}_{i,n+1}$ for all $i \in I$ and $n \in \mathbf{Z}$. Deduce that σ is then a mixing transformation. (Use the criterion of Section **15.11**, Problem 16(b).)
(c) Show that the set I is necessarily infinite. (Argue by contradiction, by first noting that if there is only a finite number of orbits, then for each $\hat{x} \neq \hat{e}$ there exist $n > 1$ and $k > 1$ such that $\prod_{j=1}^{n} ({}^t\sigma^j(\hat{x})) = {}^t\sigma^k\left(\prod_{j=1}^{n} {}^t\sigma^j(\hat{x})\right)$, and hence there exist two integers $p < q$ such that the subgroup of $\hat{G}$ generated by ${}^t\sigma^p(\hat{x}), {}^t\sigma^{p+1}(\hat{x}), \ldots, {}^t\sigma^q(\hat{x})$ is stable under ${}^t\sigma$ and ${}^t\sigma^{-1}$. We are thus reduced to the case where $\hat{G}$ is finitely generated, and therefore isomorphic to a product $\mathbf{Z}^m \times F$, where F is a finite group. The hypothesis that the orbits are infinite implies that $F = \{\hat{e}\}$. Now observe that in $\hat{G} = \mathbf{Z}^m$ if the element $\hat{x}$ is such that the equation $n\hat{y} = \hat{x}$ has no solution in $\hat{G}$ for $n > 1$, the same is true for all the elements in the orbit of $\hat{x}$.
(d) Take $G = \mathbf{T}^n$, so that the group of automorphisms of $\hat{G}$ is $\mathbf{GL}(n, \mathbf{Z})$. For σ to be ergodic, it is necessary and sufficient that ${}^t\sigma$ has no root of unity as an eigenvalue.

9. Let G be an infinite compact metrizable commutative group, written additively.
(a) Let $a \in G$. For the translation $x \mapsto a + x$ to be ergodic with respect to Haar measure, it is necessary and sufficient that $\langle a, \hat{x}\rangle \neq 1$ for all $\hat{x} \in \hat{G}$ except $\hat{x} = \hat{e}$; an equivalent condition is that the subgroup of G generated by a should be dense (which implies that G is monothetic (Problem 4)). (Use (**22.7.1.6**).)
(b) If $U: \tilde{f} \mapsto (\gamma(a)f)^{\sim}$ is the unitary operator in $L^2_{\mathbf{C}}(G)$ corresponding to the translation $x \mapsto a + x$ (Section **13.11**, Problem 10) the eigenvectors of U are the characters $\hat{x} \in \hat{G}$, and the corresponding eigenvalues are the numbers $\langle a, \hat{x}\rangle$. If the translation $x \mapsto a + x$ is ergodic, show that it is not weakly mixing (Section **15.11**, Problem 16(d)).
(c) Let X be a compact metrizable space, μ a positive measure on X, and let u: X $\to$ X be a μ-measurable mapping which leaves μ invariant. Suppose that u is ergodic, and that the

eigenvectors of the corresponding unitary operator U on $L^2_{\mathbf{C}}(X, \mu)$ form a Hilbert basis of this space. Show that u is *conjugate* to a translation $x \mapsto a + x$ in a compact metrizable group G (Section **13.17**, Problem 11). (Take G to be the dual of the at most denumerable discrete group $L \subset \mathbf{U}$ formed by the eigenvalues of U (Section **15.11**, Problem 16(c)), and consider the transpose of the canonical injection $j: L \to \mathbf{U}$. Then use Section **15.11**, Problem 16(h).)

10. Let A_n be the vector space of polynomial functions $P: x \mapsto \sum_\alpha c_\alpha x^\alpha$ on $\mathbf{R}^n$, with *complex* coefficients. Let $\bar{P}$ denote the function $x \mapsto \sum_\alpha \bar{c}_\alpha x^\alpha$, and $P(D)$ the differential operator $\sum_\alpha c_\alpha D^\alpha$.

(a) Show that the mapping $(P, Q) \mapsto (P(D) \cdot \bar{Q})(0) = (P \mid Q)$ is a positive definite Hermitian form on A_n: for $P = \sum_\alpha c_\alpha x^\alpha$ we have $(P \mid P) = \sum_\alpha \alpha! \, |c_\alpha|^2$. We have $(x^\alpha \mid x^\beta) = 0$ if $\alpha \neq \beta$.

(b) For each polynomial $P \in A_n$, put

$$\mathscr{H}P(x) = 2^{n/2} \int_{\mathbf{R}^n} e^{-2\pi|y|^2} P(x + iy) \, dy.$$

Show that $\mathscr{H}P$ is a polynomial such that if P has degree d, $\mathscr{H}P - P$ has degree $\leqq d - 2$. If the coefficients of P are real, so are those of $\mathscr{H}P$. The linear mapping $\mathscr{H}: A_n \to A_n$ is bijective. We have $\mathscr{H}D^\alpha = D^\alpha \mathscr{H}$ for all derivations D^α.

(c) Show that

$$\mathscr{H}(x_j P(x)) = x_j \mathscr{H}P(x) - \frac{1}{4\pi} D_j(\mathscr{H}P)(x)$$

for $1 \leqq j \leqq n$. Deduce that

$$\mathscr{H}P(x) = e^{2\pi|x|^2} \cdot \left(P\left(-\frac{1}{4\pi} D\right)(e^{-2\pi|x|^2})\right).$$

(Induction on the degree of P.)

(d) For each polynomial $P \in A_n$, put

$$\mathscr{W}P(x) = e^{-\pi|x|^2} \mathscr{H}P(x).$$

The functions $\mathscr{W}P$ belong to $\mathscr{L}^2_{\mathbf{C}}(\mathbf{R}^n)$. For each *homogeneous* polynomial P of degree d we have

$$(\mathscr{W}P \mid \mathscr{W}Q) = 2^{-n/2}(4\pi)^{-d}(P \mid Q),$$

the scalar product on the right-hand side being that defined in (a). (We may assume that Q is homogeneous. Show first that if $\deg(Q) < d$, we have $\int e^{-2\pi|x|^2} \mathscr{H}P(x) \cdot \bar{Q}(x) \, dx = 0$ by using (c) and integrating by parts. Using (b), deduce that $(\mathscr{W}P \mid \mathscr{W}Q) = 0$, and hence that also $(\mathscr{W}P \mid \mathscr{W}Q) = 0$ if $\deg(Q) > d$. Finally use (b) to show that when $\deg(Q) = d$ we have $(\mathscr{W}P \mid \mathscr{W}Q) = \int e^{-2\pi|x|^2} \mathscr{H}P(x) \cdot \bar{Q}(x) \, dx$, and calculate this integral as above.)

(e) Show that if P is a homogeneous polynomial of degree d, then

$$(*) \qquad \int e^{-\pi|y|^2} \mathscr{H}P(y + ix) \, dy = i^d \mathscr{H}P(x).$$

(By considering the derivatives with respect to x, reduce to proving the result when $x = 0$. Do this by induction on d, using (c).)

Deduce that

$$\mathscr{F}(\mathscr{W}\mathrm{P}) = (-i)^d \mathscr{W}\mathrm{P}.$$

(Use Cauchy's theorem to transform (*).)

(f) Show that as P runs through A_n the functions $\mathscr{W}\mathrm{P}$ form a total set in $\mathscr{L}^2_{\mathbf{C}}(\mathbf{R}^n)$. (Put $k_\zeta(x) = \zeta^{-n} \exp(-\pi|x|^2\zeta^{-2})$ for $\mathscr{R}\zeta > 0$; show that $\mathrm{G}(\zeta, x) = (k_\zeta * g)(x)$ is analytic for $\mathscr{R}\zeta > 0$ and $x \in \mathbf{R}^n$, for all functions $g \in \mathscr{L}^2_{\mathbf{C}}(\mathbf{R}^n)$. If g is orthogonal to all the functions $\mathscr{W}\mathrm{P}$, show that all the derivatives of G at the point (1, 0) vanish and hence that $\mathrm{G} = 0$; then use **(14.11.1)**.)

(g) Show that the function $e^{-\pi|x|^2}\mathscr{W}\mathrm{P}(x/\sqrt{2})$ is the Fourier transform of the function $e^{-\pi|\xi|^2}\mathrm{P}(i\xi/\sqrt{2})$; it is of positive type if and only if $\mathrm{P}(ix) \geqq 0$ for all $x \in \mathbf{R}^n$.

11. With the notation of Problem 10, when $n = 1$, the functions

$$\begin{aligned}\mathrm{W}_m(x) &= (-1)^m 2^{-m+\frac{1}{4}}\pi^{-m/2}\sqrt{m!}\,\mathscr{W}(x^m)\\ &= (-1)^m 2^{-m+\frac{1}{4}}\pi^{-m/2}\sqrt{m!}\,e^{\pi x^2}\mathrm{D}^m(e^{-2\pi x^2})\end{aligned}$$

are called the *Hermite–Weber functions.*

(a) Show that the W_m for $m \geqq 0$ form a *Hilbert basis* of $\mathrm{L}^2_{\mathbf{C}}(\mathbf{R})$, namely the basis obtained by orthonormalization **(6.6.1)** from the sequence of functions $x^m e^{-\pi x^2}$, which is dense in $\mathrm{L}^2_{\mathbf{C}}(\mathbf{R})$; also that $\mathscr{F}\mathrm{W}_m = (-i)^m \mathrm{W}_m$. The functions W_m are even or odd according as m is even or odd. For each integer $n \geqq 0$, the function $(-1)^n e^{-\pi x^2}\mathrm{W}_{2n}(x)$ is of positive type.

(b) For each function $f \in \mathscr{L}^2_{\mathbf{C}}(\mathbf{R})$, put $a_m(f) = (f \mid \mathrm{W}_m)$, so that $\tilde{f} = \sum_{m=0}^{\infty} (f \mid \mathrm{W}_m)\mathrm{W}_m$, the series being convergent in $\mathrm{L}^2_{\mathbf{C}}(\mathbf{R})$, and $\sum_{m=0}^{\infty} |(f \mid \mathrm{W}_m)|^2 = (\mathrm{N}_2(f))^2$. Show that if in addition f is of positive type, then $(-1)^n a_{2n}(f(x)e^{-\pi x^2}) \geqq 0$ for each integer $n \geqq 0$. (Use **(22.10.10.7)**.)

(c) Show that an *even* function $f \in \mathscr{L}^2_{\mathbf{C}}(\mathbf{R})$ is of positive type if and only if, for each $t > 0$, the numbers

$$c_{2n}(t) = (-1)^n \int_{-\infty}^{+\infty} f(tx)e^{-\pi x^2}\mathrm{W}_{2n}(x)\,dx$$

are $\geqq 0$. (Using (b), show that this condition implies that for each $t > 0$ the function $f(x)\exp(-\pi x^2/t^2)$ is of positive type, and let $t \to +\infty$.)

15. CONTINUOUS UNITARY REPRESENTATIONS OF LOCALLY COMPACT COMMUTATIVE GROUPS

(22.15.1) *Let* G *be a (separable, metrizable) locally compact commutative group. Every* topologically cyclic **(22.1.1)** *continuous unitary representation of* G *on a separable Hilbert space is equivalent to a representation* $s \mapsto M_\mu(s)$ *defined as follows: consider on the dual group* $\hat{\mathrm{G}}$ *a bounded positive measure* μ, *the corresponding Hilbert space* $L^2_{\mathbf{C}}(\hat{G}, \mu)$, *and for each* $s \in G$ *let* $M_\mu(s)$ *denote the unitary operator on* $L^2_{\mathbf{C}}(\hat{G}, \mu)$ *such that for each function* $g \in \mathscr{L}^2_{\mathbf{C}}(\hat{G}, \mu)$, $M_\mu(s) \cdot \tilde{g}$ *is the class of the function* $\hat{x} \mapsto \langle s, \hat{x}\rangle g(\hat{x})$ *in* $L^2_{\mathbf{C}}(\hat{G}, \mu)$.

We know from (22.1.3) that the equivalence classes of the topologically cyclic continuous unitary representations of G are in one-one correspondence with the *continuous functions of positive type* on G. By virtue of Bochner's theorem, for each continuous function p of positive type on G, the bounded positive measure μ on $\hat{G}$ which is the Plancherel transform of $p \cdot m_G$ is such that

$$p(x) = \int_{\hat{G}} \langle x, \hat{x}\rangle \, d\mu(\hat{x})$$

for $x \in G$. If E is the Hilbert space of the topologically cyclic continuous unitary representation U of G corresponding to p (22.1.3) and π the canonical mapping of the algebra $A = L^1_{\mathbf{C}}(G) + \mathbf{C}\varepsilon_e$ into E, we have $U(s) \cdot \pi(\tilde{f}) = \pi((\varepsilon_s * f)^\sim)$ for $f \in \mathscr{L}^1_{\mathbf{C}}(G)$, and the extension of U to $L^1_{\mathbf{C}}(G)$ is given by $U(g) \cdot \pi(\tilde{f}) = \pi((g * f)^\sim)$ for $f, g \in \mathscr{L}^1_{\mathbf{C}}(G)$; furthermore, there is an isomorphism T of the Hilbert space E onto $L^2_{\mathbf{C}}(\hat{G}, \mu)$ such that the composite mapping $f \mapsto \pi(\tilde{f}) \mapsto T(\pi(\tilde{f}))$ is just the Fourier cotransform $f \mapsto \bar{\mathscr{F}} f$ (22.7.4). By applying (22.7.1.6), we obtain the theorem.

(22.15.2) When G is a *compact* commutative group, the application of (22.15.1) and the decomposition of a continuous unitary representation as a Hilbert sum of topologically cyclic representations gives us again the description (21.4.1) of the continuous unitary representations of G, bearing in mind that the group $\hat{G}$ here is discrete and that the irreducible representations of G are one-dimensional.

When $G = \mathbf{R}$, there is another description of the continuous unitary representations:

(22.15.3) (Stone) *Every continuous unitary representation of the additive group* $\mathbf{R}$ *on a separable Hilbert space* E *is of the form* $t \mapsto \exp(itA)$, *where* A *is a not necessarily bounded self-adjoint operator on* E ((15.12.7) *and* (15.12.13)); *conversely, for each not necessarily bounded self-adjoint operator* A *on* E, $t \mapsto \exp(itA)$ *is a continuous unitary representation of* $\mathbf{R}$ *on* E.

If U is a *topologically cyclic* continuous unitary representation of $\mathbf{R}$ on a separable Hilbert space E, it is (up to equivalence) of the form M_μ, where μ is a bounded measure on $\mathbf{R}$, by virtue of (22.15.1). For each $n \in \mathbf{Z}$, let ψ_n be the characteristic function of the interval $]n, n+1]$ in $\mathbf{R}$; the classes of the functions in $\mathscr{L}^2_{\mathbf{C}}(\mathbf{R}, \mu)$ which vanish outside this interval form a closed subspace F_n of $\mathrm{L}^2_{\mathbf{C}}(\mathbf{R}, \mu)$, and it is clear that $\mathrm{L}^2_{\mathbf{C}}(\mathbf{R}, \mu)$ is the *Hilbert sum* of the F_n for $n \in \mathbf{Z}$, since each function $f \in \mathscr{L}^2_{\mathbf{C}}(\mathbf{R}, \mu)$ can be written in the form $f = \sum_{n \in \mathbf{Z}} f\psi_n$, the series being convergent in $\mathscr{L}^2_{\mathbf{C}}(\mathbf{R}, \mu)$. Each of the subspaces F_n is *stable* under M_μ; the restriction of M_μ being defined as in (22.15.1), $M_\mu(t) \cdot \tilde{f}$, for $\tilde{f} \in \mathrm{F}_n$, is the class of the function $x \mapsto \exp(2\pi itx)f(x)$. But since the interval $]n, n+1]$ is bounded, for each $\tilde{f} \in \mathrm{F}_n$ the class of the function $x \mapsto 2\pi x\tilde{f}(x)$ also belongs to F_n, and if we denote this class by $A_n \cdot \tilde{f}$, it is clear that A_n is a continuous self-adjoint operator on F_n, and that the restriction of $U(t)$ to F_n is equal to $\exp(itA_n)$ (15.11.1). If A is the (not necessarily bounded) self-adjoint operator on $\mathrm{L}^2_{\mathbf{C}}(\mathbf{R}, \mu)$ whose restriction to each F_n is A_n (15.12.8), it is clear that $U(t) = \exp(itA)$ for all $t \in \mathbf{R}$.

If now U is arbitrary, E is the Hilbert sum of a sequence (E_n) of closed subspaces stable under U, the restriction of U to each E_n being topologically cyclic. The argument above then shows that E_n is the Hilbert sum of closed subspaces F_{nm} stable under U, the restriction of U to each F_{nm} being of the form $t \mapsto \exp(itA_{nm})$, where A_{nm} is a continuous self-adjoint operator on F_{nm}. The proof is then completed as above by invoking (15.12.8).

Conversely, let E be a separable Hilbert space, A a not necessarily bounded self-adjoint operator on E; then E is the Hilbert sum of a sequence (E_n) of closed subspaces stable under A, the restriction A_n of A to E_n being *continuous* (15.12.8); for each $t \in \mathbf{R}$, $\exp(itA_n)$ is therefore a unitary operator on E_n (15.11.7), and there exists one and only one unitary operator $U(t)$ on E whose restriction to each E_n is $\exp(itA_n)$ (15.10.8.1); it is immediate that $t \mapsto U(t)$ is a continuous unitary representation of $\mathbf{R}$ on E.

PROBLEMS

1. Let G be a (separable, metrizable) locally compact commutative group. For each continuous unitary representation U of G on a separable Hilbert space E, show that there exists a unique representation L of the involutory Banach algebra $\mathscr{U}_{\mathbf{C}}(\hat{\mathrm{G}})$ (with the usual product) on E such that $L(\langle s, \cdot \rangle) = U(s)$ for all $s \in \mathrm{G}$.

2. (a) Let G be a (separable, metrizable) locally compact commutative group and let U (resp. V) be a continuous unitary representation of G (resp. $\hat{\mathrm{G}}$) on a separable Hilbert space E. Suppose that for $x \in \mathrm{G}$ and $\hat{x} \in \hat{\mathrm{G}}$ we have

(1) $$U(x)V(\hat{x}) = \langle x, \hat{x} \rangle^{-1} V(\hat{x})U(x).$$

Show that there exists a Hilbert space F and an isometry T of $\mathrm{L}^2_{\mathrm{F}}(m_{\mathrm{G}})$ onto E such that

$$T(\gamma(s)\tilde{\mathbf{g}}) = U(s) \cdot T(\tilde{\mathbf{g}})$$

and

$$T(\hat{s}\tilde{\mathbf{g}}) = V(\hat{s}) \cdot T(\tilde{\mathbf{g}})$$

for $s \in G$, $\hat{s} \in \hat{G}$ and $\mathbf{g} \in \mathscr{L}^2_F(m_G)$ (*Stone–von Neumann theorem*). (Apply Problem 1 to $\hat{G}$ and V, and use Section **22.3**, Problem 10.)

(b) Suppose that G and $\hat{G}$ are written additively. Consider the locally compact group A(G), whose underlying topological space is $G \times \hat{G} \times \mathbf{U}$, the multiplication being defined by the formula

$$(x_1, \hat{x}_1, \zeta_1)(x_2, \hat{x}_2, \zeta_2) = (x_1 + x_2, \hat{x}_1 + \hat{x}_2, \langle x_1, \hat{x}_2\rangle^{-1}\zeta_1\zeta_2).$$

The condition (1) is necessary and sufficient for the mapping

$$(x, \hat{x}, \zeta) \mapsto \zeta V(\hat{x})U(x)$$

to be a continuous unitary representation of A(G) on the Hilbert space E. Deduce from (a) that every continuous unitary representation W of A(G) on E, such that $W(0, 0, \zeta) = \zeta I$, is equivalent to a Hilbert sum of a (finite or infinite) sequence of representations each equivalent to the unitary representation W_0 of A(G) on $L^2_C(G)$, such that $W_0(x, \hat{x}, \zeta) \cdot \tilde{g}$ is the class of the function $s \mapsto \zeta\langle \hat{x}, s\rangle g(s - x)$. Show that W_0 is an isomorphism of A(G) onto its image, endowed with the strong topology (Section **12.15**, Problem 8). (Use the fact that $\mathscr{F} W_0(x, \hat{x}, \zeta)\mathscr{F}^{-1}$ takes $\tilde{g}' \in L^2_C(\hat{G})$ to the class of the function

$$\hat{s} \mapsto \zeta\langle x, \hat{x}\rangle\langle x, \hat{s}\rangle g'(\hat{s} + \hat{x}).)$$

(c) Let m_G, $m_{\hat{G}}$ denote associated Haar measures on G, $\hat{G}$ respectively, and let m denote the normalized Haar measure on $\mathbf{U}$; then the product measure $m_G \otimes m_{\hat{G}} \otimes m$ on A(G) is a left Haar measure, and A(G) is unimodular.

(d) For each function $h \in \mathscr{K}(G \times \hat{G})$, the "partial Fourier cotransform" $\overline{\mathscr{F}}_2 h$ defined by

$$\overline{\mathscr{F}}_2 h(x, y) = \int_{\hat{G}} \langle x, \hat{u}\rangle h(y, \hat{u})\, dm_{\hat{G}}(\hat{u})$$

is a function belonging to $\mathscr{L}^2_C(G \times G)$ such that $N_2(\overline{\mathscr{F}}_2 h) = N_2(h)$; likewise, for $k \in \mathscr{K}(G \times G)$, the "partial Fourier transform" $\mathscr{F}_1 k$ defined by

$$\mathscr{F}_1 k(x, \hat{x}) = \int_G \overline{\langle y, \hat{x}\rangle} k(x, y)\, dm_G(y)$$

is a function belonging to $\mathscr{L}^2_C(G \times \hat{G})$ such that $N_2(\mathscr{F}_1 k) = N_2(k)$. Also we have $(\overline{\mathscr{F}}_2 h | k) = (h | \mathscr{F}_1 k)$. Deduce that $\overline{\mathscr{F}}_2$ and $\mathscr{F}_1$ extend to isometries of $L^2_C(G \times \hat{G})$ onto $L^2_C(G \times G)$ and of $L^2_C(G \times G)$ onto $L^2_C(G \times \hat{G})$, respectively.

(e) For each function $h \in \mathscr{K}(G \times \hat{G})$, the function $h^* \in \mathscr{K}(G \times \hat{G} \times \mathbf{U})$ defined by $h^*(x, \hat{x}, \zeta) = \zeta^{-1}h(x, \hat{x})$ is such that the operator $W_0(h^*)$ on $L^2_C(G)$ (**21.1.6**) is defined by the formula

$$((W_0(h^*)) \cdot \tilde{g})(x) = \int_G \overline{\mathscr{F}}_2 h(x, x - y) g(y)\, dm_G(y) \tag{2}$$

and is therefore a Hilbert–Schmidt operator on $L^2_C(G)$ (Section **15.4**, Problem 14). Conversely, every Hilbert–Schmidt operator associated to a kernel belonging to $\mathscr{K}(G \times G)$ is the limit (with respect to the norm defined in (**15.4.8**)) of a sequence of Hilbert–Schmidt operators of the form $W_0(h^*)$.

(f) Deduce from (e) that if a continuous operator T on $L^2_C(G)$ commutes with all the operators $W_0(h^*)$, it commutes also with Hilbert–Schmidt operators defined by kernels belonging to $\mathscr{L}^2_C(G \times G)$. By expressing that T commutes with the operator defined by the kernel $k(x, y) = f_1(x)f_2(y)$, where f_1 and f_2 are in $\mathscr{K}(G)$, conclude that T must be a homothety, and hence that the representation W_0 is (topologically) *irreducible* **(15.5.4)**.
(g) Identify G, $\hat{G}$ and **U** canonically with subgroups of A(G); **U** is the center of A(G), and A(G) is the semidirect product of the commutative normal subgroup $\hat{G} \times \mathbf{U}$ and the subgroup G. Show that the representation of A(G) induced by the one-dimensional representation Z: $(0, \hat{x}, \zeta) \mapsto \zeta$ of $\hat{G} \times \mathbf{U}$ is equivalent to the representation W_0.

3. (a) Let M, N be two locally compact commutative groups. A *bicharacter* on the group $M \times N$ is a mapping f: $M \times N \to \mathbf{U}$ such that the partial mappings $x \mapsto f(x, y_0)$ and $y \mapsto f(x_0, y)$ are characters of M and N respectively, for all $y_0 \in N$ and all $x_0 \in M$. The product of two bicharacters is a bicharacter; the inverse of a bicharacter is a bicharacter. Show that for each bicharacter f there exists a unique continuous homomorphism ρ: $N \to \hat{M}$ such that $f(x, y) = \langle x, \rho(y)\rangle = \langle y, {}^t\rho(x)\rangle$ for all $x \in M$ and $y \in N$. If $M = N$, we have $f(x, y) = f(y, x)$ for all $x, y \in M$ if and only if ${}^t\rho = \rho$ (in which case ρ and f are said to be *symmetric*). An automorphism σ of M satisfies $f(\sigma(x), \sigma(y)) = f(x, y)$ for all $x, y \in M$ if and only if ${}^t\sigma\rho\sigma = \rho$.
(b) Suppose in particular that $M = N = G \times \hat{G}$, where G is a locally compact commutative group. If we denote an element of M by

$$\begin{pmatrix} x \\ \hat{x} \end{pmatrix}$$

(in place of $(x, \hat{x})$), every automorphism σ of M is uniquely of the form

$$(*) \qquad \begin{pmatrix} x \\ \hat{x} \end{pmatrix} \mapsto \begin{pmatrix} \alpha(x) + \beta(\hat{x}) \\ \gamma(x) + \delta(\hat{x}) \end{pmatrix}$$

where α: $G \to G$, β: $\hat{G} \to G$, γ: $G \to \hat{G}$, δ: $\hat{G} \to \hat{G}$ are continuous homomorphisms. We shall write

$$\sigma = \begin{pmatrix} \alpha & \beta \\ \gamma & \delta \end{pmatrix}$$

so that the formula $(*)$ is obtained by the usual rule for multiplying matrices. Show that we have

$${}^t\sigma = \begin{pmatrix} {}^t\alpha & {}^t\gamma \\ {}^t\beta & {}^t\delta \end{pmatrix}.$$

Put $F((x, \hat{x}), (y, \hat{y})) = \langle x, \hat{y}\rangle\langle -y, \hat{x}\rangle$, so that F is a bicharacter on M. The group of automorphisms σ of M such that $F(\sigma(z), \sigma(z')) = F(z, z')$ for all $z, z' \in M$ is called the *symplectic group* over G, denoted by **Sp**(G). If we denote by η the canonical isomorphism $(x, \hat{x}) \mapsto (-\hat{x}, x)$ of M onto $\hat{M} = \hat{G} \times G$, the relation $\sigma \in \mathbf{Sp}(G)$ is equivalent to ${}^t\sigma\eta\sigma = \eta$, or also (in the above notation) to

$$\begin{pmatrix} \alpha & \beta \\ \gamma & \delta \end{pmatrix}\begin{pmatrix} {}^t\delta & -{}^t\beta \\ -{}^t\gamma & {}^t\alpha \end{pmatrix} = \begin{pmatrix} 1_G & 0 \\ 0 & 1_{\hat{G}} \end{pmatrix}.$$

Show that $\mathrm{mod}(\sigma) = 1$.

(c) If we take $G = \mathbf{R}^n$, we have $F(z, z') = \exp(2\pi i\Phi(z, z'))$, where Φ is a nondegenerate alternating form on $\mathbf{R}^{2n}$, and hence $\mathbf{Sp}(G)$ is in this case the real symplectic group $\mathbf{Sp}(\Phi) \subset \mathbf{GL}(2n, \mathbf{R})$ (Section **16.11**, Problem 6).

4. (a) Let G be a locally compact commutative group, written additively. A *character of the second degree* on G is any continuous mapping f: $G \to \mathbf{U}$ such that the mapping $(x, y) \mapsto f(x + y)f(x)^{-1}f(y)^{-1}$ is a bicharacter on $G \times G$ (Problem 3); hence there exists a continuous homomorphism ρ_f: $G \to \hat{G}$ such that

$$(*) \qquad f(x + y) = f(x)f(y)\langle x, \rho_f(y)\rangle$$

and we have ${}^t\rho_f = \rho_f$. We say that f is *nondegenerate* if ρ_f is an isomorphism of G onto $\hat{G}$.
(b) The characters of the second degree on G form a multiplicative group $\mathbf{X}_2(G)$. The mapping $f \mapsto \rho_f$ is a homomorphism of the multiplicative group $\mathbf{X}_2(G)$ into the additive group of symmetric homomorphisms of G into $\hat{G}$, and its kernel is $\hat{G}$.
(c) Suppose that the mapping $x \mapsto 2x$ is an isomorphism of G onto itself (cf. Section **12.12**, Problem 3), and denote the inverse isomorphism by $x \mapsto 2^{-1}x$. Then for each symmetric homomorphism ρ: $G \to \hat{G}$, the function f_ρ: $x \mapsto \langle x, 2^{-1}\rho(x)\rangle$ is a character of the second degree on G, and $\rho \mapsto f_\rho$ is an isomorphism of the additive group of symmetric homomorphisms of G into $\hat{G}$, onto a subgroup $\mathbf{X}_2^0(G)$ of $\mathbf{X}_2(G)$, and $\mathbf{X}_2(G)$ is isomorphic to $\mathbf{X}_2^0(G) \times \hat{G}$.
(d) If $G = \mathbf{R}^n$, the characters of the second degree are the functions $x \mapsto \exp(2\pi i Q(x))$, where Q is a quadratic form on $\mathbf{R}^n$. Such a character is nondegenerate if and only if the quadratic form Q is nondegenerate.

5. With the notation of Problem 2, every automorphism s of the group A(G) leaves stable the center **U**, and its restriction to **U** is one of its two automorphisms, $\zeta \mapsto \zeta$ or $\zeta \mapsto \zeta^{-1}$. Let N(G) be the group of automorphisms s of A(G) whose restriction to **U** is the identity. On passing to the quotient, s defines an automorphism σ of the quotient group $A(G)/\mathbf{U} = G \times \hat{G}$. For each $(z, \zeta) \in (G \times \hat{G}) \times \mathbf{U}$ we may write

$$s(z, \zeta) = (\sigma(z), f(z)\zeta)$$

where f: $G \times \hat{G} \to \mathbf{U}$ is continuous.
(a) Show that

$$(1) \qquad f(z_1 + z_2)f(z_1)^{-1}f(z_2)^{-1} = F(\sigma(z_1), \sigma(z_2))F(z_1, z_2)^{-1}.$$

In particular, f is a character of the second degree on $G \times \hat{G}$, and σ is a symplectic automorphism. (Express that the right-hand side of (1) is symmetric.) If we put $g(x) = f(x, 0)$ and $h(\hat{x}) = f(0, \hat{x})$, show that in the notation introduced in Problem 3 we have

$$g(x_1 + x_2) = g(x_1)g(x_2)\langle x_1, {}^t\gamma(\alpha(x_2))\rangle,$$
$$h(\hat{x}_1 + \hat{x}_2) = h(\hat{x}_1)h(\hat{x}_2)\langle {}^t\delta(\beta(\hat{x}_1)), \hat{x}_2\rangle$$

(which implies that g and h are characters of the second degree on G and $\hat{G}$, respectively) and

$$(2) \qquad f(x, \hat{x}) = g(x)h(\hat{x})\langle \beta(\hat{x}), \gamma(x)\rangle.$$

(b) Suppose that $x \mapsto 2x$ is an automorphism of G. Then, for each automorphism $\sigma \in \mathbf{Sp}(G)$, if we take

$$g(x) = \langle x, 2^{-1} \cdot {}^t\gamma(\alpha(x))\rangle, \qquad h(\hat{x}) = \langle 2^{-1} \cdot {}^t\delta(\beta(\hat{x})), \hat{x}\rangle$$

the formula (2) defines a character of the second degree such that $(z, \zeta) \mapsto (\sigma(z), f(z)\zeta)$ is an automorphism of A(G) which fixes each element of **U**.
(c) Denote by (σ, f) the automorphism s of A(G) such that $s(z, \zeta) = (\sigma(z), f(z)\zeta)$. Show that $(\sigma, f) \mapsto \sigma$ is a homomorphism of N(G) into **Sp**(G), whose kernel is isomorphic to $\hat{G} \times G$. If $x \mapsto 2x$ is an automorphism of G, deduce from (b) that N(G) is the semidirect product of the normal subgroup $\hat{G} \times G$ and a group isomorphic to **Sp**(G).
(d) For each automorphism $s = (\sigma, f) \in N(G)$, the mapping $(z, \zeta) \mapsto W_0(\sigma(z), f(z)\zeta)$ is a continuous unitary representation of A(G) on $L^2_{\mathbf{C}}(G)$, in the notation of Problem 2. Deduce from Problem 2(b) that there exists an isometry $T(s)$ of $L^2_{\mathbf{C}}(G)$ onto itself such that $W_0(\sigma(z), f(z)\zeta) = T(s)W_0(z, \zeta)T(s)^{-1}$, and that $T(s)$ is unique up to multiplication by a nonzero scalar.

6. (a) On $\mathbf{C}^n$, let $(z \mid w) = \sum_{j=1}^{n} z_j \bar{w}_j$ denote the usual Hermitian scalar product, and put $B(z, w) = \sum_{j=1}^{n} z_j w_j$; also put $|z|^2 = (z \mid z)$ and $Q(z) = B(z, z)$. If λ is Lebesgue measure on $\mathbf{C}^n = \mathbf{R}^{2n}$, let μ denote the measure which has density $\exp(-\pi |z|^2)$ with respect to λ. The monomials

$$M_\nu(z) = (\nu!)^{-1/2}(\pi^{1/2}z)^\nu$$

(with the usual notation for multi-indices (19.5)) form a *Hilbert basis* of the subspace F_n of $L^2_{\mathbf{C}}(\mu)$ consisting of the *entire functions* on $\mathbf{C}^n$ which are square-integrable with respect to μ; furthermore, the function $\exp(\pi(z \mid w))$ is a *reproducing kernel* in F_n (Section 6.3, Problem 4): in other words, for all functions $f \in F_n$ we have

$$f(z) = \int_{\mathbf{C}^n} \exp(\pi(z \mid w)) f(w)\, d\mu(w).$$

(b) The Hermite–Weber functions

$$W_\nu(x) = W_{\nu_1}(x_1) \cdots W_{\nu_n}(x_n)$$

form a Hilbert basis of the space $L^2_{\mathbf{C}}(\mathbf{R}^n)$ (Section 22.14, Problem 11). Hence we may define an isometry F of $L^2_{\mathbf{C}}(\mathbf{R}^n)$ onto F_n by putting

$$F \cdot \bar{W}_\nu = M_\nu$$

for each multi-index ν. Show that if we define

$$K(x, z) = 2^{n/4} \exp(-\pi Q(x) + \tfrac{1}{2}\pi Q(z) + 2\pi i B(x, z))$$

for $x \in \mathbf{R}^n$ and $z \in \mathbf{C}^n$, then we have

$$(F \cdot g)(z) = \int_{\mathbf{R}^n} K(x, z) g(x)\, dx$$

and

$$K(x, z) = \sum_{\nu \in \mathbf{N}^n} M_\nu(z) W_\nu(x).$$

(Reduce to the case $n = 1$.)
(c) The nilpotent Lie group $A(\mathbf{R}^n)$ (Problem 2) is called the *Heisenberg group*, and its irreducible representation W_0 on $L^2_{\mathbf{C}}(\mathbf{R}^n)$ is called the *Schrödinger representation*: if u,

$u' \in \mathbf{R}^n$ and $\zeta \in \mathbf{U}$, then $W_0(u, u', \zeta)$ transforms the class of a function g to that of the function $x \mapsto \zeta \cdot \exp(2\pi i(u \mid x))g(x - u)$.

The representation $V_0 = FW_0F^{-1}$ of $\mathrm{A}(\mathbf{R}^n)$ on the Hilbert space F_n is called the *Fock representation*: for each function $f \in \mathrm{F}_n$, $V_0(u, u', \zeta) \cdot f$ is the entire function

$$z \mapsto \zeta \cdot \exp(\pi \mathrm{B}(z + \tfrac{1}{2}v, \bar{v}) + \tfrac{1}{2}i\pi \mathscr{I}\mathrm{Q}(v)) \cdot f(x + v)$$

where $v = u' + iu$, $\bar{v} = u' - iu$.

7. The group $\mathbf{Sp}(2n, \mathbf{R})$ consists of the matrices

$$\sigma = \begin{pmatrix} \alpha & \beta \\ \gamma & \delta \end{pmatrix}$$

(where $\alpha, \beta, \gamma, \delta$ are real $n \times n$ matrices) such that ${}^t\sigma\eta\sigma = \eta$, where

$$\eta = \begin{pmatrix} 0 & 1_n \\ -1_n & 0 \end{pmatrix}.$$

(a) For each pair of matrices $\xi_1, \xi_2 \in \mathbf{M}_n(\mathbf{C})$, the $n \times n$ matrix

$$\frac{1}{2i}({}^t\xi_1\ {}^t\xi_2)\eta\begin{pmatrix} \bar{\xi}_1 \\ \bar{\xi}_2 \end{pmatrix} = \frac{1}{2i}({}^t\xi_1\bar{\xi}_2 - {}^t\xi_2\bar{\xi}_1) \tag{1}$$

is Hermitian. Show that if it is positive definite, ξ_2 is invertible. (Observe that otherwise there would exist a matrix $\xi \neq 0$ such that $\xi_2\xi = 0$; obtain a contradiction by multiplying the right-hand side of (1) on the left by ${}^t\xi$ and on the right by $\bar{\xi}$.)

(b) Let P_n be the subset of $\mathbf{M}_n(\mathbf{C})$ consisting of the symmetric matrices τ such that $\mathscr{I}\tau$ is positive definite. Show with the aid of (a) that if $\tau \in \mathrm{P}_n$ and $\sigma \in \mathbf{Sp}(2n, \mathbf{R})$, the matrix $\gamma\tau + \delta$ is invertible, and $\tau' = (\alpha\tau + \beta)(\gamma\tau + \delta)^{-1}$ belongs to P_n. The set P_n, endowed with the structure of a complex manifold as an open subset of a complex vector subspace of $\mathbf{M}_n(\mathbf{C})$, is called the *Siegel half-space* of index n. The group $\mathbf{Sp}(2n, \mathbf{R})$ acts analytically on P_n by

$$(\sigma, \tau) \mapsto \sigma \cdot \tau = (\alpha\tau + \beta)(\gamma\tau + \delta)^{-1},$$

the mapping $\tau \mapsto \sigma \cdot \tau$ being holomorphic on P_n. Show that the stabilizer K of the matrix $i1_n \in \mathrm{P}_n$ is the group of matrices

$$\begin{pmatrix} \delta & -\gamma \\ \gamma & \delta \end{pmatrix}$$

such that $\gamma \cdot {}^t\delta = \delta \cdot {}^t\gamma$ and $\gamma \cdot {}^t\gamma + \delta \cdot {}^t\delta = 1_n$, and show that this group is isomorphic to the group of matrices $i\gamma + \delta$, with γ and δ satisfying the same relations, hence is isomorphic to the unitary group $\mathbf{U}(n)$. Show that $\mathbf{Sp}(2n, \mathbf{R})$ acts transitively on P_n; more precisely, the group of matrices

$$\begin{pmatrix} {}^t\delta^{-1} & \lambda\delta \\ 0 & \delta \end{pmatrix},$$

where δ is invertible and λ symmetric, acts transitively on P_n, and we have $\mathrm{S}_0 \cap \mathrm{K} = \mathbf{O}(n)$. (Relate these facts to the general theory of **(21.18)**.)

8. Let $t \mapsto e^{itA} = U(t)$ be a continuous unitary representation of **R** on a separable Hilbert space E (22.15.3).

(a) Show that dom(A) is the set of $x \in$ E such that $(U(t) \cdot x - x)/it$ tends to a limit in E as $t \to 0$ in **R**; this limit is then equal to $A \cdot x$. (Use the Hilbert sum decomposition of E introduced in the proof of (22.15.3).)

(b) The operator $(A - \zeta I)^{-1}$ is defined and continuous on E for all $\zeta \notin \mathbf{R}$. Show that (with the notation of (21.1.4))

$$(A - \zeta I)^{-1} = \begin{cases} i \displaystyle\int_{-\infty}^{0} e^{-i\zeta t} U(t)\, dt & \text{if } \mathscr{I}\zeta > 0, \\ -i \displaystyle\int_{0}^{\infty} e^{-i\zeta t} U(t)\, dt & \text{if } \mathscr{I}\zeta < 0, \end{cases}$$

(same method).

9. Let P and Q be two not necessarily bounded self-adjoint operators on a separable Hilbert space E (15.12.7). Suppose that for each complex number $\zeta \notin \mathbf{R}$,

$$\operatorname{dom} Q(P - \zeta I)) \subset \operatorname{dom}(PQ)$$

(15.12.5) and that for all $x \in \operatorname{dom}(PQ) \cap \operatorname{dom}(QP)$, *Heisenberg's commutation relation* holds:

$$(PQ) \cdot x - (QP) \cdot x = ix.$$

(a) Put $R(\zeta) = (P - \zeta I)^{-1}$ for all $\zeta \notin \mathbf{R}$. For $x \in \operatorname{dom}(Q)$, show that $R(\zeta) \cdot x$ belongs to dom(P) and to $\operatorname{dom}(PQ) \subset \operatorname{dom}(Q)$, and deduce that

$$QR(\zeta) \cdot x - R(\zeta)Q \cdot x = iR(\zeta)^2 \cdot x.$$

If G is the (closed) graph of Q in E × E, then G is stable under the operators $S(\zeta)$ on E × E defined by

$$S(\zeta) \cdot (x, y) = (R(\zeta) \cdot x, iR(\zeta)^2 \cdot x + R(\zeta) \cdot y)$$

for $\zeta \notin \mathbf{R}$.

(b) Put $U(t) = e^{itP}$, $V(t) = e^{2\pi i t Q}$, and let $W(t)$ be the continuous operator on E × E defined by

$$W(t) \cdot (x, y) = (U(t) \cdot x, tU(t) \cdot x + U(t) \cdot y).$$

Show that, in the notation of (21.1.4), we have

$$S(\zeta) = \begin{cases} i \displaystyle\int_{-\infty}^{0} e^{-i\zeta t} W(t)\, dt & \text{if } \mathscr{I}\zeta > 0, \\ -i \displaystyle\int_{0}^{\infty} e^{-i\zeta t} W(t)\, dt & \text{if } \mathscr{I}\zeta < 0, \end{cases}$$

for $\zeta \notin \mathbf{R}$. (Use Problem 8(b).) Deduce that G is stable under the $W(t)$, and that $U(t)QU(t)^{-1} = Q - tI$. Hence show that

$$U(t)V(s) = e^{-2\pi i s t} V(s)U(t)$$

for all $s, t \in \mathbf{R}$ (*H. Weyl's commutation relation*).

(c) Deduce from (b) and from Problem 2(a) that there exists a separable Hilbert space F and an isometry T of $L^2_F(m_{\mathbf{R}})$ onto E such that (1) $\operatorname{dom}(T^{-1}PT)$ contains the space of continuously differentiable functions with compact support on **R** with values in F, and the restriction of the operator $T^{-1}PT$ to this subspace is the operator id/dt; (2) $\operatorname{dom}(T^{-1}QT)$ contains the same subspace, and the restriction of $T^{-1}QT$ to this subspace is the operator of multiplication by the function $1_{\mathbf{R}}: t \mapsto t$.

10. Let $t \mapsto U(t)$ be a continuous unitary representation of **R** on a separable Hilbert space E, and let E_0 be a closed vector subspace of E which is stable under the $U(t)$ *such that* $t \geqq 0$.
(a) Show that the relation $s \leqq t$ in **R** implies that $U(t) \cdot E_0 \subset U(s) \cdot E_0$. Let E_∞ denote the intersection of the closed subspaces $U(t) \cdot E_0$ for $t \in \mathbf{R}$, and $E_{-\infty}$ the closure of the union of the $U(t) \cdot E_0$, $t \in \mathbf{R}$; also let E′ be the orthogonal supplement of E_∞ in $E_{-\infty}$. Then E_0 is the Hilbert sum of E_∞ and $E'_0 = E_0 \cap E'$. The subspaces $E_{-\infty}$, E_∞ and E′ are stable under all $U(t)$; E'_0 is stable under the $U(t)$ with $t \geqq 0$.
(b) Let Q be the orthogonal projection of E′ onto E'_0, and let $V(t)$ be the restriction of $U(t)$ to E′. Show that, in the Hilbert space E′, the operators

$$P(t) = 1_{E'} - V(t)QV(t)^{-1}$$

satisfy the conditions of Section 22.3, Problem 11. Deduce that there exists a Hilbert space F and an isometry T_0 of $L^2_F(m_{\mathbf{R}})$ onto E′ such that $T_0^{-1}V(t)T_0$ is the linear mapping $\tilde{\mathbf{g}} \mapsto (\gamma(t)\mathbf{g})^{\sim}$, and such that E'_0 is the image under T_0 of the subspace of classes of functions in $L^2_F(m_{\mathbf{R}})$ which are *zero for* $t \leqq 0$. (Use Section 22.3, Problems 10 and 11, and consider the projection $L(Y)$, where Y is the Heaviside function (17.5.7).)

11. Let G be a simply connected Lie group, $\mathfrak{g}$ its Lie algebra. A continuous one-dimensional unitary representation of G is a homomorphism of G into **U**; for such a homomorphism V, there exists a unique linear form f on $\mathfrak{g}$ such that $V(\exp \mathbf{x}) = e^{if(\mathbf{x})}$ for all $\mathbf{x} \in \mathfrak{g}$; moreover, we have $f([\mathbf{u}, \mathbf{v}]) = 0$ for all $\mathbf{u}, \mathbf{v} \in \mathfrak{g}$ (in other words, f may be considered as an element of the dual of $\mathfrak{g}/[\mathfrak{g}, \mathfrak{g}]$). Conversely, for each linear form f on $\mathfrak{g}$ satisfying this condition, there exists a continuous homomorphism V of G into **U** such that $V(\exp \mathbf{x}) = e^{if(\mathbf{x})}$ for all $\mathbf{x} \in \mathfrak{g}$ (19.7.6).

12. Let H be the simply connected nilpotent group consisting of the matrices

$$\begin{pmatrix} 1 & x & z \\ 0 & 1 & y \\ 0 & 0 & 1 \end{pmatrix}.$$

For brevity, let (x, y, z) denote the matrix just written. Then the center Z of H consists of the matrices $(0, 0, z)$, and the sub-groups L_1 (resp. L_2) consisting of the matrices $(x, 0, z)$ (resp $(0, y, z)$) are isomorphic to $\mathbf{R}^2$ and normal in H.

The Lie algebra $\mathfrak{h}$ of H has as a basis the matrices E_{12}, E_{23}, and E_{13} (19.4.2). Show that every irreducible continuous unitary representation of H on a separable Hilbert space is equivalent to one of the following representations:

(1) a representation $U^{(\mu, \nu)}$ of dimension 1 such that

$$U^{(\mu, \nu)}(\exp(uE_{12} + vE_{23} + wE_{13})) = \exp(2\pi i(\mu u + \nu v))$$

for $u, v, w \in \mathbf{R}$, where μ and ν are arbitrary real numbers;

(2) a representation U^λ on the Hilbert space $L^2_{\mathbf{C}}(\mathbf{R})$ such that, for $\tilde{f} \in L^2_{\mathbf{C}}(\mathbf{R})$, $U^\lambda(\exp(uE_{12} + vE_{23} + wE_{13})) \cdot \tilde{f}$ is the class of the function

$$t \mapsto \exp(2\pi i\lambda(vt + w + \tfrac{1}{2}uv))f(t - u)$$

for $u, v, w \in \mathbf{R}$, where λ is any nonzero real number. (Distinguish two cases, according as the restriction of the representation to the center Z of H is trivial or not. In the second case, consider the restrictions of the representation to the matrices $(x, 0, 0)$ and to the matrices $(0, y, 0)$, and use Problem 2(a). Recall that $\exp: \mathfrak{h} \to \mathrm{H}$ is a diffeomorphism (Section **19.14**, Problem 6(a)).)

Let f be a linear form on $\mathfrak{h}$ such that $f(E_{23}) \neq 0$. Show that if V_1 (resp. V_2) is the one-dimensional representation of L_1 (resp. L_2) defined by f (Problem 11), then the representations V_1^{ind} and V_2^{ind} of G are equivalent to the same representation U^λ. Two representations U^λ and $U^{\lambda'}$ are inequivalent unless $\lambda = \lambda'$.

13. Let G be a simply connected nilpotent Lie group, $\mathfrak{g}$ its Lie algebra. The mapping $\exp: \mathfrak{g} \to \mathrm{G}$ is a diffeomorphism which transforms each Lie subalgebra of $\mathfrak{g}$ into the corresponding connected subgroup (which is necessarily closed) of G (Section **19.14**, Problem 6). The center Z of G is connected and not equal to $\{e\}$ (Section **19.14**, Problem 7); let $\mathfrak{z}$ be its Lie algebra, the center of $\mathfrak{g}$. *We shall assume that* $\dim \mathrm{Z} = 1$, *and* $\dim \mathrm{G} \geqq 3$.

(a) There exists an ideal $\mathfrak{a} \subset \mathfrak{g}$ of dimension 2 containing $\mathfrak{z}$ and such that $[\mathfrak{a}, \mathfrak{g}] \subset \mathfrak{z}$. If $\mathfrak{g}_1$ is the centralizer of $\mathfrak{a}$ in $\mathfrak{g}$, then $\mathfrak{g}_1$ is an ideal in $\mathfrak{g}$ of codimension 1. If $\mathbf{c} \in \mathfrak{z}$, $\mathbf{c} \neq 0$, and $\mathbf{b} \in \mathfrak{a} \cap \complement\mathfrak{z}$, there exists $\mathbf{a} \in \mathfrak{g}_1$ such that $[\mathbf{a}, \mathbf{b}] = \mathbf{c}$, and if $\mathfrak{n} = \mathbf{Ra} + \mathbf{Rb} + \mathbf{Rc}$, then $\mathrm{N} = \exp(\mathfrak{n})$ is a subgroup of G isomorphic to the group H of Problem 10 (cf. Section **19.12**, Problem 4). If $\mathrm{G}_1 = \exp(\mathfrak{g}_1)$, then G is the semidirect product of the normal subgroup G_1 and the one-parameter subgroup $\exp(\mathbf{Ra})$, so that G/G_1 may be identified with $\mathbf{R}$.

(b) Let U be a continuous unitary representation of G on a separable Hilbert space F; suppose that the restriction of U to the center Z is nontrivial. Show that we may assume that F is of the form $L^2_{\mathrm{E}}(m_{\mathbf{R}})$ and that, for all $s \in \mathrm{G}_1$, $U(s)$ commutes with multiplication by any character of $\mathbf{R}$ (use Problem 10 and the fact that $\mathbf{b}$ lies in the center of $\mathfrak{g}_1$). Deduce first that if $\mathbf{f} \in \mathscr{L}^2_{\mathrm{E}}(m_{\mathbf{R}})$ is zero almost everywhere in the complement of a compact subset S of $\mathbf{R}$, the same is true of every function in the class $U(s) \cdot \tilde{\mathbf{f}}$ for $s \in \mathrm{G}_1$ (argue by contradiction, using Section **22.10**, Problem 6). Then use Section **22.3**, Problem 12, and Section **22.10**, Problem 6 to show that there exists an $(m_{\mathrm{G}_1} \otimes m_{\mathbf{R}})$-measurable mapping $(s, t) \mapsto K(s, t)$ of $\mathrm{G}_1 \times \mathbf{R}$ into $\mathscr{L}(\mathrm{E})$ such that, for each $t \in \mathbf{R}$, $s \mapsto K(s, t)$ is a continuous unitary representation of G_1 on E, and for each function $\mathbf{g} \in \mathscr{L}^2_{\mathrm{E}}(m_{\mathbf{R}})$ and each $s \in \mathrm{G}_1$, the class $U(s) \cdot \tilde{\mathbf{g}}$ is that of the function $t \mapsto K(s, t) \cdot \mathbf{g}(t)$.

(c) Deduce from (a) and (b) that, for $s = s_1 \exp(\mathbf{a}u) \in \mathrm{G}$, with $s_1 \in \mathrm{G}_1$ and $u \in \mathbf{R}$, the class $U(s) \cdot \tilde{\mathbf{g}}$ is that of the function $t \mapsto K(s_1, t) \cdot \mathbf{g}(t - u)$ (cf. Problem 2). Conclude that, for each $t_0 \in \mathbf{R}$, U is equivalent to the representation induced by the representation $V: s_1 \mapsto K(s_1, t_0)$ of G_1; this representation is such that $V(\exp(\mathbf{b}v + \mathbf{c}w)) = \exp(2\pi i\lambda(t_0 v + w))1_{\mathrm{E}}$ for $v, w \in \mathbf{R}$, with $\lambda \neq 0$ (cf. (**22.3.6**)). We may therefore always choose t_0 so that $V(\exp(\mathbf{b}v + \mathbf{c}w)) = \exp(2\pi i\lambda w)1_{\mathrm{E}}$ with $\lambda \neq 0$.

(d) Let V, V' be two continuous unitary representations of G_1 on separable Hilbert spaces E, E′ respectively, such that $V(\exp(\mathbf{b}v + \mathbf{c}w)) = \exp(2\pi i\lambda w)1_{\mathrm{E}}$ and $V'(\exp(\mathbf{b}v + \mathbf{c}w)) = \exp(2\pi i\lambda' w)1_{\mathrm{E}}$ for $v, w \in \mathbf{R}$, with $\lambda' \neq 0$ and $\lambda \neq 0$. Let U (resp. U') be the representation of G on $L^2_{\mathrm{E}}(m_{\mathbf{R}})$ (resp. $L^2_{\mathrm{E}'}(m_{\mathbf{R}})$) induced by V (resp. V'). In order that there should exist an isometry T of $L^2_{\mathrm{E}}(m_{\mathbf{R}})$ onto $L^2_{\mathrm{E}}(m_{\mathbf{R}})$ such that $U' = TUT^{-1}$, it is necessary and sufficient that $\lambda' = \lambda$ and that there exist an isometry T_0 of E onto E′ such that $T \cdot \tilde{\mathbf{f}}$ is the class of the function $t \mapsto T_0 \cdot \mathbf{f}(t)$ (express that $TU(s) = U'(s)T$ for

$s = \exp(\mathbf{c}w)$, $s = \exp(\mathbf{b}v)$ and $s = \exp(\mathbf{a}u)$: use Problem 12 to calculate $U(s)$ and $U'(s)$ for these values of s). Conclude that U and U' are equivalent if and only if $\lambda' = \lambda$ and V, V' are equivalent, and that U is irreducible if and only if V is irreducible.

14. With the notation of Problem 13, *we now drop the restriction on the dimension of* Z (it is always $\geqq 1$). For each nonzero linear form f on $\mathfrak{g}$, and each Lie subgroup $\mathrm{H} = \exp \mathfrak{h}$ of G, such that $f([\mathbf{x}, \mathbf{y}]) = 0$ for $\mathbf{x}, \mathbf{y} \in \mathfrak{h}$ (in which case $\mathfrak{h}$ is said to be *subordinate* to f) we denote by $V_{f,\mathfrak{h}}$ the one-dimensional continuous representation of H defined by $V_{f,\mathfrak{h}}(\exp \mathbf{x}) = \exp(2\pi i f(\mathbf{x}))$ (Problem 11), and by $U_{f,\mathfrak{h}}$ the representation of G *induced* by $V_{f,\mathfrak{h}}$. Show that every *irreducible* continuous unitary representation of G is equivalent to some $U_{f,\mathfrak{h}}$ (*Dixmier–Kirillov theorem*). (Proceed by induction on dim G. If U is an irreducible representation of G on F, consider the restriction of U to Z, which is such that $U(\exp \mathbf{z}) = \exp(2\pi i f(\mathbf{z}))1_{\mathrm{F}}$, where f is a linear form on $\mathfrak{z}$. Distinguish two cases, according as $\mathrm{Ker}(f) \neq \{0\}$ or $\mathrm{Ker}(f) = \{0\}$. In the first case, use the inductive hypothesis; in the second, which implies that $\dim \mathrm{Z} = 1$, use Problem 11 and the inductive hypothesis, together with Section **22.3**, Problem 8.)

15. (a) With the notation of Problems 13 and 14, *suppose again* that $\dim \mathrm{Z} = 1$. Let f be a linear form on $\mathfrak{g}$ whose restriction to $\mathfrak{z}$ is not identically zero, and let $\mathfrak{h}$ be a Lie subalgebra of $\mathfrak{g}$ subordinate to f which contains $\mathfrak{z}$ and is not contained in $\mathfrak{g}_1$. We may then assume that the elements **a**, **b**, **c** of Problem 12(a) are such that $\mathbf{a} \in \mathfrak{h}$; show that $\mathbf{b} \notin \mathfrak{h}$. Put $\mathfrak{h}' = (\mathfrak{h} \cap \mathfrak{g}_1) \oplus \mathbf{Rb}$; then $\mathfrak{h}'$ is a Lie subalgebra of $\mathfrak{g}$ subordinate to f, and $\dim \mathfrak{h}' = \dim \mathfrak{h}$.
(b) Put $\mathfrak{w} = (\mathrm{Ker} f) \cap \mathfrak{h} \cap \mathfrak{g}_1$. Show that $\mathfrak{h} + \mathfrak{h}'$ is a Lie algebra, that $\mathfrak{w}$ is an ideal in $\mathfrak{h} + \mathfrak{h}'$, and that $\mathfrak{h} + \mathfrak{h}'$ is the semidirect product of $\mathfrak{w}$ and the subalgebra $\mathfrak{n}$ defined in Problem 12(a).
(c) Show that the representations $U_{f,\mathfrak{h}}$ and $U_{f,\mathfrak{h}'}$ of G (Problem 14) are equivalent. (Prove that the representations $W_{f,\mathfrak{h}}$ and $W_{f,\mathfrak{h}'}$ of the group $\exp(\mathfrak{h} + \mathfrak{h}')$, induced, respectively, by $V_{f,\mathfrak{h}}$ and $V_{f,\mathfrak{h}'}$, are equivalent; by observing that these representations are trivial on $\exp \mathfrak{w}$, reduce to Problem 12.)

16. The notation and assumptions are those of Problem 14. The *coadjoint representation* of G is the representation $s \mapsto \mathrm{Coad}(s) = {}^t\mathrm{Ad}(s)^{-1}$ of G on the vector space $\mathfrak{g}^*$ dual to $\mathfrak{g}$, i.e., it is the contragredient of the adjoint representation. We propose to prove *Kirillov's theorems*:

(1) The representation $U_{f,\mathfrak{h}}$ of G is irreducible if and only if $\mathfrak{h}$ is of maximum dimension among the Lie subalgebras of $\mathfrak{g}$ subordinate to f.

(2) The irreducible representations $U_{f,\mathfrak{h}}$ and $U_{f_1,\mathfrak{h}_1}$ of G are equivalent if and only if f and f_1 belong to the same orbit of G in $\mathfrak{g}^*$ under the coadjoint representation.

(a) If $\mathfrak{h}$ is of maximal dimension among the subalgebras of $\mathfrak{g}$ subordinate to f, then $\mathfrak{z} \subset \mathfrak{h}$.
(b) If $U_{f,\mathfrak{h}}$ is irreducible, then $\mathfrak{z} \subset \mathfrak{h}$. (Otherwise $\mathfrak{h} + \mathfrak{z} = \mathfrak{h}_1$ would be the direct sum of $\mathfrak{h}$ and a nonzero subalgebra $\mathfrak{z}_1$ of $\mathfrak{z}$. Show that the representation of $\exp \mathfrak{h}_1$ induced by $V_{f,\mathfrak{h}}$ would be reducible.)
(c) If $\mathfrak{z} \subset \mathfrak{h}$ and $\mathfrak{z} \subset \mathfrak{h}_1$, the representations $U_{f,\mathfrak{h}}$ and $U_{f_1,\mathfrak{h}_1}$ cannot be equivalent unless f and f_1 have the same restriction to $\mathfrak{z}$.
(d) To prove (1) and (2), we proceed by induction on dim G. Suppose to start with that $\mathfrak{z}_0 = (\mathrm{Ker} f) \cap \mathfrak{z} \neq \{0\}$. If $\mathrm{Z}_0 = \exp \mathfrak{z}_0$, reduce to (1) and (2) for the group G/Z_0 by observing that if f and f_1 agree on $\mathfrak{z}$, and if f' and f'_1 are the linear forms on $\mathfrak{g}/\mathfrak{z}_0$ defined by f and f_1, then f and f_1 are in the same orbit under the coadjoint representation of G if and only if f' and f'_1 are in the same orbit under the coadjoint representation of G/Z_0.
(e) If on the other hand $\mathfrak{z}_0 = \{0\}$, we must have $\dim \mathrm{Z} = 1$. We may then assume that $\mathfrak{h} \subset \mathfrak{g}_1$ and $\mathfrak{h}_1 \subset \mathfrak{g}_1$ by virtue of Problem 15. Show also that we may assume that $f(\mathbf{b}) =$

$f_1(\mathbf{b}) = 0$, by replacing f and f_1 by other representatives of their orbits (consider an automorphism $\mathrm{Coad}(\exp(\mathbf{a}u))$ of $\mathfrak{g}^*$ for a suitable choice of $u \in \mathbf{R}$). Now let f' and f'_1 be the restrictions of f and f_1 to $\mathfrak{g}_1$, and show that if $f_1 = \mathrm{Coad}(s) \cdot f$, then $s \in G_1 = \exp \mathfrak{g}_1$ (use the fact that G is the semidirect product of G_1 and $\exp(\mathbf{Ra})$ (Problem 13(a))). Conversely, if $f'_1 = \mathrm{Coad}(s') \cdot f'$, where $s' \in G_1$, show that $f_1 = \mathrm{Coad}(s' \exp(\mathbf{b}v)) \cdot f$ for some $v \in \mathbf{R}$.
(f) Complete the proof by using the inductive hypothesis and by showing that if $\mathfrak{h}$ has maximal dimension, then in the notation of Problem 11(d) we have

$$U_{f', \mathfrak{h}}(\exp(\mathbf{b}v + \mathbf{c}w)) = \exp(2\pi i\lambda w)1_{\mathrm{E}}$$

by virtue of the relation $f'(\mathbf{b}) = 0$.

17. With the hypotheses and notation of Problem 16, let f be a nonzero linear form on $\mathfrak{g}$. Let S_f be the subgroup of G consisting of the $s \in G$ such that $\mathrm{Coad}(s) \cdot f = f$. Show that the Lie algebra $\mathfrak{s}_f$ of S_f is the set of $\mathbf{x} \in \mathfrak{g}$ such that $\langle f, [\mathbf{x}, \mathbf{y}]\rangle = 0$ for all $\mathbf{y} \in \mathfrak{g}$, and deduce that the dimension of G/S_f (which is in one-one correspondence with the orbit of f) is equal to the rank of the alternating form B_f: $(\mathbf{x}, \mathbf{y}) \mapsto \langle f, [\mathbf{x}, \mathbf{y}]\rangle$, and hence is an even number.
If $\mathfrak{h}$ is a Lie subalgebra of $\mathfrak{g}$ of maximal dimension subordinate to f, show that $\dim \mathfrak{h} = \dim \mathfrak{g} - \frac{1}{2}\dim(G/S_f)$. (Argue by induction on dim G, distinguishing the two cases $\mathfrak{z}_0 \neq \{0\}$ and $\mathfrak{z}_0 = \{0\}$ as in Problem 16. In the second case we may again assume that $\mathfrak{h} \subset \mathfrak{g}_1$.)

16. DECLINING FUNCTIONS ON $\mathbf{R}^n$

(22.16.1) The Fourier transformation on the additive group $\mathbf{R}^n$ relates *differentiability properties* of an integrable function to properties of *decrease at infinity* of its Fourier transform.

For $x = (x_1, \ldots, x_n) \in \mathbf{R}^n$, we shall write

$$r(x) = (x_1^2 + \cdots + x_n^2)^{1/2},$$

the Euclidean distance from 0 to x. For each complex number α, the function $(1 + r^2)^\alpha$ is analytic on $\mathbf{R}^n$.

(22.16.2) *Let $f \in \mathscr{L}^1_{\mathrm{C}}(\mathbf{R}^n)$ be such that fr^m is integrable, for some integer $m \geqq 1$. Then $\mathscr{F}f$ is of class C^m on $\mathbf{R}^n$.*

Consider first the case $m = 1$. Since $|x_j| \leqq r(x)$ for $1 \leqq j \leqq n$, the function $x \mapsto x_j f(x)$ is integrable **(13.9.13)**. We may therefore apply to the integral

$$\mathscr{F}f(\xi) = \int \exp(-2\pi i(x\,|\,\xi))f(x)\,dx$$

the formula for differentiating under the integral sign (13.8.6.1), and obtain

(22.16.2.1) $$D_j(\mathscr{F}f)(\xi) = -2\pi i \int \exp(-2\pi i(x \mid \xi))x_j f(x)\, dx;$$

in other words, $-(1/2\pi i)D_j(\mathscr{F}f)$ is the Fourier transform of the integrable function $x \mapsto x_j f(x)$, hence is continuous. The proof is completed by induction on m.

(22.16.3) *Let f be a function of class C^m on $\mathbf{R}^n$ which is integrable, together with its derivatives of order $\leq m$. Then the function $\xi \mapsto r^m(\xi)\mathscr{F}f(\xi)$ tends to 0 at infinity.*

By induction on m, it is again sufficient to consider the case $m = 1$, and to show that

(22.16.3.1) $$\mathscr{F}(D_j f)(\xi) = 2\pi i \xi_j \mathscr{F}f(\xi) \qquad (1 \leq j \leq n).$$

Write $x = (x', x_j)$, where $x' = (x_1, \ldots, x_{j-1}, x_{j+1}, \ldots, x_n) \in \mathbf{R}^{n-1}$. By virtue of the Lebesgue–Fubini theorem (13.21.7), for almost all $x' \in \mathbf{R}^{n-1}$, the partial functions $x_j \mapsto f(x', x_j)$ and $x_j \mapsto D_j f(x', x_j)$ are integrable over $\mathbf{R}$. Since

$$f(x', b) - f(x', a) = \int_a^b D_j f(x', x_j)\, dx_j,$$

the integrability of $D_j f$ implies that $f(x', x_j)$ tends to a finite limit as x_j tends to $\pm\infty$ (13.9.14), and this limit can only be 0 because $x_j \mapsto f(x', x_j)$ is integrable. Integrating by parts, we obtain for $N > 0$

$$\int_{-N}^{N} \exp(-2\pi i(x \mid \xi))D_j f(x)\, dx_j$$

$$= (\exp(-2\pi i(x \mid \xi))f(x', x_j))\Big|_{-N}^{+N} + 2\pi i \xi_j \int_{-N}^{N} \exp(-2\pi i(x \mid \xi))f(x)\, dx_j$$

and therefore, on passing to the limit (13.9.14)

$$\int_{-\infty}^{+\infty} \exp(-2\pi i(x \mid \xi))D_j f(x)\, dx_j = 2\pi i \xi_j \int_{-\infty}^{+\infty} \exp(-2\pi i(x \mid \xi))f(x)\, dx_j;$$

if we now apply the Lebesgue–Fubini theorem, we shall obtain (22.16.3.1).

(22.16.4) A complex-valued measurable function f on $\mathbf{R}^n$ is said to be *rapidly decreasing* if, for each polynomial P in n variables with complex coefficients, the function $x \mapsto \mathrm{P}(x)f(x)$ tends to 0 at infinity. An equivalent condition is that for each integer $m > 0$ the function $r^m f$ (or $(1 + r^2)^m f$) tends to 0 at infinity (or is bounded for sufficiently large $r(x)$, because $r^m f = r^{-1}(r^{m+1}f)$). It is clear that a bounded, rapidly decreasing function f belongs to $\mathscr{L}_{\mathbf{C}}^1(\mathbf{R}^n)$ and to $\mathscr{L}_{\mathbf{C}}^2(\mathbf{R}^n)$, and so do all the functions $|f|^\alpha$ for $\alpha > 0$. The functions in $\mathscr{K}(\mathbf{R}^n)$ are obviously rapidly decreasing, and so are the functions $x \mapsto \exp(-c(r(x))^\alpha)$ for $c > 0$ and $\alpha > 0$. All sums and products of rapidly decreasing functions are rapidly decreasing.

(22.16.5) A complex-valued function f on $\mathbf{R}^n$ is said to be *declining* if f is of class C^∞ and if *f and all its derivatives* are rapidly decreasing. Leibniz's formula **(8.13.2)** shows that the set of declining functions on $\mathbf{R}^n$, which is denoted by $\mathscr{S}(\mathbf{R}^n)$ or just $\mathscr{S}$, is an *algebra* over $\mathbf{C}$. It is clear that $\mathscr{S}(\mathbf{R}^n)$ contains the space $\mathscr{D}(\mathbf{R}^n)$ of complex-valued C^∞-functions of compact support on $\mathbf{R}^n$. The functions $\exp(-c(r(x))^{2m})$, where $c > 0$ and m is a positive integer, and $\exp(-c(1 + r^2(x))^\alpha)$, where $c > 0$ and $\alpha > 0$, belong to $\mathscr{S}$, because every derivative of such a function is majorized in absolute value by the product of the function and the absolute value of a polynomial, as is easily verified by induction.

(22.16.6) We endow $\mathscr{S}(\mathbf{R}^n)$ with the topology of a *metrizable* locally convex space by means of the sequence of *seminorms*

(22.16.6.1)
$$q_{s,m}(f) = \sup_{x \in \mathbf{R}^n,\ |\nu| \leq s} (1 + r^2(x))^m |\mathrm{D}^\nu f(x)|$$

for $s \geqq 0$ and $m \geqq 0$. The fact that these numbers are finite for each function $f \in \mathscr{S}(\mathbf{R}^n)$ follows from the definition, and it is immediately seen that the $q_{s,m}$ are seminorms such that $q_{s,m} \leqq q_{s',m'}$ whenever $s \leqq s'$ and $m \leqq m'$; since moreover the relation $q_{0,m}(f) = 0$ implies that f is identically zero, the topology defined by the $q_{s,m}$ is Hausdorff and therefore metrizable. Since on each compact subset K of $\mathbf{R}^n$ the function $1 + r^2$ is bounded above and below by positive numbers depending only on K, the topology induced on the subspace $\mathscr{D}(\mathbf{R}^n; \mathrm{K})$ of C^∞ functions with support contained in K, by the topology of $\mathscr{S}(\mathbf{R}^n)$, is that defined in **(17.3.1)**.

(22.16.7) (i) *The space $\mathscr{S}(\mathbf{R}^n)$ is complete* (in other words, is a *Fréchet space* **(12.14.5)**).

(ii) *The space $\mathscr{D}(\mathbf{R}^n)$ is dense in $\mathscr{S}(\mathbf{R}^n)$, and $\mathscr{S}(\mathbf{R}^n)$ is separable.*

(i) If (f_k) is a Cauchy sequence in $\mathscr{S}(\mathbf{R}^n)$, it is clear that (f_k) is also a Cauchy sequence in $\mathscr{E}(\mathbf{R}^n)$, hence converges *in this space* to a function f (17.1.2). Since $D^\nu f_k(x)$ tends to $D^\nu f(x)$ for each $x \in \mathbf{R}^n$, and since by hypothesis the sequence $(q_{s,m}(f_k))_{k \geq 0}$ is bounded, for each pair of integers $s \geqq 0$, $m \geqq 0$, it follows from the definition (22.16.6.1) that $f \in \mathscr{S}(\mathbf{R}^n)$ and that, if $q_{s,m}(f_p - f_q) \leqq \varepsilon$ for all $p, q \geqq k_0$, then also $q_{s,m}(f_p - f) \leqq \varepsilon$ for all $p \geqq k_0$; in other words, f is the limit of the sequence (f_k) *in the space* $\mathscr{S}(\mathbf{R}^n)$.

(ii) Let $h \in \mathscr{D}(\mathbf{R}^n)$ take its values in $[0, 1]$ and be such that $h(x) = 1$ for $r(x) \leqq 1$ (16.4.2). For each function $f \in \mathscr{S}(\mathbf{R}^n)$ and each integer $k \geqq 1$, put $f_k(x) = f(x)h(x/k)$; clearly $f_k \in \mathscr{D}(\mathbf{R}^n)$, and for each multi-index α, it follows from Leibniz's formula (8.13.2) that we may write

$$D^\alpha f_k(x) = h(x/k)D^\alpha f(x) + \sum_{\substack{\beta+\gamma=\alpha \\ \gamma \neq 0}} \frac{c_{\beta\gamma}}{k^{|\gamma|}} D^\beta f(x) D^\gamma h(x/k)$$

where the $c_{\beta\gamma}$ are constants independent of f and of k. We have $D^\alpha f(x) = D^\alpha f_k(x)$ for $r(x) < k$; for each $m > 0$, the functions $(1 + r^2)^m D^\beta f$ and $D^\gamma h$ are bounded for $r(x) \geqq k$; moreover, the function $(1 + r^2)^m D^\alpha f$ is arbitrarily small for all $r(x) \geqq k$ provided that k is sufficiently large; hence we see that for k sufficiently large, $q_{s,m}(f - f_k)$ is arbitrarily small. This establishes the first assertion.

If B_k is the ball $r(x) \leqq k$, the argument above shows that each function in $\mathscr{S}(\mathbf{R}^n)$ is the limit of a sequence of functions $g_k \in \mathscr{D}(\mathbf{R}^n; B_k)$; since each of the subspaces $\mathscr{D}(\mathbf{R}^n; B_k)$ is separable (17.12), the same is true of $\mathscr{S}(\mathbf{R}^n)$.

(22.16.8) A complex-valued function f on $\mathbf{R}^n$ is said to be *slowly increasing* if there exists an integer $m \geqq 0$ such that $|f(x)| \leqq (1 + r^2(x))^m$ for all sufficiently large $r(x)$. A complex-valued function f on $\mathbf{R}^n$ is said to be *tempered* if it is *of class* C^∞ and if *f and all its derivatives* are slowly increasing. By virtue of Leibniz's formula (8.13.2), the product of two tempered functions is tempered. The function $\exp(icr^2)$ is tempered, for c real, because it is bounded, and each derivative is the product of $\exp(icr^2)$ by a polynomial.

(22.16.9) (i) *For each multi-index ν, the derivative $D^\nu f$ of a function $f \in \mathscr{S}$ belongs to $\mathscr{S}$, and the linear mapping $f \mapsto D^\nu f$ of $\mathscr{S}$ into itself is continuous.*

(ii) *For each tempered function g, the product of g with a function $f \in \mathscr{S}$ belongs to $\mathscr{S}$, and the linear mapping $f \mapsto gf$ of $\mathscr{S}$ into itself is continuous.*

(iii) *The bilinear mapping $(f, g) \mapsto fg$ of $\mathscr{S} \times \mathscr{S}$ into $\mathscr{S}$ is continuous.*

(i) This follows from the definition and the inequality $q_{s,m}(D^\nu f) \leqq q_{s+|\nu|,m}(f)$.

(ii) By hypothesis, for each integer $s \geqq 0$ there exists an integer $k > 0$ and a real number $c > 0$ such that $|D^\nu g(x)| \leqq c(1 + r^2(x)^k)$ for all $x \in \mathbf{R}^n$ and

$|\nu| \leqq s$. By virtue of Leibniz's formula, there exists a number $a_{s,m}$ depending only on s and m, such that $q_{s,m}(fg) \leqq c a_{s,m} q_{s,m+k}(f)$, which proves the result.

(iii) Likewise, Leibniz's formula shows that there exists a number $b_{s,m}$ depending only on s and m, such that $q_{s,m}(fg) \leqq b_{s,m} q_{s,m}(f) q_{s,0}(g)$; hence the bilinear mapping $(f, g) \mapsto fg$ is continuous **(12.14.11.2)**.

(22.16.10) *The Fourier transformation $\mathscr{F}$ is an isomorphism of the locally convex space $\mathscr{S}(\mathbf{R}^n)$ onto itself, and the Fourier cotransformation $\bar{\mathscr{F}}$ is the inverse isomorphism.*

Since $r^{2m} f \in \mathscr{S}$ for all $f \in \mathscr{S}$ and all $m \geqq 0$, it follows from **(22.16.2)** that $\mathscr{F}f$ is of class C^∞; and since all the derivatives of f are integrable, it follows from **(22.16.3)** that $\mathscr{F}f$ is rapidly decreasing. Furthermore, again by **(22.16.2)**, for each multi-index ν the function $D^\nu(\mathscr{F}f)$ is the Fourier transform of the function $x \mapsto (-2\pi i x)^\nu f(x)$, which belongs to $\mathscr{S}$, and hence $\mathscr{F}f \in \mathscr{S}$. Next, by **(22.16.2)** and **(22.16.3)**, the function $\xi \mapsto \xi^\beta D^\alpha(\mathscr{F}f)(\xi)$ is the Fourier transform of the function $x \mapsto (2\pi i)^{|\alpha| - |\beta|} D^\beta((-x)^\alpha f(x))$, so that we have

(22.16.10.1) $$\xi^\beta D^\alpha(\mathscr{F}f)(\xi) = (2\pi i)^{|\alpha| - |\beta|} \int \exp(-2\pi i (x \mid \xi)) D^\beta((-x)^\alpha f(x))\, dx.$$

From this we shall deduce that for each pair of integer $s \geqq 0$, $m \geqq 0$, there exists a constant $c_{s,m}$ depending only on s and m, such that the relation $q_{2m,s+n+1}(f) \leqq 1$ implies $q_{s,m}(\mathscr{F}f) \leqq c_{s,m}$; this will establish the continuity of the mapping $f \mapsto \mathscr{F}f$ *in* $\mathscr{S}$ **(12.14.11)**. The hypothesis implies that for $|\beta| \leqq 2m$ and $|\alpha| \leqq s$ we have $|D^\beta((-x)^\alpha f(x))| \leqq a_{s,m}(1 + r^2(x))^{-n}$, where the constant $a_{s,m}$ depends only on s and m; it follows then from **(22.16.10.1)** that $|\xi^\beta D^\alpha(\mathscr{F}f)(\xi)| \leqq b_{s,m}$ for these values of α and β, where the constant $b_{s,m}$ depends only on s and m, in view of the integrability of the function $(1 + r^2)^{-n}$ on $\mathbf{R}^n$ **(16.24.9.6)**; by the definition of the seminorms $q_{s,m}$, this proves our assertion. The argument for $\bar{\mathscr{F}}$ is analogous, and the fact that $\mathscr{F}$ is an isomorphism follows from the relations $\bar{\mathscr{F}}\mathscr{F}f = f$ and $\mathscr{F}\bar{\mathscr{F}}f = f$ **(22.10.3)**.

(22.16.11) Poisson's formula **(22.14.4.5)** is applicable to all functions $f \in \mathscr{S}(\mathbf{R})$: for the conditions (1), (2), and (4) are obviously satisfied, and the series $\sum_{n \in \mathbf{Z}} f(x + n)$ is normally convergent on each compact subset of $\mathbf{R}$, which shows that condition (3) of **(22.14.4)** is also satisfied.

PROBLEMS

1. Let $f(x) = e^{-|x|}$ for $x \in \mathbf{R}$. Show that $\mathcal{F}f(\xi) = 2/(1 + 4\pi^2\xi^2)$. Deduce that in the statement of **(22.16.2)**, the condition that $r^m f$ should be integrable cannot be replaced by the condition that $r^m f$ tends to 0 with $1/r$.

2. Let $g \in \mathcal{K}(\mathbf{R})$ be a continuous function whose support is contained in the interval $[-\frac{1}{2}, \frac{1}{2}]$; suppose that its Fourier series does not converge at the point 0 (Section **11.6**, Problem 2). Show that $\mathcal{F}g$ is not Lebesgue-integrable. (Use the fact that the Fourier series of the function $x \mapsto g(x)(e^{2\pi i x} - 1 - 2\pi i x)/(2\pi i x)$ converges at $x = 0$ (Section **(22.19**, Problem 3).) Deduce that in the statement of **(22.16.3)**, the conclusion that the function $\xi \mapsto r^m(\xi)\mathcal{F}f(\xi)$ tends to 0 at infinity cannot be replaced by the conclusion that this function is integrable.

3. (a) Let (f_n) be a sequence of rapidly decreasing measurable functions on $\mathbf{R}^n$. Show that there exists a rapidly decreasing measurable function $h \geqq 0$ such that for each integer m there exists a number a_m with the property that $|f_m(x)| \leqq |h(x)|$ whenever $r(x) \geqq a_m$. (By hypothesis, for each integer k, the number $b_{mk} = \sup_x r^k(x)|f_m(x)|$ is finite; let $c_m = \sup_{k \leq m} (1 + b_{mk})$, and show that the function

$$h = \sum_{m=1}^{\infty} |f_m|/2^m c_m$$

has the required properties.)
(b) Let f be a rapidly decreasing measurable function on $\mathbf{R}^n$. Show that there exists a declining function g such that $|f| \leqq g$. (Let I be the cube in $\mathbf{R}^n$ defined by $0 \leqq x_j \leqq 1$ for $1 \leqq j \leqq n$, I_ν the cube $\mathrm{I} + \nu$ for each $\nu \in \mathbf{Z}^n$, and a_ν the least upper bound of $|f(x)|$ in I_ν. The family (a_ν) is rapidly decreasing **(22.19.3)**. If h is a function $\geqq 0$ belonging to $\mathcal{D}(\mathbf{R}^n)$ and equal to 1 on I, show that the function $g: x \mapsto \sum_{\nu \in \mathbf{Z}^n} a_\nu h(x - \nu)$ has the required properties.)

4. (a) In order that a subset $\mathrm{B} \subset \mathcal{S}(\mathbf{R}^n)$ should be bounded in the Fréchet space $\mathcal{S}(\mathbf{R}^n)$, i.e., that each of the seminorms $q_{s,m}$ **(22.16.6)** should be bounded in B, it is necessary and sufficient that there should exist a function $h \geqq 0$ belonging to $\mathcal{S}(\mathbf{R}^n)$ such that for each multi-index α there is a constant c_α such that $|\mathrm{D}^\alpha f| \leqq c_\alpha \cdot h$ for *all* functions $f \in \mathrm{B}$. (Observe that if B is bounded, then for each α the upper envelope of the functions $|\mathrm{D}^\alpha f|$, $f \in \mathrm{B}$, is rapidly decreasing, and use Problem 3.)
(b) Show that every bounded subset of $\mathcal{S}(\mathbf{R}^n)$ is relatively compact. (Argue as in **(17.1.2)**.)

5. With the notation of Section **22.14**, Problem 11, for each function $f \in \mathcal{E}(\mathbf{R})$ define

$$(T_\pm \cdot f)(x) = \pm \mathrm{D}f(x) + 2\pi x f(x).$$

(a) For the Hermite–Weber functions, show that

$$T_+ \cdot \mathrm{W}_m = 2\sqrt{\pi m}\,\mathrm{W}_{m-1}, \qquad T_- \cdot \mathrm{W}_m = 2\sqrt{\pi(m+1)}\,\mathrm{W}_{m+1}.$$

(b) Deduce from (a) that if a function $f \in \mathcal{L}^2_{\mathbf{C}}(\mathbf{R})$ is such that the series $\sum_{m=0}^{\infty} m^k |a_m(f)|^2$

converges for some integer $k \geqq 1$, then f is equal almost everywhere to a function of class C^{k-1}, and the derivatives $D^h f$, $0 \leqq h \leqq k$ belong to $\mathscr{L}^2_{\mathbf{C}}(\mathbf{R})$; also the functions $x^h D^{k-h} f(x)$ belong to $\mathscr{L}^2_{\mathbf{C}}(\mathbf{R})$ for $0 \leqq h \leqq k$. Consider the converse.
(c) Show that the Fréchet space $\mathscr{S}(\mathbf{R})$ is isomorphic to the space $\mathscr{S}(\mathbf{N})$ of rapidly decreasing sequences (22.19.3). (Use (b).)
(d) Generalize the preceding results to $\mathbf{R}^n$. (Consider the products of Hermite-Weber functions $W_\nu(x_1, \ldots, x_n) = W_{\nu_1}(x_1) \cdots W_{\nu_n}(x_n)$ for each multi-index $\nu = (\nu_1, \ldots, \nu_n) \in \mathbf{N}^n$.)

17. TEMPERED DISTRIBUTIONS

(22.17.1) We shall denote by $\mathscr{S}'(\mathbf{R}^n)$, or simply $\mathscr{S}'$, the *dual* of the locally convex space $\mathscr{S}(\mathbf{R}^n)$, that is to say the space of continuous linear forms on $\mathscr{S}$ (12.15). If T is such a form, its restriction to each of the subspaces $\mathscr{D}(\mathbf{R}^n; K)$ of $\mathscr{S}(\mathbf{R}^n)$, for each compact subset K of $\mathbf{R}^n$, is by virtue of (22.16.6) continuous on this subspace with respect to its topology defined in (17.3.1); the restriction of T to the space $\mathscr{D}(\mathbf{R}^n)$ is therefore a *distribution.* By definition, this distribution is continuous in the topology induced on $\mathscr{D}(\mathbf{R}^n)$ by that of $\mathscr{S}(\mathbf{R}^n)$. Conversely, every distribution T_0 on $\mathbf{R}^n$ with this property has a unique continuous extension to a continuous linear form T on $\mathscr{S}(\mathbf{R}^n)$, because $\mathscr{D}(\mathbf{R}^n)$ is dense in $\mathscr{S}(\mathbf{R}^n)$ ((22.16.7) and (12.9.4)). We may therefore *identify* the elements of $\mathscr{S}'(\mathbf{R}^n)$ with their restrictions to $\mathscr{D}(\mathbf{R}^n)$, and we shall speak of them as *tempered distributions* on $\mathbf{R}^n$; they are therefore (12.14.11) the distributions T with the property that there exist integers $s \geqq 0$, $m \geqq 0$ and a constant $a > 0$ such that, for each function $f \in \mathscr{D}(\mathbf{R}^n)$ (or $f \in \mathscr{S}(\mathbf{R}^n)$)

(22.17.1.1) $$|\langle T, f \rangle| \leqq a \cdot q_{s,m}(f).$$

We give $\mathscr{S}'(\mathbf{R}^n)$ the *weak topology* of the dual (12.15) defined by the seminorms $T \mapsto |\langle T, f \rangle|$, where f runs through $\mathscr{S}(\mathbf{R}^n)$. Notice that this topology *is not identical* with the topology induced by the weak topology on $\mathscr{D}'(\mathbf{R}^n)$ (17.8); it can be shown that it is strictly finer than the latter (Section 12.15, Problem 2).

(22.17.2) *Examples.* The distributions with *compact support* on $\mathbf{R}^n$ (17.7.1) are tempered. For if T is such a distribution, K a compact neighborhood of its support, and $h \in \mathscr{D}(\mathbf{R}^n)$ a function equal to 1 on K, then we have $T(f) = T(hf)$ for all functions $f \in \mathscr{D}(\mathbf{R}^n)$, and since $f \mapsto hf$ is a continuous mapping of $\mathscr{S}(\mathbf{R}^n)$ into $\mathscr{D}(\mathbf{R}^n; L)$, where $L = \operatorname{Supp}(h)$ (22.16.9), it follows that $T \in \mathscr{S}'(\mathbf{R}^n)$.

A measure μ on $\mathbf{R}^n$ is said to be *slowly increasing* if there exists an integer $m \geqq 0$ such that the measure $(1 + r^2)^{-m} \cdot \mu$ is *bounded*, or equivalently if

$|\mu|(B(0; \rho)) = O(\rho^m)$ for some integer m as $\rho \to +\infty$. A slowly increasing measure (and in particular a bounded measure) is a *tempered distribution.* For if the measure $\nu = (1 + r^2)^{-m} \cdot \mu$ is bounded, there exists a constant $c > 0$ such that for each function $f \in \mathscr{K}(\mathbf{R}^n)$ we have $||\nu|(f)| \leq c\|f\|$. Since $|\mu|(f) = |\nu|((1 + r^2)^m f)$, it follows that $||\mu|(f)| \leq c \cdot q_{0,m}(f)$, which proves our assertion.

If as usual we identify a locally integrable function with the measure of which it is the density relative to Lebesgue measure **(17.5.3)**, we see that the slowly increasing functions **(22.16.8)**, and more generally the functions f such that $(1 + r^2)^{-m} f$ is integrable for some integer $m > 0$, are tempered distributions. This shows already that we have $L^1_{\mathbf{C}}(\mathbf{R}^n) \subset \mathscr{S}'(\mathbf{R}^n)$ and $L^\infty_{\mathbf{C}}(\mathbf{R}^n) \subset \mathscr{S}'(\mathbf{R}^n)$. We have also $L^2_{\mathbf{C}}(\mathbf{R}^n) \subset \mathscr{S}'(\mathbf{R}^n)$, because $(1 + r^2)^{-n/2}$ is integrable **(16.24.9.6)**, hence $(1 + r^2)^{-n/4}$ belongs to $\mathscr{L}^2$, and if the same is true of f, then their product belongs to $\mathscr{L}^1$.

More generally, it can be shown that the measure spaces $\mathbf{P}(\mathbf{R}^n)$ and $\mathbf{P}'(\mathbf{R}^n)$ are contained in $\mathscr{S}'(\mathbf{R}^n)$ (Problem 5).

(22.17.3) (i) *For each multi-index* ν, *the derivative* $D^\nu T$ *of a tempered distribution is tempered, and the linear mapping* $T \mapsto D^\nu T$ *of* $\mathscr{S}'$ *into itself is continuous.*

(ii) *For each tempered function* g, *the product* $g \cdot T$ *of* g *with a tempered distribution* T *is a tempered distribution, and the linear mapping* $T \mapsto g \cdot T$ *of* $\mathscr{S}'$ *into itself is continuous.*

(iii) *The space* $\mathscr{D}(\mathbf{R}^n)$ *is dense in* $\mathscr{S}'(\mathbf{R}^n)$; *more precisely, every tempered distribution is the limit in* $\mathscr{S}'(\mathbf{R}^n)$ *of a sequence of functions belonging to* $\mathscr{D}(\mathbf{R}^n)$.

In view of the definitions (**(17.5.5)** and **(17.3.5)**), (i) and (ii) are particular cases of the theorem on the continuity of transposes **(12.15.3)**, by virtue of **(22.16.9)**. To prove (iii), we shall first show that $\mathscr{E}'(\mathbf{R}^n)$ is dense in $\mathscr{S}'(\mathbf{R}^n)$. Consider the function $h \in \mathscr{D}(\mathbf{R}^n)$ defined in the proof of **(22.16.7(ii))**; it is enough to show that for each tempered distribution T, if $h_k(x) = h(x/k)$, the sequence of compactly supported distributions $h_k \cdot T$ converges to T in $\mathscr{S}'(\mathbf{R}^n)$, or equivalently that, for each function $f \in \mathscr{S}(\mathbf{R}^n)$, the sequence $(\langle T, h_k f\rangle)$ converges to $\langle T, f\rangle$; but we have seen in the proof of **(22.16.7(ii))** that the sequence $(h_k f)$ tends to f in $\mathscr{S}(\mathbf{R}^n)$, which proves our assertion. Moreover, since a distribution with compact support is the limit, for the weak topology of $\mathscr{D}'(\mathbf{R}^n)$, of a sequence of functions in $\mathscr{D}(\mathbf{R}^n)$ having their supports contained in a *fixed* compact set **(17.12.3)**, it is immediately seen that this sequence converges also for the weak topology of $\mathscr{S}'(\mathbf{R}^n)$.

(22.17.4) *Example.* The bounded function $f(x) = \cos(e^x)$ is a tempered distribution, hence so is its derivative $-\sin(e^x) \cdot e^x$. This shows on the one hand that there exist measures which are tempered distributions although they are *not slowly increasing*; and on the other hand that there exist C^∞ functions f which are slowly increasing (and hence are *tempered distributions*) but whose derivative is not slowly increasing (and therefore f is not a *tempered function*).

(22.17.5) Since the Fourier transformation $\mathcal{F}$ and the Fourier cotransformation $\bar{\mathcal{F}}$ are mutually inverse isomorphisms of $\mathcal{S}(\mathbf{R}^n)$ onto itself, their *transposes* **(12.15.3)** are (mutually inverse) *isomorphisms* of the space $\mathcal{S}'(\mathbf{R}^n)$ onto itself. We denote these isomorphisms by the same symbols $\mathcal{F}$ and $\bar{\mathcal{F}}$, and refer to them as the *Fourier transformation* and *Fourier cotransformation*. They are therefore defined by the relations

(22.17.5.1) $$\langle \mathcal{F}\mathrm{T}, f\rangle = \langle \mathrm{T}, \mathcal{F}f\rangle, \qquad \langle \bar{\mathcal{F}}\mathrm{T}, f\rangle = \langle \mathrm{T}, \bar{\mathcal{F}}f\rangle$$

for $f \in \mathcal{S}(\mathbf{R}^n)$. We have

(22.17.5.2) $$\bar{\mathcal{F}}\mathrm{T} = \mathcal{F}\check{\mathrm{T}} = (\mathcal{F}\mathrm{T})\check{}, \qquad \bar{\mathcal{F}}\mathrm{T} = \overline{\mathcal{F}\bar{\mathrm{T}}}$$

for $\mathrm{T} \in \mathcal{S}'(\mathbf{R}^n)$, in view of the definitions of $\bar{\mathrm{T}}$ **(17.6.1)** and $\check{\mathrm{T}}$ **(17.11.7)**. These relations follow immediately from the corresponding relations for functions in $\mathcal{S}(\mathbf{R}^n)$ **(22.10.6.7)**.

Again, for each $s \in \mathbf{R}^n$ we have

(22.17.5.3) $$\mathcal{F}(\gamma(s)\mathrm{T}) = \exp(-2\pi i(s\,|\,\cdot)) \cdot \mathcal{F}\mathrm{T},$$

(22.17.5.4) $$\mathcal{F}(\exp(2\pi i(s\,|\,\cdot)) \cdot \mathrm{T}) = \gamma(s)\mathcal{F}\mathrm{T};$$

these relations again follow from the definition **(22.17.5.1)** and the analogous relations for functions in $\mathcal{S}(\mathbf{R}^n)$ **(22.7.1.6)**.

Finally, for each multi-index ν, we have†

(22.17.5.5) $$\mathcal{F}(\mathrm{D}^\nu \mathrm{T}) = (2\pi i)^{|\nu|}\xi^\nu \cdot \mathcal{F}\mathrm{T},$$
$$\bar{\mathcal{F}}(\mathrm{D}^\nu \mathrm{T}) = (-2\pi i)^{|\nu|}\xi^\nu \cdot \bar{\mathcal{F}}\mathrm{T},$$

(22.17.5.6) $$\mathrm{D}^\nu(\mathcal{F}\mathrm{T}) = (-2\pi i)^{|\nu|}\mathcal{F}(x^\nu \cdot \mathrm{T}),$$
$$\mathrm{D}^\nu(\bar{\mathcal{F}}\mathrm{T}) = (2\pi i)^{|\nu|}\bar{\mathcal{F}}(x^\nu \cdot \mathrm{T}),$$

which are proved in the same way from **(22.16.3.1)** and **(22.16.2.1)**.

† It seems to be impossible here to avoid abuses of notation: $x^\nu \cdot \mathrm{T}$ is an abuse of notation for the distribution $p_\nu \cdot \mathrm{T}$ **(17.3.5)**, where p_ν is the continuous function $x \mapsto x^\nu$.

Notice in particular the formulas for the Laplacian $\Delta = D_1^2 + \cdots + D_n^2$:

$$\text{(22.17.5.7)} \quad \mathcal{F}(\Delta T) = -4\pi^2 r^2 \cdot \mathcal{F}T, \qquad \Delta(\mathcal{F}T) = -4\pi^2 \mathcal{F}(r^2 \cdot T).$$

All these formulas may also be proved by *continuous extension* from $\mathcal{S}(\mathbf{R}^n)$ to $\mathcal{S}'(\mathbf{R}^n)$, since $\mathcal{S}(\mathbf{R}^n)$ is dense in $\mathcal{S}'(\mathbf{R}^n)$ **(22.17.3)**.

(22.17.6) Let $u: \mathbf{R}^n \to \mathbf{R}^n$ be a *linear* bijection. Recall that the image $u(T)$ of a distribution T on $\mathbf{R}^n$ is defined by the relation $\langle u(T), f\rangle = \langle T, f \circ u\rangle$ for $f \in \mathcal{D}(\mathbf{R}^n)$ **(17.3.7)**. The definition **(22.16.6.1)** shows immediately that there exists a constant $c_{s,m}(u)$ depending only on s, m, and u, such that for each function $f \in \mathcal{S}(\mathbf{R}^n)$ we have $q_{s,m}(f \circ u) \leqq c_{s,m}(u) q_{s,m}(f)$; hence if T is a tempered distribution, so also is $u(T)$. Moreover, we have

$$\text{(22.17.6.1)} \qquad \mathcal{F}(u(T)) = |\det u|^{-1}(u^*)^{-1}(\mathcal{F}T)$$

where u^* denotes the adjoint of u **(11.5.1)** with respect to the Euclidean scalar product on $\mathbf{R}^n$. For if $f \in \mathcal{S}(\mathbf{R}^n)$, we have by definition

$$\langle \mathcal{F}(u(T)), f\rangle = \langle u(T), \mathcal{F}f\rangle = \langle T, (\mathcal{F}f) \circ u\rangle.$$

But also by definition

$$\begin{aligned}
\mathcal{F}f(u(\xi)) &= \int \exp(-2\pi i(x \mid u(\xi))) f(x)\, dx \\
&= \int \exp(-2\pi i(u^*(x) \mid \xi)) f(x)\, dx \\
&= |\det u|^{-1} \int \exp(-2\pi i(y \mid \xi)) f((u^*)^{-1}(y))\, dy
\end{aligned}$$

by virtue of **(14.3.6.1)** and **(14.3.9)**. This shows that we have

$$\begin{aligned}
\langle T, (\mathcal{F}f) \circ u\rangle &= |\det u|^{-1} \langle T, \mathcal{F}(f \circ (u^*)^{-1})\rangle \\
&= |\det u|^{-1} \langle \mathcal{F}T, f \circ (u^*)^{-1}\rangle \\
&= |\det u|^{-1} \langle (u^*)^{-1}(\mathcal{F}T), f\rangle
\end{aligned}$$

which proves **(22.17.6.1)**.

(22.17.7) Let μ be a measure on $\mathbf{R}^n$ which belongs to one of the spaces $M_{\mathbf{C}}^1(\mathbf{R}^n)$, $L_{\mathbf{C}}^2(\mathbf{R}^n)$, or $\mathcal{P}(\mathbf{R}^n)$. The duality relation **(22.10.10.7)**, with ν' replaced by a measure having as density with respect to Lebesgue measure a function $f \in \mathscr{D}(\mathbf{R}^n)$, gives the relation

(22.17.7.1) $$\langle \mu^{\Delta}, f \rangle = \langle \mu, \mathscr{F}f \rangle.$$

It follows from (22.17.7.1) that $\langle \mathscr{F}\mu, f \rangle = \langle \mu^{\Delta}, f \rangle$, where $\mathscr{F}\mu$ is the *distributional* Fourier transform of μ; by (17.3.2) we see that this distribution is a *measure* equal to μ^{Δ}. When $\mu \in \mathbf{M}^1_{\mathbf{C}}(\mathbf{R}^n)$, the notation $\mathscr{F}\mu$ as the *density* of the measure μ^{Δ} is concordant with the notation $\mathscr{F}\mu$ for μ^{Δ}, in view of the canonical identification of locally integrable functions with the corresponding distributions (17.5.3.5). Likewise if $\mu = f \cdot m_{\mathbf{R}^n}$, with $f \in \mathscr{L}^2_{\mathbf{C}}(\mathbf{R}^n)$, the measure $\mathscr{F}\mu$ has density $\mathscr{F}f \in \mathscr{L}^2_{\mathbf{C}}(\mathbf{R}^n)$, and is therefore identified with $\mathscr{F}f$.

(22.17.8) It follows immediately from the Banach–Steinhaus theorem applied to the Fréchet space $\mathscr{S}(\mathbf{R}^n)$ (12.16.5) that if a sequence (T_k) of tempered distributions is such that, for each function $f \in \mathscr{S}(\mathbf{R}^n)$, the sequence $(\langle \mathrm{T}_k, f \rangle)$ tends to a limit, then the linear form $f \mapsto \lim_{k \to \infty} \langle \mathrm{T}_k, f \rangle$ is a tempered distribution, the limit of the sequence (T_k). Likewise (12.16.6), if $z \mapsto \mathrm{T}_z$ is a mapping of an open set in $\mathbf{R}$ (resp $\mathbf{C}$) into $\mathscr{S}'(\mathbf{R}^n)$ which is *weakly differentiable* (resp. *weakly analytic*), then the weak derivative T'_z is a tempered distribution; moreover $z \mapsto \mathscr{F}\mathrm{T}_z$ is weakly differentiable (resp. weakly analytic) and the weak derivative is given by

$$(\mathscr{F}\mathrm{T}_z)' = \mathscr{F}(\mathrm{T}'_z)$$

as follows immediately from (22.17.5.1).

(22.17.9) *Example.* Consider the function r^{ζ}, defined for $x \neq 0$ and all complex numbers ζ. We have seen in (17.9.2) that this function has an analytic continuation to a distribution T_{ζ} on $\mathbf{R}^n$, for each $\zeta \neq -(n + 2k)$, k an integer $\geqq 0$. For these values of ζ, T_{ζ} is a *tempered* distribution. For if h is a function in $\mathscr{D}(\mathbf{R}^n)$ which is equal to 1 on a neighborhood of 0, the distribution $h \cdot \mathrm{T}_{\zeta}$ is compactly supported and therefore tempered, and $(1 - h) \cdot \mathrm{T}_{\zeta}$ is (identified with) the function $(1 - h)r^{\zeta}$, which is slowly increasing. We shall calculate the Fourier transform $\mathscr{F}\mathrm{T}_{\zeta}$. If $\mathscr{R}\zeta > -n$, the function $r^{\zeta}h$ is integrable (17.9.2), whereas $r^{\zeta}(1 - h)$ belongs to $\mathscr{L}^2_{\mathbf{C}}(\mathbf{R}^n)$ *provided that* $2\mathscr{R}\zeta + n - 1 < -1$ (16.24.9.6), i.e., $\mathscr{R}\zeta < -\frac{1}{2}n$. Since the Fourier transforms of functions in $\mathscr{L}^1$ and $\mathscr{L}^2$ are functions, it follows that for $-n < \mathscr{R}\zeta < -\frac{1}{2}n$ the Fourier transform $\mathscr{F}\mathrm{T}_{\zeta}$ is a locally integrable *function* Φ_{ζ}. In fact, this function is *continuous on* $\mathbf{R}^n - \{0\}$. For we already know that the Fourier transform of $r^{\zeta}h$ is continuous. On the other hand, if k is an integer > 0 such that $\mathscr{R}\zeta - 2k + n - 1 < -1$, i.e., $\mathscr{R}\zeta < 2k - n$, then the function

$r^{\zeta-2k}(1-h)$ is integrable **(16.24.9.6)**, and its Fourier transform is therefore a continuous function; but for $r(x)$ sufficiently large the function $r^{\zeta-2k}(1-h)$ coincides with $r^{\zeta-2k}$ and hence, up to a constant factor, with $\Delta^k(r^{\zeta}(1-h))$ **(17.9.2.8)**. Since $\mathscr{F}(\Delta^k(r^{\zeta}(1-h))) = (-4\pi^2 r^2)^k \cdot \mathscr{F}(r^{\zeta}(1-h))$, and since the difference $\mathscr{F}(\Delta^k(r^{\zeta}(1-h))) - \mathscr{F}(r^{\zeta-2k}(1-h))$ is the Fourier transform of a function in $\mathscr{D}(\mathbf{R}^n)$ and therefore is continuous, our assertion is proved.

This being so, since T_ζ is invariant under all orthogonal transformations u, the same is true of $\mathscr{F}T_\zeta$ **(22.17.6)**; consequently the functions Φ_ζ and $\Phi_\zeta \circ u$ are equal almost everywhere, and therefore equal because they are continuous functions. In other words, Φ_ζ is a *continuous function of r alone* for $r > 0$ **(12.10.6)**. Suppose in addition that ζ is *real* and satisfies $-n < \zeta < -\frac{1}{2}n$. Then, for each real number $t > 0$, if h_t denotes the homothety with ratio t, we have for each function $f \in \mathscr{K}(\mathbf{R}^n)$

$$\begin{aligned}\langle h_t(T_\zeta), f\rangle &= \langle T_\zeta, f \circ h_t\rangle \\ &= \int r^{\zeta}(x) f(tx)\, dx \\ &= \frac{1}{t^{n+\zeta}} \int r^{\zeta}(y) f(y)\, dy \\ &= \frac{1}{t^{n+\zeta}} \langle T_\zeta, f\rangle,\end{aligned}$$

so that $h_t(T_\zeta) = t^{-\zeta-n} T_\zeta$. It follows **(22.17.6.1)** that $h_t(\mathscr{F}T_\zeta) = t^{\zeta}\mathscr{F}T_\zeta$; bearing in mind that Φ_ζ is continuous on $\mathbf{R}^n - \{0\}$, we see as above that Φ_ζ is *homogeneous of degree* $-(n+\zeta)$, which is possible for a function of r only if $\Phi_\zeta(r) = C(\zeta) r^{-n-\zeta}$ (for $r > 0$). It remains to determine the constant $C(\zeta)$, which we shall do by applying the formula of definition **(22.17.5.1)** with a particular choice of $f \in \mathscr{S}(\mathbf{R}^n)$.

We shall take the function $f(x) = \exp(-\pi r^2(x))$, which indeed belongs to $\mathscr{S}(\mathbf{R}^n)$, and we shall show that $\mathscr{F}f = f$. Namely, we have

$$\begin{aligned}\mathscr{F}f(\xi) &= \int_{\mathbf{R}^n} \exp(-\pi r^2 - 2\pi i (x \mid \xi))\, dx \\ &= \prod_{k=1}^{n} \int_{-\infty}^{+\infty} \exp(-\pi x_k^2 - 2\pi i x_k \xi_k)\, dx_k.\end{aligned}$$

Now, for each real number u,

$$\int_{-\infty}^{+\infty} \exp(-\pi t^2 - 2\pi i t u)\, dt = \exp(-\pi u^2) \int_{-\infty}^{+\infty} \exp(-\pi (t + iu)^2)\, dt;$$

by applying Cauchy's theorem (9.6.3) to the entire function $e^{-\pi z^2}$ and the rectangle with vertices $\pm R$ and $\pm R + iu$, and using the fact that $\int_0^u \exp(-\pi(R + iv)^2)\, dv$ tends to 0 as $R \to +\infty$, we obtain

$$\int_{-\infty}^{+\infty} \exp(-\pi(t + iu)^2)\, dt = \int_{-\infty}^{+\infty} \exp(-\pi t^2)\, dt = 1 \ (*),$$

which proves our assertion. The formula (22.17.5.1) then gives

$$\int r^{\zeta} \exp(-\pi r^2)\, dx = C(\zeta) \int r^{-\zeta-n} \exp(-\pi r^2)\, dx.$$

If we now apply the formula for integration in polar coordinates (16.24.9.1), this equation becomes

$$\int_0^{\infty} u^{\zeta+n-1} \exp(-\pi u^2)\, du = C(\zeta) \int_0^{\infty} u^{-\zeta-1} \exp(-\pi u^2)\, du;$$

making the change of variable $\pi u^2 = t$, we obtain(*)

$$C(\zeta) = \pi^{-\zeta-\frac{1}{2}n} \frac{\Gamma(\frac{1}{2}(\zeta+n))}{\Gamma(-\frac{1}{2}\zeta)};$$

and the result we have obtained may also be written in the form

(22.17.9.1) $$\mathscr{F}\left(\frac{\pi^{\zeta/2}}{\Gamma(\frac{1}{2}(n+\zeta))} T_{\zeta}\right) = \frac{\pi^{-(n+\zeta)/2}}{\Gamma(-\frac{1}{2}\zeta)} T_{-\zeta-n}.$$

This formula has been established only for ζ *real* and $-n < \zeta < -\frac{1}{2}n$; however, both sides, in view of (22.17.8), are *weakly analytic* functions for *all* $\zeta \in \mathbf{C}$ (17.9.2); the theorem of analytic continuation ((9.4.4) and (12.16.6)) shows therefore that (22.17.9.1) is valid for *all* $\zeta \in \mathbf{C}$.

PROBLEMS

1. For a *positive* measure μ on $\mathbf{R}^n$ to be a tempered distribution, it is necessary (and sufficient) that μ should be slowly increasing. (Express that the relation (22.17.1.1) is satisfied by taking

$$f(x) = \varepsilon(1 + r^2(x))^{-m} h(x/k)$$

(*) See my book, *Infinitesimal Calculus*, Boston (Houghton-Mifflin), 1971.

with ε fixed and sufficiently small, and k an arbitrary positive integer, the function h being defined as in the proof of (22.16.7(ii)); then let $k \to +\infty$). In particular, the function e^x on **R** is not a tempered distribution.

2. (a) Show that every summable distribution (Section 17.11, Problem 1) is tempered.
(b) Show that every tempered distribution is of finite order.

3. (a) Let T be a distribution on $\mathbf{R}^n$. For each complex number ζ, T is said to be *homogeneous of degree* ζ if for each real number $t > 0$ we have

$$h_t(\mathrm{T}) = t^{-(\zeta+n)}\mathrm{T},$$

where h_t denotes the homothety with ratio t. With this definition, for each locally integrable function f such that $f(tx) = t^{\zeta} f(x)$ for $x \in \mathbf{R}^n$ and $t > 0$, the distribution T_f is homogeneous of degree ζ (17.5.3.4). The Dirac measure ε_0 is homogeneous of degree $-n$.

If T is homogeneous of degree ζ, $\mathrm{D}^\alpha\mathrm{T}$ is homogeneous of degree $\zeta - |\alpha|$, for each multi-index α.

(b) Show that a distribution T is homogeneous of degree ζ if and only if it satisfies *Euler's equation*

$$\zeta\mathrm{T} = \sum_{j=1}^{n} x_j \mathrm{D}_j \mathrm{T}$$

(consider the function $t \mapsto \langle h_t(\mathrm{T}), u\rangle$ for $u \in \mathscr{D}(\mathbf{R}^n)$, and take its derivative).

(c) Show that every homogeneous distribution T is tempered. (Show first that if T is homogeneous of order ζ, there exists an integer m_0 and a constant C such that, for each function $u \in \mathscr{D}(\mathbf{R}^n)$ with support contained in the ball $\mathrm{B}(0; \mathrm{R})$ with radius $\mathrm{R} > 1$, we have

$$|\mathrm{T}(u)| \leqq \mathrm{C}\mathrm{R}^{n+|\zeta|+m} \sup_{|\nu| \leq m} \|\mathrm{D}^\nu u\|$$

for $m \geqq m_0$. Remark next that if $u \in \mathscr{D}(\mathbf{R}^n)$ has support contained in a ball of radius $\frac{1}{2}$ and center x_0, where $|x_0| = \mathrm{R} > 4$, then for each m we have

$$(\mathrm{R}-1)^{2m} \sup_{|\nu| \leq m} \|\mathrm{D}^\nu u\| \leqq \sup_{|\nu| \leq m,\, x \in \mathbf{R}^n} (1 + r^2(x))^m |\mathrm{D}^\nu u(x)|.$$

Show that there exists a partition of unity $(g_\nu)_{\nu \in \mathbf{Z}^n}$ consisting of C^∞-functions, such that $g_\nu(x) = g_0(x - c\nu)$ for each $\nu \in \mathbf{Z}^n$ (with $c > 0$ a small constant) and such that the support of g_0 is contained in the ball $\mathrm{B}(0; \frac{1}{2})$. Finally take m sufficiently large.)

(d) Show that for each distribution T which is homogeneous of order ζ, the distribution $\mathscr{F}\mathrm{T}$ is homogeneous of order $-n-\zeta$.

4. Calculate the Fourier transforms of the distributions P.V. $(1/x)$ (Section 17.9, Problem 1) and P.V. $(1/z^m)$ (Section 17.9, Problem 2).

5. A distribution T on $\mathbf{R}^n$ is said to be *of positive type* if, for each function $u \in \mathscr{D}(\mathbf{R}^n)$, we have $\langle \mathrm{T}, \check{u} * u\rangle \geqq 0$. For measures on $\mathbf{R}^n$, this definition agrees with that of (22.2.1). (Use the fact that if a compact set K is contained in the interior of a compact set K′, the closure of $\mathscr{D}(\mathbf{R}^n; \mathrm{K}')$ in $\mathscr{K}(\mathbf{R}^n; \mathrm{K}')$ contains $\mathscr{K}(\mathbf{R}^n; \mathrm{K})$.)

(a) Show that for each function $u \in \mathcal{D}(\mathbf{R}^n)$, $\mathrm{T} * u * \check{u}$ is a continuous function of positive type; hence there exists a bounded measure $\mu_u \geqq 0$ such that, for each function $v \in \mathcal{D}(\mathbf{R}^n)$, we have

$$\langle \mathrm{T} * u * \check{u}, v * \check{v} \rangle = \int |\mathcal{F}v(\xi)|^2 \, d\mu_u(\xi).$$

Deduce that $|\mathcal{F}u|^2 \cdot \mu_v = |\mathcal{F}v|^2 \cdot \mu_u$.
(b) For each function $u \neq 0$ in $\mathcal{K}(\mathbf{R}^n)$, show that the Fourier transform of $(u * \check{u})^2$ is > 0 at each point of $\mathbf{R}^n$. (Use the fact that $\mathcal{F}u$ cannot vanish at every point of a nonempty open set in $\mathbf{R}^n$ (22.18.1)).
(c) Deduce from (a) and (b) that there exists a measure $\mu \geqq 0$ on $\mathbf{R}^n$ such that $\langle \mathrm{T}, f * \tilde{f} \rangle = \int |\mathcal{F}f(\xi)|^2 \, d\mu(\xi)$ for all $f \in \mathcal{D}(\mathbf{R}^n)$. By writing down the condition of continuity of T (17.3.1) for the functions $x \mapsto kg(kx)$, where $g = f * \tilde{f}$ with $f(0) \neq 0$, show that μ is a slowly increasing measure. Conclude that T is a *tempered* distribution, and with the help of a regularizing sequence show that $\langle \mathrm{T}, f \rangle = \int \mathcal{F}f(\xi) \, d\mu(\xi)$ for all $f \in \mathcal{D}(\mathbf{R}^n)$, and hence that the Fourier transform $\mathcal{F}\mathrm{T}$ is a *slowly increasing* positive measure. Consider the converse.
(d) Deduce from (c) that the spaces $\mathrm{P}(\mathbf{R}^n)$ and $\mathrm{P}'(\mathbf{R}^n)$ (22.10.10) are contained in $\mathcal{S}'(\mathbf{R}^n)$, and that $\mathcal{F}\mu = \mu^{\wedge}$ for $\mu \in \mathrm{P}(\mathbf{R}^n)$.
(e) If T is a distribution of positive type, $-\Delta\mathrm{T}$ (where Δ is the Laplacian) is also a distribution of positive type. Hence construct examples of distributions of positive type and arbitrary order.
(f) Lebesgue measure $m_{\mathbf{R}}$ is equal to $\varepsilon_0^{\wedge}$, but the measure $\mathrm{Y} \cdot m_{\mathbf{R}}$ (where Y is the Heaviside function) is not the Plancherel transform of a measure of positive type. (Use Problem 4.)

6. Let f be a continuous bounded real-valued function on $\mathbf{R}$, such that $f(-x) = f(x) \geqq 0$ for all $x \geqq 0$, and such that the restriction of f to $[0, +\infty[$ is a convex function (Section 8.5, Problem 8). We propose to show that f is a positive type.
(a) If $(x_i)_{1 \leq i \leq m}$ is a finite sequence of real numbers, show that there exist a finite number of affine-linear functions $x \mapsto l_k(x) = a_k x + b_k$ $(1 \leqq k \leqq \mathrm{N})$ such that, if we define $g(x) = \sup(0, l_1(x), \ldots, l_{\mathrm{N}}(x))$ for $x \geqq 0$, and $g(x) = g(-x)$ for $x \leqq 0$, then the function g has the following properties: (1) $g(x) \leqq f(x)$ for all $x \in \mathbf{R}$; (2) $g(x_i - x_j) = f(x_i - x_j)$ for $i \neq j$. It is sufficient to show that g is of positive type.
(b) When g is considered as a distribution, we have $\mathrm{D}^2 g = -m\varepsilon_0 + \mu$, where μ is a positive measure of mass m. Deduce that $-4\pi^2\xi^2\mathcal{F}g(\xi) = \mathcal{F}\mu(\xi) - m$, and by using the fact that $\mathcal{F}\mu$ is a continuous function of positive type, conclude that $\mathcal{F}g(\xi) \geqq 0$ for all $\xi \in \mathbf{R}$.

7. Let T be a distribution of positive type on $\mathbf{R}^n$, and suppose that the restriction of T to a sufficiently small neighborhood of 0 is a function of class C^{2k} for some integer k. Put $\mathrm{S} = (1 - \Delta)^k \mathrm{T}$, so that if $\mu = \mathcal{F}\mathrm{T}$ we have $\mathcal{F}\mathrm{S} = (1 + 4\pi^2 r^2)^k \cdot \mu$, a slowly increasing positive measure. The hypothesis implies that the restriction of S to a neighborhood of 0 is a continuous function; if $g = f * \tilde{f}$, where $f \geqq 0$ belongs to $\mathcal{D}(\mathbf{R}^n)$, and if g_k denotes the function $x \mapsto k^n g(kx)$, then $\mathrm{S} * g_k$ is a locally integrable function of positive type, and there exists $\mathrm{M} > 0$ such that $|(\mathrm{S} * g_k)(0)| \leqq \mathrm{M}$ for sufficiently large k. Deduce that $\mathcal{F}\mathrm{S}$ is a bounded measure, and hence that the same is true of $x^\alpha \cdot \mu$ for each multi-index α such that $|\alpha| \leqq 2k$. Conclude with the help of (22.17.5.5) that T is a bounded function of class C^{2k}.

8. (a) In order that a bounded continuous complex-valued function f on **R** should be almost periodic (Section 22.10, Problem 7) it is necessary and sufficient that, for each $\varepsilon > 0$, there should exist a number $T > 0$ with the property that in each interval of **R** of length T there exists a number s such that $|f(t) - f(t-s)| \leqq \varepsilon$ for all $t \in \mathbf{R}$. (To prove necessity, use Section 22.10, Problem 9. To prove that the condition is sufficient, show that it implies that f is uniformly continuous, and then use the fact that as s runs through a compact interval in **R** the set of functions $\gamma(s)f$ is relatively compact in $\mathscr{C}_{\mathbf{C}}^{\infty}(\mathbf{R})$.)
(b) Show that if f is almost periodic on **R**, its mean (Section 22.10, Problem 9) is given by

$$M(f) = \lim_{T\to\infty} \frac{1}{2T}\int_{-T}^{T} f(x)\,dx.$$

(Integrate the inequality $M(f) - \sum_j c_j f(x - b_j) \leqq \varepsilon$ for $c_j \geqq 0$ and $\sum_j c_j = 1$.)
(c) Suppose that f is almost periodic on **R**, and that there exists $a > 0$ such that $|f(t)| \geqq a$ for all $t \in \mathbf{R}$; then we may write $f = e^g$, where g is a uniformly continuous mapping of **R** into **C** (16.28.9). Show that $g(t) = ict + h(t)$, where $c \in \mathbf{R}$ and h is almost periodic. (Reduce first to the case where f is a linear combination of a finite number of characters, by observing that if $0 < \varepsilon < \frac{1}{2}a$ and $\|f - f_1\| \leqq \varepsilon$, we may write $f_1 = e^{g_1}$ with $\|g - g_1\| \leqq 2\varepsilon/a$, and then using the definition of almost periodic functions. Suppose then that $f(t) = \sum_{j=1}^{r} b_j \exp(2\pi i\lambda_j t)$; we may write $f = F \circ \psi$, where ψ is the homomorphism $t \mapsto (\lambda_j t)_{1\leq j\leq r}$ of **R** into $\mathbf{R}^r$, and F is the periodic mapping $(x_1, \dots, x_r) \mapsto \sum_{j=1}^{r} b_j \exp(2\pi i x_j)$ of $\mathbf{R}^r$ into **C**. If p: $(x_1, \dots, x_r) \mapsto (e^{2\pi i x_1}, \dots, e^{2\pi i x_r})$ is the canonical homomorphism of $\mathbf{R}^r$ onto $\mathbf{U}^r$, the closure of the image of the homomorphism $p \circ \psi$: $\mathbf{R} \to \mathbf{U}^r$ is isomorphic to a torus $\mathbf{T}^s$ with $s \leqq r$. There exists a basis $(\mathbf{u}_j)_{1\leq j\leq r}$ of $\mathbf{Z}^r$ such that if V is the vector subspace of $\mathbf{R}^r$ generated by $\mathbf{u}_1, \dots, \mathbf{u}_s$, we have $p(V) = \overline{p(\psi(\mathbf{R}))}$; if F_1 is the restriction of F to V, then F_1 is periodic and $|F_1(z)| \geqq a$ for all $z \in V$. Deduce that $F_1(z) = \exp(G_1(z))$, where G_1 is a mapping of V into **C**, of the form

$$z = \sum_{j=1}^{s} z_j \mathbf{u}_j \mapsto 2\pi i \sum_{j=1}^{s} m_j z_j + H(z)$$

where H is periodic on V and the m_j are integers.)

9. (a) Let $\mathscr{A}_2$ be the set of complex-valued Lebesgue-measurable functions on **R** which are square-integrable on each compact interval, and such that the number

$$M_2(f) = \limsup_{T\to\infty}\left(\frac{1}{2T}\int_{-T}^{T} |f(t)|^2\,dt\right)^{1/2}$$

is finite. If $f, g \in \mathscr{A}_2$ and $\lambda \in \mathbf{C}$ we have $f + g \in \mathscr{A}_2$, $\lambda f \in \mathscr{A}_2$ and

$$M_2(f+g) \leqq M_2(f) + M_2(g), \qquad M_2(\lambda f) \leqq |\lambda| M_2(f);$$

in other words, M_2 is a seminorm on $\mathscr{A}_2$. Also we have

$$M_2(\gamma(s)f) = M_2(f)$$

for all $s \in \mathbf{R}$, and

$$\limsup_{T\to\infty} \frac{1}{2T}\int_{-T}^{T} |f(t)g(t)|\,dt \leqq M_2(f)M_2(g).$$

(b) The spaces $\mathscr{L}^p_{\mathbf{C}}(\mathbf{R})$, $2 \leqq p \leqq +\infty$, are contained in $\mathscr{A}_2$ (Section **13.11**, Problem 12; observe that $\mathscr{L}^p \subset \mathscr{L}^2 + \mathscr{L}^\infty$ for $2 \leqq p \leqq +\infty$); for $f \in \mathscr{L}^2_{\mathbf{C}}(\mathbf{R})$ we have $\mathrm{M}_2(f) = 0$; for $f \in \mathscr{L}^\infty_{\mathbf{C}}(\mathbf{R})$, we have $\mathrm{M}_2(f) \leqq \mathrm{N}_\infty(f)$.
(c) For each function $f \in \mathscr{A}_2$ and each real number $\alpha > 1$, show that the function $x \mapsto |f(x)|^2/(1 + |x|^\alpha)$ is integrable. (Majorize each of the integrals

$$\int_n^{n+1} \frac{|f(x)|^2}{1 + |x|^\alpha}\, dx$$

for $n \in \mathbf{Z}$, and use Abel's partial summation formula.)
(d) Let f be a measurable function which is square-integrable on each compact interval. For each $\mathrm{T} > 0$, define $f_\mathrm{T}(x)$ to be $f(x)$ if $|x| \leqq \mathrm{T}$, and 0 otherwise. Then $f \in \mathscr{A}_2$ if and only if the total masses of the positive measures $\mu_\mathrm{T} = (1/2\mathrm{T})|\mathscr{F}f_\mathrm{T}|^2 \cdot m_\mathbf{R}$ are bounded as $\mathrm{T} \to +\infty$.

(e) Let $g \in \mathscr{D}(\mathbf{R})$ be $\geqq 0$ and such that $\int g(x)\, dx = 1$. Show that if $f \in \mathscr{A}_2$, then for each $k > 1$ we have

$$\left| \frac{1}{2\mathrm{T}} \int_{-\mathrm{T}}^{\mathrm{T}} f(t)\, dt \right|^2 \leqq \frac{\mathrm{T} + k}{\mathrm{T}} \int |\mathscr{F}g(k\xi)|^2\, d\mu_\mathrm{T}(\xi).$$

(Observe that if we put $g_{1/k}(x) = k^{-n}g(x/k)$, we have

$$\int_{-\mathrm{T}-k}^{\mathrm{T}+k} (f_\mathrm{T} * g_{1/k})(t)\, dt = \int_{-\mathrm{T}}^{\mathrm{T}} f(t)\, dt;$$

then use Cauchy–Schwarz and Plancherel's theorem.)

10. With the notation of Problem 9, two functions $f, g \in \mathscr{A}_2$ are said to be *correlated* if, for each $s \in \mathbf{R}$, the limit

$$\mathrm{C}_{f,g}(s) = \lim_{\mathrm{T} \to \infty} \frac{1}{2\mathrm{T}} \int_{-\mathrm{T}}^{\mathrm{T}} f(t - s)\overline{g(t)}\, dt$$

exists. The function $\mathrm{C}_{f,g}$ is called the *correlation function* of f and g. A function $f \in \mathscr{A}_2$ is said to be *autocorrelated* if $\mathrm{C}_{f,f}$ exists, and $\mathrm{C}_{f,f}$ is called the *autocorrelation function* of f. If $\mathrm{C}_{f,1}$ exists, it is a constant function, whose value $\mathrm{M}(f)$ is also called the *mean* of f (cf. Problem 8). For each character $\chi\colon x \mapsto e^{2\pi i\lambda x}$ of $\mathbf{R}$, if f and g are correlated, then so also are χf and χg, and $\mathrm{C}_{\chi f, \chi g}(s) = \chi(-s)\mathrm{C}_{f,g}(s)$.
(a) If $\mathrm{C}_{f,g}$ exists, we have $|\mathrm{C}_{f,g}(s)| \leqq \mathrm{M}_2(f)\mathrm{M}_2(g)$ for all $s \in \mathbf{R}$. If f is autocorrelated, $\mathrm{C}_{f,f}(0) = (\mathrm{M}_2(f))^2$ and $|\mathrm{C}_{f,f}(s)| \leqq \mathrm{C}_{f,f}(0)$ for all $s \in \mathbf{R}$.
(b) If

$$\lim_{\mathrm{T} \to \infty} \frac{1}{2\mathrm{T}} \int_{-\mathrm{T}}^{\mathrm{T}} |f(t)|^2\, dt$$

exists, show that

$$\lim_{\mathrm{T} \to \infty} \frac{1}{\mathrm{T}} \int_{\mathrm{T}}^{\mathrm{T}+a} |f(t)|^2\, dt = 0$$

for all $a \in \mathbf{R}$. (Compare

$$\frac{1}{2T}\int_{-T}^{T} |f(t)|^2\,dt \quad \text{and} \quad \frac{1}{2(T+a)}\int_{-T-a}^{T+a} |f(t)|^2\,dt$$

for large T.) Deduce that, if f is autocorrelated, $C_{f,f}(s)$ is the limit, for each $s \in \mathbf{R}$, of $(1/2T)(f_T * \tilde{f}_T)(s)$, in the notation of Problem 9(d). Show that the function $C_{f,f}$ is locally integrable, bounded and of positive type, and therefore (Section 22.2, Problem 1) equal almost everywhere to a continuous function of positive type; its Fourier transform, which is a bounded positive measure μ, is the vague limit of the measures $\mu_T = (1/2T)|\mathscr{F}f_T|^2 \cdot m_{\mathbf{R}}$. Moreover, for each $h \in \mathbf{R}$ we have

$$\limsup_{T\to\infty} \left| \frac{1}{2T}\int_{-T}^{T} f(t)e^{2\pi iht}\,dt \right|^2 \leqq \mu(\{h\})$$

(*Van der Corput's theorem*). (Use Problem 9(e).) In particular,

$$\lim_{T\to\infty} \frac{1}{2T}\int_{-T}^{T} f(t)e^{i\lambda t}\,dt = 0$$

except for an at most denumerable set of values of $\lambda \in \mathbf{R}$.

(c) Let (f_n), (g_n) be two sequences of functions in $\mathscr{A}_2$. Show that if f_n and g_n are correlated for each n, and if the sequence (f_n) (resp. (g_n)) converges uniformly on $\mathbf{R}$ to f (resp. g), then f and g belong to $\mathscr{A}_2$ and are correlated, and $C_{f,g}$ is the uniform limit of the sequence of functions C_{f_n, g_n}.

11. (a) Let f, g be two functions in $\mathscr{P}(\mathbf{R})$, and let $\mu = \mathscr{F}f$, $\nu = \mathscr{F}g$ be the bounded measures which are their Fourier transforms. Show that for all $a \in \mathbf{R}$ we have

$$\begin{aligned} C_{f,g}(s) &= \lim_{T\to\infty} \frac{1}{T}\int_{T}^{a+T} f(t-s)\overline{g(t)}\,dt \\ &= \sum_x e^{-2\pi isx}\mu(\{x\})\overline{\nu(\{x\})} \end{aligned}$$

summed over the at most denumerable set of $x \in \mathbf{R}$ such that $\mu(\{x\}) \neq 0$ and $\nu(\{x\}) \neq 0$. (Replace f by $\bar{\mathscr{F}}\mu$ and g by $\bar{\mathscr{F}}\nu$ in the integral.) In particular,

$$C_{f,f}(s) = \sum_x e^{-2\pi isx}|\mu(\{x\})|^2$$

so that the function $C_{f,f}$ is *almost periodic* and *of positive type*. We have $M(f) = \mu(\{0\})$, and

$$M(|f|)^2 = \sum_x |\mu(\{x\})|^2.$$

Deduce that

$$\lim_{T\to\infty} \frac{1}{T}\int_{T}^{a+T} f(t-s)\overline{g(t)}\,dt$$

exists for all functions $f, g \in \overline{\mathscr{P}(\mathbf{R})}$, and is equal to $C_{f,g}(s)$; $C_{f,g}$ is an almost periodic function belonging to $\overline{\mathscr{P}(\mathbf{R})}$ (use Problem 10); $\mathscr{F}C_{f,g}$ is therefore a bounded measure, necessarily atomic (Problem 15). If we put $\chi_x(s) = e^{2\pi isx}$, the measure $\mathscr{F}C_{f,g}$ is concentrated on the at most denumerable set of $x \in \mathbf{R}$ such that $\mathrm{M}(f\chi_x) \neq 0$ and $\mathrm{M}(g\chi_x) \neq 0$, and we have

$$C_{f,g}(s) = \sum_x e^{-2\pi i s x}\mathrm{M}(f\chi_x)\overline{\mathrm{M}(g\chi_x)}.$$

(b) Show that if $f \in \overline{\mathscr{P}(\mathbf{R})}$ is such that $\mathrm{M}(|f|^2) > 0$, then for each $a > 0$ there exists a measurable set $\mathrm{A} \subset [a, +\infty[$ of infinite outer Lebesgue measure, such that

$$|f(x)| \geqq \tfrac{1}{2}\mathrm{M}(|f|^2)$$

for all $x \in \mathrm{A}$.

12. (a) Show that the function $f(t) = e^{it}$ is correlated to all functions in $\mathscr{P}(\mathbf{R})$ with an identically zero correlation function; f is also autocorrelated and we have $C_{f,f}(0) = 1$, $C_{f,f}(s) = 0$ for $s \neq 0$.
(b) Show that the function $f(x) = \exp(i \log(1 + |x|))$ is autocorrelated, but that its mean does not exist. (Integrate by parts.)

13. Let $(a_k)_{k \geq 1}$ be a sequence of real numbers. For each integer $h > 0$, put $b_k^{(h)} = a_{k+h} - a_k$ for $k \geqq 1$. For each integer $n > 0$, let f_n denote the function which is equal to 0 for $t < 0$, and to $\exp(2\pi i n a_k)$ for $k - 1 \leqq t < k$, where $k \geqq 1$.
(a) Suppose that for each integer $h > 0$ the sequence $(b_k^{(h)} - [b_k^{(h)}])_{k \geq 1}$ is equipartitioned on $[0, 1]$ with respect to Lebesgue measure (Section 13.4, Problem 7). Show that for each n the function f_n is autocorrelated and that its autocorrelation function is equal to $\frac{1}{2}$ for $s = 0$, is zero for $|s| \geqq 1$, and is affine-linear in each of the intervals $[-1, 0]$ and $[0, 1]$. (Observe that it is enough to prove the convergence of $(1/2\mathrm{T})\int_0^{\mathrm{T}} f_n(t + h)\overline{f_n(t)}\,dt$ as $\mathrm{T} \to +\infty$, where h is a positive integer.) Deduce that, for each $n > 0$, we have

$$\lim_{\mathrm{T}\to\infty} \frac{1}{\mathrm{T}}\int_0^{\mathrm{T}} f_n(t)\,dt = 0$$

(use Problem 11(b)), and conclude that the sequence $(a_k - [a_k])$ is equipartitioned on $[0, 1]$ (*Van der Corput's theorem*).
(b) Deduce from (a) that if $a_k = \alpha_0 k^m + \alpha_1 k^{m-1} + \cdots + \alpha_m$, where α_0 is irrational, the sequence $(a_k - [a_k])$ is equipartitioned (*H. Weyl's theorem*). (Argue by induction on m.)

14. A *pseudomeasure* on $\mathbf{R}$ is a tempered distribution whose Fourier transform is a function in $\mathscr{L}_{\mathbf{C}}^{\infty}(\mathbf{R})$. Every bounded measure is a pseudomeasure, but Lebesgue measure is not. It can be shown that a pseudomeasure is a distribution of order $\leqq 1$ (Section 22.19, Problem 5(e)).
(a) Let T be a pseudomeasure, A its support, and suppose that A is the union of a compact set A_1 and a disjoint closed set A_2. Show that the restriction of T to the complement of A_2 is a pseudomeasure with support A_1. (Use (22.10.4).)
(b) Show that a pseudomeasure with finite support is a measure (use (a) and (17.7.3)).
(c) Show that for $2 \leqq q \leqq +\infty$, the Fourier transform of a function in $\mathscr{L}_{\mathbf{C}}^{q}(\mathbf{R})$ is a distribution of order $\leqq 1$ which is the sum of a function in $\mathscr{L}_{\mathbf{C}}^{2}(\mathbf{R})$ and a pseudomeasure. (Compare with Section 22.10, Problem 17.)

15. Let f be an almost periodic function on $\mathbf{R}$, and let (λ_n) be the (finite or infinite) sequence of numbers belonging to the set Ξ_f (Section **22.10**, Problem 9), i.e. the numbers such that $c_n = \mathrm{M}(f\chi_{\lambda_n}) \neq 0$, where $\chi_\lambda(x) = \exp(-2\pi i\lambda x)$.

(a) Show that the pseudomeasure $\mathscr{F}f$ is the weak limit in $\mathscr{D}'(\mathbf{R})$ of a sequence of measures of finite support carried by Ξ_f, and that the support of $\mathscr{F}f$ is the closure of Ξ_f.

(b) Show that if $\mathscr{F}f$ is a bounded measure, it is the atomic measure (**13.18**) defined by the mass c_n placed at the point λ_n for each n (**13.1.3**), the series $\sum_n |c_n|$ being convergent. (If $f = \mathscr{F}\mu$, where $\mu = \overline{\mathscr{F}}f$ is a bounded measure, and if ν is the atomic part of μ (**13.18.6**), show that if $g = \mathscr{F}\nu$ we have $\mathrm{M}(|f - g|^2) = 0$.) Conversely, if the series $\sum_n |c_n|$ converges, $\mathscr{F}f$ is a bounded measure and $f(x) = \sum_n c_n \exp(2\pi i\lambda_n x)$ for all $x \in \mathbf{R}$.

(c) For each integer $\mathrm{N} > 0$ and each integer $k > 0$, we may write

$$\left(\sum_{n=1}^{\mathrm{N}} c_n \exp(2\pi i\lambda_n x)\right)^k = \sum_{m=1}^{\mathrm{H}(k)} c_m^{(k)} \exp(2\pi i\lambda_m^{(k)} x)$$

where the $\lambda_m^{(k)}$ are all distinct, and $\mathrm{H}(k) \leqq (k+1)^{\mathrm{N}}$. Show that

$$\left|\sum_{n=1}^{\mathrm{N}} c_n\right|^{2k} \leqq (k+1)^{\mathrm{N}} \sum_{m=1}^{\mathrm{H}(k)} |c_m^{(k)}|^2 \leqq (k+1)^{\mathrm{N}}\left(\sum_{n=1}^{\mathrm{N}} |c_n|\right)^{2k}.$$

(Use Section **22.10**, Problem 9(c).)

(d) Deduce from (c) that if $c_n \geqq 0$ for all n (so that $c_m^{(k)} \geqq 0$ for all k and all $m \leqq \mathrm{H}(k)$), we have

$$\sum_{m=1}^{\mathrm{H}(k)} (c_m^{(k)})^2 \leqq \|f\|^{2k}.$$

(By using Section **22.10**, Problem 9(f), show that if (μ_n) is the sequence of numbers in Ξ_{f^k}, and $\mathrm{C}_n^{(k)} = \mathrm{M}(f^k\chi_{\mu_n})$, then

$$\sum_{m=1}^{\mathrm{H}(k)} (c_m^{(k)})^2 \leqq \sum_{n=1}^{\infty} (\mathrm{C}_n^{(k)})^2.)$$

Deduce that the series $\sum_n c_n$ is convergent, and hence that in this case $\mathscr{F}f$ is a *bounded measure*.

16. (a) Show that there exists a number $\mathrm{C} > 0$ such that

$$\left|\frac{\sin t}{1} + \frac{\sin 2t}{2} + \cdots + \frac{\sin \mathrm{N}t}{\mathrm{N}}\right| \leqq \mathrm{C}$$

for each integer $\mathrm{N} \geqq 1$ and each $t \in \mathbf{R}$. (Observe that the improper integral $\int_0^\infty \frac{\sin x}{x}\,dx$ converges and that the function $(1/x) - (1/\sin x)$ is continuous, and use Section **22.19**, Problem 3.)

(b) For each integer $m > 0$ put

$$g_m(t) = \frac{\sin 2\pi t}{1} + \frac{\sin 4\pi t}{2} + \cdots + \frac{\sin 2m\pi t}{m}$$

and choose real numbers α_m such that $0 < m\alpha_m < 1$, and such that the α_m, $m \geqq 1$, are linearly independent over the field **Q** of rational numbers. Put $h_m(t) = g_m(\alpha_m t)$. Also let (ε_m) be a sequence of positive real numbers such that $\sum_m \varepsilon_m < +\infty$ and $\sum_m \varepsilon_m \log m = +\infty$. Deduce from (a) that the series $f = \sum_m \varepsilon_m h_m$ is uniformly convergent, and hence that f is an almost periodic function such that Ξ_f is the set of numbers $\pm j\alpha_m$ for $m \geqq 1$ and $1 \leqq j \leqq m$, and $\mathrm{M}(f\chi_{\pm j\alpha_m}) = \pm\varepsilon_m/2ij$. Show that $\mathscr{F}f$ is not a measure. (Otherwise $\mathscr{F}f$ would be a bounded measure, and the family formed by the $\mathrm{M}(f\chi_{\pm j\alpha_m})$ would be absolutely summable.)

(c) Deduce from (b) that there exists an almost periodic function g for which the pseudomeasure $\mathscr{F}g$ is such that there exists a nonempty open interval I having the property that the restriction of $\mathscr{F}g$ to every open interval which meets I is a distribution of order 1. (Consider the product of f by a function of the form $x \mapsto \sum_n a_n \exp(i\lambda_n x)$, where the a_n are sufficiently small and the λ_n are chosen so that the set of them is dense in an interval I.)

17. Let μ be a bounded measure on **R**. Show that if $a < b$ in **R** are such that $\mu(\{a\}) = \mu(\{b\}) = 0$, we have

$$\mu([a, b]) = \lim_{\mathrm{T}\to\infty} \frac{1}{2\mathrm{T}} \int_{-\mathrm{T}}^{\mathrm{T}} \frac{e^{2\pi ib\xi} - e^{2\pi ia\xi}}{i\xi} \mathscr{F}\mu(\xi)\, d\xi$$

(*P. Lévy's inversion formula*). (Use the Lebesgue–Fubini theorem.) How must this formula be modified when $\mu(\{a\})$ and $\mu(\{b\})$ are arbitrary?

18. Let $f \in \mathscr{L}^1_{\mathrm{C}}(\mathbf{R})$. In order that $x \in \mathbf{R}$ should be such that the integral

$$(*) \qquad \int_{-\mathrm{T}}^{\mathrm{T}} e^{2\pi ix\xi} \mathscr{F}f(\xi)\, d\xi$$

tends to a finite limit as $\mathrm{T} \to +\infty$, it is necessary and sufficient that, if we put $g(t) = \frac{1}{2}(f(x+t) + f(x-t))$, the integral

$$\int_0^a \frac{\sin \mathrm{T}u}{u} g(u)\, du$$

tends to a finite limit, for any positive real number a. (Use the Lebesgue–Fubini theorem.) In particular, this is the case if, in a neighborhood of the point x, the function f is equal almost everywhere to a function f_0, continuous and differentiable at x, and the limit of the integral (*) is then $f_0(x)$. (Use the Riemann–Lebesgue theorem.) More particularly, if the restriction of f to some open interval $\mathrm{I} \subset \mathbf{R}$ is negligible, the integral (*) tends to zero at each point of I. (This result expresses the *local character* of Fourier's reciprocity formula.)

19. (a) Let $\Phi \geqq 0$ be a function in $\mathscr{L}^1_{\mathrm{C}}(\mathbf{R})$, such that $\int_{\mathbf{R}} \Phi(t)\, dt = 1$ and $\mathscr{F}\Phi$ is integrable. Show that, for each function $f \in \mathscr{L}^1_{\mathrm{C}}(\mathbf{R})$, the function

$$(*) \qquad x \mapsto \int_{-\infty}^{+\infty} e^{2\pi ix\xi} \mathscr{F}\Phi(\xi/\mathrm{T}) \mathscr{F}f(\xi)\, d\xi$$

belongs to $\mathscr{L}^1_{\mathrm{C}}(\mathbf{R})$ and converges in mean to f as $\mathrm{T} \to +\infty$ (cf. **(14.11.1)**).

This result applies in particular in the following cases:

$$\Phi(x) = \frac{\sin^2 \pi x}{\pi^2 x^2}, \qquad \mathscr{F}\Phi(\xi) = (1 - |\xi|)^+ \qquad \text{(Cesàro–Féjer summation)}$$

$$\Phi(x) = \frac{2}{1 + 4\pi^2 x^2}, \qquad \mathscr{F}\Phi(\xi) = e^{-|\xi|} \qquad \text{(Abel summation)}$$

$$\Phi(x) = \sqrt{2\pi} \cdot e^{-2\pi^2 x^2}, \qquad \mathscr{F}\Phi(\xi) = e^{-\xi^2/2} \qquad \text{(Weierstrass summation).}$$

(b) Suppose in addition that $\Phi(-x) = \Phi(x)$ and that, on the complement of each interval $[-\alpha, \alpha]$ (with $\alpha > 0$), the functions $x \mapsto \mathrm{T}\Phi(\mathrm{T}x)$ tend uniformly to 0 as $\mathrm{T} \to +\infty$. Show that if f is continuous at a point x_0, the integral (∗), for $x = x_0$, tends to $f(x_0)$ as $\mathrm{T} \to +\infty$. These conditions are satisfied in the three examples in (a) above.
(c) In order that a continuous function g on **R** should be the Fourier transform of a bounded measure, it is necessary and sufficient that for each $\mathrm{T} > 0$ the function

$$x \mapsto \int_{-\mathrm{T}}^{\mathrm{T}} e^{2\pi i x \xi} g(\xi) \mathscr{F}\Phi(\xi/\mathrm{T})\, d\xi$$

should belong to $\mathscr{L}^1_{\mathbf{C}}(\mathbf{R})$ and that the norm of this function in $\mathscr{L}^1_{\mathbf{C}}(\mathbf{R})$ should remain bounded as $\mathrm{T} \to +\infty$. (Use **(13.4.2)** and Section **22.10**, Problem 10(h).)

20. Let f be an almost periodic function on **R**, and let $(\lambda_1, \ldots, \lambda_n, \ldots)$ be the sequence (in any order) of the $\lambda \in \mathbf{R}$ such that $\mathrm{M}(f\chi_\lambda) \neq 0$, where $\chi_\lambda(t) = \exp(-2\pi i \lambda t)$. Let $a_n = \mathrm{M}(f\chi_{\lambda_n})$. Suppose in addition that the "exponents" λ_n are *linearly independent* over the field **Q** of rational numbers (and hence in particular that $\lambda_n \neq 0$ for each n).
(a) For each integer $\mathrm{N} > 1$ put

$$\mathrm{K_N}(t) = (1 + \cos 2\pi(\lambda_1 t + \omega_1)) \cdots (1 + \cos 2\pi(\lambda_\mathrm{N} t + \omega_\mathrm{N}))$$

where the $\omega_n \in \mathbf{R}$ are arbitrary. Show that we have

$$\mathrm{M}(f\mathrm{K_N}) = \tfrac{1}{2}(a_1 e^{-2\pi i \omega_1} + \cdots + a_\mathrm{N} e^{-2\pi i \omega_\mathrm{N}}).$$

(Use the independence of the λ_n.)
(b) Deduce that

$$|a_1| + \cdots + |a_\mathrm{N}| \leq 2\|f\|$$

and hence that $f(t) = \sum_{n=1}^{\infty} a_n \exp(2\pi i \lambda_n t)$, this series being absolutely convergent for all $t \in \mathbf{R}$, so that $\overline{\mathscr{F}}f$ is the atomic measure defined by the masses a_n at the points λ_n.

21. Let $\mathscr{P}_0 \subset \mathscr{P}_+$ be the set of continuous functions of positive type on **R** such that $f(0) = 1$, or equivalently the set of Fourier transforms of positive measures μ on **R** such that $\mu(\mathbf{R}) = 1$ ("probability measures").
(a) Show that each function $f \in \mathscr{P}_0$ satisfies the following inequalities:

(α) $1 - \mathscr{R}f(2t) \leq 4(1 - \mathscr{R}f(t))$ for all $t \in \mathbf{R}$;

(β) $|\mathscr{I}f(t)| \leq (\tfrac{1}{2}(1 - \mathscr{R}f(2t)))^{1/2}$ for all $t \in \mathbf{R}$;

(γ) $\left| \dfrac{1}{2h} \displaystyle\int_{t-h}^{t+h} f(u)\, du \right|^2 \leq \tfrac{1}{2}(1 + \mathscr{R}f(h))$ for all $t \in \mathbf{R}$ and all $h > 0$.

(If $f(t) = \mathscr{F}\mu(t)$, observe that

$$\mathscr{R}f(t) = \int_{-\infty}^{+\infty} \cos(2\pi t x)\, d\mu(x), \qquad \mathscr{I}f(t) = \int_{-\infty}^{+\infty} \sin(2\pi t x)\, d\mu(x),$$

and use Cauchy–Schwarz for (β) and (γ).)

(b) Show that if $f \in \mathscr{P}_0$ is such that $(f(t) - 1)/t^2 \to 0$ as $t \to 0$, then f is constant. (If $1 - \mathscr{R}f(t) = t^2 g(t)$, show with the help of (a) that the function g satisfies the inequality $0 \leqq g(2t) \leqq g(t)$ for all $t \in \mathbf{R}$.)

(c) Suppose that $f \in \mathscr{P}_0$ is such that there exist two constants $0 < \alpha < 1$, $\beta > 0$ such that $|f(t)| \leqq \alpha$ whenever $|t| \geqq \beta$. Then we have

$$f(t) \leqq 1 - \frac{1 - \alpha^2}{8\beta^2} t^2$$

for $|t| < \beta$. (Observe that $|f|^2 \in \mathscr{P}_0$ and deduce from (a) that

$$1 - |f(2t)|^2 \leqq 4(1 - |f(t)|^2).)$$

Deduce that if $f \in \mathscr{P}_0$ is such that $\mathscr{F}f \in \mathscr{L}^1_{\mathbf{C}}(\mathbf{R})$, then for each $\varepsilon > 0$ there exists $c < 1$ such that $|f(t)| \leqq c$, whenever $|t| \geqq \varepsilon$.

(d) If $f = \mathscr{F}\mu \in \mathscr{P}_0$, then for $t > 0$ we have

$$1 - \mathscr{R}f(t) \geqq \tfrac{1}{3}t^2 \int_{|x| < 1/t} x^2\, d\mu(x),$$

$$\int_0^t (1 - \mathscr{R}f(u))\, du \geqq \tfrac{1}{7}t \int_{|x| \geq 1/t} d\mu(x).$$

(Minorize $1 - \cos u$ and $1 - u^{-1} \sin u$.)

(e) If a sequence (f_n) of functions in $\mathscr{P}_0$ is uniformly bounded and converges almost everywhere to a function f, then f is equal almost everywhere to a function in $\mathscr{P}_0$.

22. Let (μ_n) be a sequence of *positive* bounded measures on $\mathbf{R}$. For each n, let θ_n be the increasing function, continuous on the right, such that $\theta_n(0) = 0$ and $\mu_n(]a, b]) = \theta_n(b) - \theta_n(a)$ (Section **13.18**, Problem 6).

(a) In order that the sequence (μ_n) should be vaguely convergent, it is necessary and sufficient that there should exist a denumerable subset D of $\mathbf{R}$ such that, for each $x \notin$ D, the sequence $(\theta_n(x))$ converges in $\mathbf{R}$. For $x \in$ D, the sequence $(\theta_n(x))$ is then bounded; show by an example that it need not converge. The limit μ of the sequence (μ_n) is a not necessarily bounded positive measure.

(b) Let D′ be the union of D and the sets of points of discontinuity of the functions θ_n (i.e., the points of measure > 0 for μ_n). There exists a unique increasing function θ, continuous on the right, such that $\theta(x) = \lim_{n \to \infty} \theta_n(x)$ for all $x \notin$ D′, and we have $\mu(]a, b]) = \theta(b) - \theta(a)$ for all $a < b$ in $\mathbf{R}$. Show that for $x \in$ D′ the sequence $(\theta_n(x))$ may tend to a limit other than $\theta(x)$.

(c) Suppose that $\|\mu_n\| = 1$ for each n ("probability measures"). In order that the sequence (μ_n) should converge vaguely to a measure μ such that $\|\mu\| = 1$, it is necessary and sufficient that the sequence $(\mathscr{F}\mu_n)$ should converge pointwise in $\mathbf{R}$ to a function g which is continuous in a neighborhood of 0; we have then $g = \mathscr{F}\mu$ (cf. Section **22.10**, Problem 10). An equivalent condition is that the sequence $(\mathscr{F}\mu_n)$ converges uniformly on each compact subset of $\mathbf{R}$.

(d) Let (λ_n), (μ_n) be two sequences of probability measures, such that the sequences (λ_n) and $(\lambda_n * \mu_n)$ both converge vaguely to the *same* probability measure. Show that the sequence (μ_n) converges vaguely to the Dirac measure ε_0.

23. Let $(\mu_n)_{n \geq 1}$ be an infinite sequence of probability measures on $\mathbf{R}$. For each integer $n \geqq 1$, put

$$\mathop{\ast}_{k=1}^{n} \mu_k = \mu_1 * \mu_2 * \cdots * \mu_n.$$

The *infinite* (convolution) product $\mathop{\ast}\limits_{n=1}^{\infty} \mu_n$ is said to exist (or to *converge*) if the sequence $\left(\mathop{\ast}\limits_{k=1}^{n} \mu_k\right)_{n \geq 1}$ converges vaguely to a probability measure, denoted by $\mathop{\ast}\limits_{n=1}^{\infty} \mu_n$.

(a) The infinite product $\mathop{\ast}\limits_{n=1}^{\infty} \mu_n$ converges if and only if the sequence of products $\prod\limits_{k=1}^{n} \mathscr{F}\mu_k$ converges pointwise in $\mathbf{R}$ and uniformly on some neighborhood of 0. When this is so, for each mapping $n \mapsto p(n)$ of the set of integers $n \geqq 1$ into itself, the sequence of products

$$\prod_{k=1}^{p(n)} \mathscr{F}\mu_{n+k}$$

converges uniformly to 1 on each compact subset of $\mathbf{R}$ (use Problem 22(d)), and the sequence of probability measures

$$\mu_{n+1} * \mu_{n+2} * \cdots * \mu_{n+p(n)}$$

converges vaguely to the Dirac measure ε_0. Consider the converse.

(b) Show that if $\mu = \mathop{\ast}\limits_{n=1}^{\infty} \mu_n$ exists, $\operatorname{Supp}(\mu)$ is the set of limits of convergent sequences (x_n) such that $x_n \in \operatorname{Supp}(\mu_1) + \cdots + \operatorname{Supp}(\mu_n)$. (Use (a) and the fact that for two bounded positive measures μ, ν on $\mathbf{R}$ we have

$$(\mu * \nu)([c - r, c + r]) \geqq \mu([-\tfrac{1}{2}r, \tfrac{1}{2}r])\nu([c - \tfrac{1}{2}r, c + \tfrac{1}{2}r]).)$$

(c) Put $\lambda_n = \mu_n * \check{\mu}_n$, so that $\mathscr{F}\lambda_n = |\mathscr{F}\mu_n|^2$. Show that if the product $\mathop{\ast}\limits_{n=0}^{\infty} \mu_n$ converges, then so also does the product $\mathop{\ast}\limits_{n=0}^{\infty} \lambda_n$, and that

$$\sum_{n=0}^{\infty} \int_{-a}^{a} x^2 \, d\lambda_n(x) < +\infty$$

for all $a > 0$. (Use Problem 21(d), and the fact that the series $\sum\limits_{n} (1 - \mathscr{F}\lambda_n(t))$ converges for all $t \in \mathbf{R}$.)

(d) With the same notation and assumptions, put $\nu_n = \lambda_1 * \lambda_2 * \cdots * \lambda_n$. Show that for $\alpha > 0$ the sequence of integrals

$$\int \left(1 - \frac{|x|}{\alpha}\right)^{+} d\nu_n(x)$$

is decreasing and converges to $\int (1 - (|x|/\alpha))^{+} \, d\lambda(x)$. (Consider the Fourier transforms.)

24. If μ is a probability measure on **R** such that the function x^2 is μ-integrable, then the function x is μ-integrable. Let $\mathrm{E}(\mu) = \int_{\mathbf{R}} x \, d\mu(x)$, then we have $\mathrm{E}(\mu)^2 \leq \int_{\mathbf{R}} x^2 \, d\mu(x)$. The number

$$\mathrm{V}(\mu) = \int_{\mathbf{R}} x^2 \, d\mu(x) - \mathrm{E}(\mu)^2$$

is called the *variance* of μ, and μ is said to be *centered* if $\mathrm{E}(\mu) = 0$.
(a) Show that if x^2 is integrable with respect to each of the probability measures μ, ν, then it is integrable with respect to $\mu * \nu$, and

$$\mathrm{E}(\mu * \nu) = \mathrm{E}(\mu) + \mathrm{E}(\nu), \qquad \mathrm{V}(\mu * \nu) = \mathrm{V}(\mu) + \mathrm{V}(\nu).$$

(b) Let (μ_n) be a sequence of centered probability measures whose supports are contained in a compact interval $\mathrm{I} \subset \mathbf{R}$. For the product $\mathop{\ast}\limits_{n=1}^{\infty} \mu_n$ to converge, it is necessary and sufficient that the series $\sum_{n=1}^{\infty} \mathrm{V}(\mu_n)$ should converge, and then we have $\mathrm{V}\left(\mathop{\ast}\limits_{n=1}^{\infty} \mu_n\right) = \sum_{n=1}^{\infty} \mathrm{V}(\mu_n)$. (To show that the condition is necessary, use Problem 23(b). To show that it is sufficient, observe that $\mathrm{D}(\mathscr{F}\mu_n)(0) = 0$ and $\mathrm{D}^2(\mathscr{F}\mu_n)(0) = -4\pi^2 \mathrm{V}(\mu_n)$, and that the function $u^{-2}(e^{iu} - 1 - iu)$ is bounded on **R**.)
(c) Let (μ_n) be a sequence of probability measures, not necessarily centered, with their supports contained in a compact interval I. For the product $\mathop{\ast}\limits_{n=1}^{\infty} \mu_n$ to converge, it is necessary and sufficient that both the series $\sum_{n=1}^{\infty} \mathrm{V}(\mu_n)$ and $\sum_{n=1}^{\infty} \mathrm{E}(\mu_n)$ should converge.

(Observe that if (s_n) is a sequence of real numbers, and if the infinite product $\mathop{\ast}\limits_{n=1}^{\infty} \mu_n$ is convergent, then the infinite product $\mathop{\ast}\limits_{n=1}^{\infty} \gamma(s_n)\mu_n$ converges if and only if the series $\sum_{n=1}^{\infty} s_n$ converges; observe also that $\gamma(-\mathrm{E}(\mu_n))\mu_n$ is centered.)

25. A probability measure μ on **R** may be considered as a measure on the completed line $\bar{\mathbf{R}}$, by identifying μ with its image (Section **13**.9, Problem 24) under the canonical injection of **R** into $\bar{\mathbf{R}}$. If $(\mu_n)_{n \geq 0}$ is a sequence of probability measures on **R**, we may therefore consider the product measure $\varpi = \bigotimes_{n=0}^{\infty} \mu_n$ on the compact space $\bar{\mathbf{R}}^{\mathbf{N}}$, which is concentrated on the measurable set $\mathbf{R}^{\mathbf{N}}$ (Section **13**.21, Problem 9). For each $\mathbf{x} = (x_n)_{n \geq 0} \in \mathbf{R}^{\mathbf{N}}$, put $x_n = \mathrm{pr}_n(\mathbf{x})$.
(a) Show that the infinite product $\mathop{\ast}\limits_{n=0}^{\infty} \mu_n$ converges if and only if the series

$$\mathrm{pr}_0(\mathbf{x}) + \mathrm{pr}_1(\mathbf{x}) + \cdots + \mathrm{pr}_n(\mathbf{x}) + \cdots$$

is *convergent in measure* with respect to the measure ϖ (Section **13**.12, Problem 2). (Let $s_n(\mathbf{x}) = \mathrm{pr}_0(\mathbf{x}) + \cdots + \mathrm{pr}_n(\mathbf{x})$. Use the criterion of Problem 23(a), together with Section **22**.10, Problem 10(e), to show that the convergence of the product $\mathop{\ast}\limits_{n=0}^{\infty} \mu_n$ is equivalent to the following property: for each $a > 0$, the set of $\mathbf{x}$ such that

$$|s_{n+p(n)}(\mathbf{x}) - s_n(\mathbf{x})| \geqq a$$

has a measure which tends to 0 as $n \to +\infty$. Then use Section 13.9, Problem 2(c).)

The measure $\mathop{\ast}\limits_{n=0}^{\infty} \mu_n$ is then the image (Section 13.9, Problem 24) of the measure ϖ under the mapping $\mathbf{x} \mapsto s(\mathbf{x}) = \sum_{n=0}^{\infty} \mathrm{pr}_n(\mathbf{x})$ (defined to within a ϖ-negligible function).

(b) For each real number $c \geqq 0$, define $u_n^c(\mathbf{x}) = \mathrm{pr}_n(\mathbf{x})$ if $|\mathrm{pr}_n(\mathbf{x})| \leqq 1$, and $u_n^c(\mathbf{x}) = c$ if $|\mathrm{pr}_n(\mathbf{x})| > 1$. If the series $\mathrm{pr}_0(\mathbf{x}) + \cdots + \mathrm{pr}_n(\mathbf{x}) + \cdots$ converges in measure with respect to ϖ, then so also does the series $u_0^c(\mathbf{x}) + \cdots + u_n^c(\mathbf{x}) + \cdots$. Deduce that if $\mathrm{I} = [-1, 1]$ and if we put $\mathrm{P}(\mu_n) = \mu_n(\complement \mathrm{I})$ and $\mu_n^c = \varphi_1 \cdot \mu_n + \mathrm{P}(\mu_n) \cdot \varepsilon_c$, then the convergence of the product $\mathop{\ast}\limits_{n=0}^{\infty} \mu_n$ implies that of the product $\mathop{\ast}\limits_{n=0}^{\infty} \mu_n^c$ (use (a)). Conclude that if the product $\mathop{\ast}\limits_{n=0}^{\infty} \mu_n$ is convergent, the series $\sum_{n=0}^{\infty} \mathrm{P}(\mu_n)$ is convergent. (Use Problem 24(c), giving c two different values.)

(c) Show that for the product $\mathop{\ast}\limits_{n=0}^{\infty} \mu_n$ to converge, it is necessary and sufficient that the three real series

$$\sum_{n=0}^{\infty} \mathrm{P}(\mu_n), \qquad \sum_{n=0}^{\infty} \mathrm{V}(\mu_n^0), \qquad \sum_{n=0}^{\infty} \mathrm{E}(\mu_n^0)$$

should converge (*Kolmogoroff's three series theorem*). (Observe that

$$|\mathscr{F}\mu_n^0(t) - \mathscr{F}\mu(t)| \leqq 2\mathrm{P}(\mu_n)$$

for all $t \in \mathbf{R}$, and use (b) above and Problems 23 and 24.)

26. (a) Let μ, ν be two bounded measures on $\mathbf{R}$. Let $\mathrm{A}(\mu)$ denote the (at most denumerable) set of $x \in \mathbf{R}$ such that $\mu(\{x\}) \neq 0$. Show that $\mathrm{A}(\mu * \nu) \subset \mathrm{A}(\mu) + \mathrm{A}(\nu)$, and that these two sets are equal if μ and ν are positive (the sum $\mathrm{A}(\mu) + \mathrm{A}(\nu)$ is the empty set if either of the two sets $\mathrm{A}(\mu)$, $\mathrm{A}(\nu)$ is empty, i.e., if either μ or ν is diffuse).

(b) Deduce from (a) that

$$\mathrm{M}(|\mathscr{F}(\mu * \nu)|^2) = \mathrm{M}(|\mathscr{F}\mu|^2)\mathrm{M}(|\mathscr{F}\nu|^2).$$

(Consider the measures $\mu * \check{\mu}$ and $\nu * \check{\nu}$.)

(c) Let $(\mu_n)_{n \geq 1}$ be a sequence of probability measures on $\mathbf{R}$, such that the infinite product $\mu = \mathop{\ast}\limits_{n=1}^{\infty} \mu_n$ is convergent. Show that

$$(*) \qquad \mathrm{M}(|\mathscr{F}\mu|^2) = \lim_{n \to \infty} \prod_{k=1}^{n} \mathrm{M}(|\mathscr{F}\mu_k|^2).$$

(Let $\lambda_n = \mu_n * \check{\mu}_n$, and $\lambda = \mathop{\ast}\limits_{n=1}^{\infty} \lambda_n$; observe that the left-hand side of $(*)$ is equal to $\lambda(\{0\})$; then use Problem 23(d), and the fact that, for each positive measure ν on $\mathbf{R}$, $\nu(\{0\})$ is the greatest lower bound of the integrals

$$\int \left(1 - \frac{|x|}{\alpha}\right)^{+} dv(x)$$

for values $\alpha > 0$.)
(d) Under the hypotheses of (c), let α_n denote the least upper bound of the numbers $\mu_n(\{x\})$, $x \in \mathbf{R}$. For the measure $\mu = \mathop{\ast}\limits_{n=1}^{\infty} \mu_n$ to be diffuse, it is necessary and sufficient that $\lim\limits_{n\to\infty} \prod\limits_{k=1}^{n} \alpha_k = 0$ (use Problem 11).

27. (a) With the notation of Section **13.21**, Problem 9, let f be a μ-integrable function on X, having the following property: for each finite subset L of **N** and each $x_{L'} \in X_{L'}$ (where $L' = N - L$), $f(y_L, x_{L'}) = f(z_L, x_{L'})$ for each pair of points $(y_L, z_L) \in X_L$. Show that f is equal almost everywhere to a constant. (Observe that $f_{L'} = f$ for all finite subsets L of **N**, and use Section **13.21**, Problem 9(d).) In particular, if f is the characteristic function φ_A of an integrable subset A of X, then either A or $\complement A$ is negligible.
(b) Let $(\mu_n)_{n \geq 0}$ be a sequence of probability measures on **R**, all of which are atomic; let D_n be the at most denumerable set of $x \in \mathbf{R}$ such that $\mu_n(\{x\}) > 0$; we may write D_n as the set of points of a sequence $(a_{nk})_{0 \leq k < \omega_n}$ of distinct terms, where ω_n is an integer $\geqq 1$ or $+\infty$. Let K_n denote the set $\{0, 1, \ldots, \omega_n - 1\}$ if ω_n is an integer, and the set $\mathbf{N} \cup \{+\infty\}$ if $\omega_n = +\infty$; endow K_n with the topology induced by that of $\bar{\mathbf{R}}$, so that K_n is compact. Let v_n be the measure on K_n defined by $v_n(\{k\}) = \mu_n(\{a_{nk}\})$ for $0 \leqq k < \omega_n$, and $v_n(\{+\infty\}) = 0$ if $\omega_n = +\infty$. On the space $K = \prod\limits_n K_n$ let v be the product measure $\bigotimes\limits_{n=0}^{\infty} v_n$, and let π be the mapping of K into $\bar{\mathbf{R}}^{\mathbf{N}}$ defined as follows: $\pi((k_n))$ is the point (a_{n, k_n}) if all the k_n are integers, and is 0 if one of the k_n is $+\infty$. Show that π is v-measurable, and that $\pi(v) = \varpi$.
(c) Suppose in addition that the product $\mu = \mathop{\ast}\limits_{n=0}^{\infty} \mu_n$ is convergent; then the set $E \subset K$ of points **k** such that the series $u(\mathbf{k}) = \sum\limits_{n=0}^{\infty} \mathrm{pr}_n(\pi(\mathbf{k}))$ converges has a v-negligible complement. Let M be the **Z**-module generated by the points $a_{nk} \in \mathbf{R}$ ($n \geqq 0$, $0 \leqq k < \omega_n$), which is a countable subset of **R**. For each Lebesgue-measurable set $B \subset \mathbf{R}$, the set $M + B$ is measurable, and if B is negligible, then $M + B$ is negligible (all with respect to Lebesgue measure). Let A be the set of $\mathbf{k} \in K$ such that $u(\mathbf{k}) \in M + B$. Show that if $\mathbf{k} = (k_n) \in A$ and if $\mathbf{k}' = (k'_n) \in K$ is such that $k'_n = k_n$ for all but finitely many indices n, then also $\mathbf{k}' \in A$. Deduce from (a) that either A or $\complement A$ is v-negligible.
(d) Show that the measure $\mu = \mathop{\ast}\limits_{n=0}^{\infty} \mu_n$ is either atomic, or diffuse and disjoint from Lebesgue measure λ, or else is a measure with base λ. (Consider the three possibilities: either there exists an at most denumerable set $B \subset \mathbf{R}$ such that $u^{-1}(M + B)$ has measure equal to 1; or this condition is not satisfied, but there exists a (nondenumerable) λ-negligible set B such that $u^{-1}(M + B)$ has measure equal to 1; or else $\mu^{-1}(M + B)$ is v-negligible for each λ-negligible set B.) (*Jessen–Wintner theorem.*)

28. (a) Let $(a_n)_{n \geq 1}$ be a sequence of positive real numbers, and put $\mu_n = \frac{1}{2}(\varepsilon_{a_n} + \varepsilon_{-a_n})$. For the infinite product $\mu = \mathop{\ast}\limits_{n=1}^{\infty} \mu_n$ to converge, it is necessary and sufficient that the series $\sum\limits_{n=1}^{\infty} a_n^2$

should converge (cf. Problem 25). Show that the measure μ is then always diffuse (Problem 26), and is therefore either disjoint from Lebesgue measure λ, or else is a measure with base λ (Problem 27). We have $\mathscr{F}\mu(t) = \prod_{n=1}^{\infty} \cos(2\pi a_n t)$ for all $t \in \mathbf{R}$.

(b) Suppose that $\sum_{n=1}^{\infty} a_n^2 < +\infty$. The support of μ is then the set of sums $\sum_{n=1}^{\infty} (\pm a_n)$, for all choices of signs which make the series converge (Problem 23(b)). Show that if $\sum_{n=1}^{\infty} a_n = +\infty$, then $\operatorname{Supp}(\mu) = \mathbf{R}$. (For each $x \in \mathbf{R}$, choose the signs of the terms in the series $\sum_{n=1}^{\infty} (\pm a_n)$ so that for some strictly increasing sequence (n_k), the partial sums $s_n = \sum_{\nu=0}^{n} (\pm a_\nu)$ are such that $s_{n_{2k}} \leqq x \leqq s_{n_{2k-1}}$, and $|x - s_{n_k}| \leqq a_{n_k}$.) If $\sum_{n=1}^{\infty} a_n < +\infty$, $\operatorname{Supp}(\mu)$ is a perfect compact set (observe that the support of a diffuse measure cannot have an isolated point).

(c) If $(r_n)_{n \geq 1}$ is the sequence of Rademacher functions (Section **13.21**, Problem 10), the series $v(t) = \sum_{n=1}^{\infty} a_n r_n(t)$ converges almost everywhere in $[0, 1]$ (Section **13.21**, Problem 12). Show that μ is the image, under the mapping v: $[0, 1] \to \mathbf{R}$, of Lebesgue measure on $[0, 1]$. (Use Problem 27(a) and Section **13.21**, Problem 10(a).)

(d) Suppose that $\sum_{n=1}^{\infty} a_n < +\infty$ and also that, putting $q_n = \sum_{k=n+1}^{\infty} a_k$, we have $q_n < a_n$ for all n (or, equivalently, $2q_n < q_{n-1}$). Show that the function v is increasing on $[0, 1]$; its points of discontinuity are the $t_{jn} = j/2^n$ for odd $j < 2^n$, and we have

$$v(t_{jn}+) - v(t_{jn}-) = 2(q_{n-1} - 2q_n).$$

Deduce that $\lambda(\operatorname{Supp}(\mu))$ is the limit of the decreasing sequence $(2^{n+1}q_n)$; hence μ is disjoint from Lebesgue measure λ if and only if this limit is 0. If on the other hand this limit is > 0, the measure μ is equal to the measure $\varphi_{\operatorname{Supp}(\mu)} \cdot \lambda$ multiplied by the constant $\lambda(\operatorname{Supp}(\mu))^{-1}$.

(e) Suppose that $\sum_{n=1}^{\infty} a_n < +\infty$ but also that $a_n \leqq q_n$ (or equivalently $2q_n \geqq q_{n-1}$) for all n. Show that $\operatorname{Supp}(\mu)$ is the interval $[-q_0, q_0]$ (argue as in (b)).

29. The notation is the same as in Problem 28.

(a) Show that if we take $a_n = 2^{-n} + 3^{-n}$, then $\operatorname{Supp}(\mu)$ is a totally disconnected perfect set, but μ is a measure with base λ.

(b) Show that if we take $a_n = 1/n!$, $\operatorname{Supp}(\mu)$ is a totally disconnected perfect set which is λ-negligible, and $\limsup_{t \to +\infty} \mathscr{F}\mu(t) = 1$, although $\mathscr{F}\mu$ is not almost periodic. (Evaluate $\mathscr{F}\mu(k!)$ by observing that $\sum_{k=n+1}^{\infty} (k!/n!)^2$ tends to 0 with $1/k$.)

(c) Let $m_0 = 0$, $m_k = 2^{1!} + 2^{2!} + \cdots + 2^{k!}$ for $k \geqq 1$, and define a_n by taking $a_n = 2^{-k!}$ for $m_{k-1} < n \leqq m_k$ and all $k \geqq 1$. Show that μ is then disjoint from λ, although $\operatorname{Supp}(\mu) = \mathbf{R}$. (Show that $\limsup_{t \to +\infty} |\mathscr{F}\mu(t)| > 0$ by arguing as in (b), and use the Riemann–Lebesgue theorem.)

(d) Take $a_n = b^n$, where $0 < b < 1$, and let ν_b denote the corresponding measure $\mu = \mathop{\ast}\limits_{n=1}^{\infty} \mu_n$. If $0 < b < \frac{1}{2}$, then $\operatorname{Supp}(\nu_b)$ is a λ-negligible totally disconnected perfect set, so

that ν_b is disjoint from λ; if $\frac{1}{2} \leqq b < 1$, then $\text{Supp}(\nu_b)$ is the interval $[-b/(1-b), b/(1-b)]$.
(e) With the notation of (d), suppose that $\limsup_{t \to +\infty} |\mathscr{F}\mu(t)| > 0$. Show that there exists a number $c \neq 0$ such that $d(\mathbf{Z}, cb^{-n}) \to 0$ as $n \to +\infty$. (The hypothesis implies that there exists a sequence (t_j) tending to $+\infty$, such that $|\mathscr{F}\mu(t_j)| \geqq a > 0$; if k_j is the unique positive integer such that $b < t_j b^{k_j} \leqq 1$, we may suppose that the sequence of numbers $u_j = t_j b^{k_j}$ tends to a limit c. We have

$$|\mathscr{F}\mu(u_j)| \cdot \prod_{m=1}^{k} |\cos(2\pi u_j b^{-m})| \geqq a$$

for all integers $k \leqq k_j$. Deduce that

$$\prod_{m=1}^{k} |\cos(2\pi c b^{-m})| \geqq a' > 0$$

for all integers k.)
(f) Show that the set of numbers b such that $0 < b < 1$ and such that there exists a number $c \neq 0$ for which $d(\mathbf{Z}, cb^{-n}) \to 0$ as $n \to +\infty$ is denumerable (in any case it contains the numbers $1/m$, for each integer $m > 1$). (Write $cb^{-n} = v_n + \varepsilon_n$, where $v_n \in \mathbf{Z}$ and $\varepsilon_n \to 0$; show that there exists an integer $n_0(b, c)$ such that $|v_{n+2} - v_{n+1}^2 v_n^{-1}| < \frac{1}{2}$ for $n \geqq n_0(b, c)$. Deduce that the set of sequences (v_n), corresponding to all pairs (b, c) for which $n_0(b, c)$ has a given value, is denumerable, and complete the proof by noting that $b = \lim_{n \to \infty} (v_n / v_{n+1})$.)
Deduce that there is only an at most denumerable set of values of b such that $0 < b < \frac{1}{2}$ and such that $\mathscr{F}\nu_b(t)$ does not tend to 0 as $t \to +\infty$, although ν_b is disjoint from Lebesgue measure.
(g) A *Pisot number* is a real number $\theta > 1$ which is an algebraic integer, all of whose conjugates $\theta^{(k)} \neq \theta$ satisfy $|\theta^{(k)}| < 1$. For example $\frac{1}{2}(\sqrt{5} + 1)$ is a Pisot number. Show that if b is such that b^{-1} is a Pisot number $\neq 2$, then $\mathscr{F}\nu_b(t)$ does not tend to 0 as $t \to +\infty$ (and consequently ν_b is a measure disjoint from Lebesgue measure, although it may happen that $\frac{1}{2} < b < 1$). (Observe that if θ is a Pisot number, there exists α such that $0 < \alpha < 1$ and $a(\mathbf{Z}, \theta^n) \leqq \alpha^n$. Minorize $\mathscr{F}\nu_b(\theta^m)$.)
(h) Let $\nu \neq 0$ be a bounded positive measure on $\mathbf{R}$, whose support is compact and totally disconnected, which is disjoint from Lebesgue measure and is such that $\mathscr{F}\nu(t) \to 0$ as $t \to +\infty$ (see (d) and (f) above). Show that for each $x \in \mathbf{R}$ not in the support of ν (and hence almost everywhere with respect to Lebesgue measure), the integral

$$\int_{-T}^{T} e^{2\pi i x \xi} \mathscr{F}\nu(\xi) \, d\xi$$

tends to 0 as $T \to +\infty$. (Argue as in Problem 18, using **(14.10.16(iv))**.)

30. A probability measure μ (or its Fourier transform $\mathscr{F}\mu$) is said to be *infinitely divisible* if, for each integer $n \geqq 1$, there exists a probability measure μ_n such that $\mu = \mu_n * \mu_n * \cdots * \mu_n$ (n factors) (or equivalently $\mathscr{F}\mu = (\mathscr{F}\mu_n)^n$).
(a) If μ is infinitely divisible, then $\mathscr{F}\mu(\xi) \neq 0$ for all $\xi \in \mathbf{R}$. (Observe that $\mathscr{F}\mu(0) = \mathscr{F}\mu_n(0) = 1$ and that if I is a neighborhood of 0 in which $\mathscr{F}\mu(\xi) \neq 0$, the relation $\mathscr{F}\mu = (\mathscr{F}\mu_n)^n$ determines $\mathscr{F}\mu_n$ uniquely in I **(16.28.9)**. From $|\mathscr{F}\mu(\xi)| \leqq 1$, deduce that $\mathscr{F}\mu_n(\xi) \to 1$ as $n \to +\infty$, for all $\xi \in$ I. Complete the proof by using Section **22.10**, Problem 10(g).)

(b) It follows from (a) that there exists a continuous function Φ on $\mathbf{R}$ such that $\exp(\Phi(\xi)) = \mathscr{F}\mu(\xi)$ for all $\xi \in \mathbf{R}$ (**16.28.9**). Show that

$$\Phi(\xi) = \lim_{n\to\infty} n(\mathscr{F}\mu_n(\xi) - 1) = \lim_{n\to\infty} n \int_{-\infty}^{+\infty} (e^{-2\pi i x\xi} - 1)\, d\mu_n(x)$$

for all $\xi \in \mathbf{R}$. We may also write $\Phi(\xi) = \lim_{n\to\infty} \Phi_n(\xi)$, where

$$(1) \qquad \Phi_n(\xi) = ia_n\xi + \int_{-\infty}^{+\infty} \left(\exp(-2\pi i x\xi) - 1 + \frac{2\pi i x\xi}{1+x^2}\right)\frac{1+x^2}{x^2}\, d\nu_n(x)$$

with

$$a_n = -2\pi n \int_{-\infty}^{+\infty} \frac{x\, d\mu_n(x)}{1+x^2},$$

ν_n being the positive bounded measure with density $nx^2/(1+x^2)$ with respect to μ_n.

(c) If ν is a bounded positive measure on $\mathbf{R}$, the function

$$G(\xi) = \int_{-\infty}^{+\infty} \left(\exp(-2\pi i x\xi) - 1 + \frac{2\pi i x\xi}{1+x^2}\right)\frac{1+x^2}{x^2}\, d\nu(x)$$

is continuous on $\mathbf{R}$ (the function under the integral sign being taken equal to its limit $-2\pi^2\xi^2$ for $x = 0$). Show that if we put

$$H(\xi) = \int_0^1 (G(\xi) - \tfrac{1}{2}(G(\xi+t) + G(\xi - t)))\, dt,$$

then $H = \mathscr{F}\lambda$, where λ is the measure with density

$$\left(1 - \frac{\sin x}{x}\right)\frac{1+x^2}{x^2}$$

with respect to ν; there exist two constants $a > 0$, $b > 0$, independent of ν, such that $a\nu \leq \lambda \leq b\nu$.

(d) Deduce from (c) that in the formula (1) the measures ν_n converge vaguely to a bounded measure ν such that $\|\nu\| = \lim_{n\to\infty} \|\nu_n\|$ (cf. Section **22.10**, Problem 10(g)); the sequence (a_n) then converges to a real limit a, and we have

$$(2) \qquad \Phi(\xi) = ia\xi + \int_{-\infty}^{+\infty} \left(\exp(-2\pi i x\xi) - 1 + \frac{2\pi i x\xi}{1+x^2}\right)\frac{1+x^2}{x^2}\, d\nu(\xi).$$

(e) For each real number $a > 0$, and each probability measure μ, show that the function $\exp(a(\mathscr{F}\mu - 1))$ is the Fourier transform of an infinitely divisible probability measure. (Observe that this is the case for $(1 + (a/n)(\mathscr{F}\mu - 1))^n$ for sufficiently large n.)

(f) Deduce from (e) that for each positive bounded measure ν on $\mathbf{R}$ and each real number a, the function e^{Φ}, where Φ is given by the formula (2), is the Fourier transform of an infinitely divisible measure (*Lévy–Khintchine theorem*).

31. (a) Let g be a continuous real-valued function on $\mathbf{R}$, with support contained in $[0, 1]$, and such that $\mathscr{F}g \in \mathscr{L}^1_{\mathbf{C}}(\mathbf{R})$. For each integer $m > 1$, put

$$g_m(x) = \frac{1}{m+1} \sum_{j=-m}^{m} \left(1 - \frac{|j|}{m+1}\right) g(x - j).$$

Show that $N_1(\mathscr{F}g_m) \to 0$ as $m \to +\infty$.

(b) Show that there exists a sequence $(g^{(j)})_{j \geq 1}$ of continuous nonnegative functions on $\mathbf{R}$, with supports contained in $[0, 1]$, such that $\mathscr{F}g^{(j)} \in \mathscr{L}^1_{\mathbf{C}}(\mathbf{R})$ for all $j \geqq 1$, and $\sum_{j=1}^{\infty} g^{(j)}(x) = 1$ for $0 < x < 1$, and $= 0$ otherwise.

(c) Let (m_j) be a sequence of integers tending to $+\infty$ so rapidly that $\mathrm{N}_1(g^{(j)}_{m_j}) \leqq 2^{-j}$ for all $j \geqq 1$. Show that the function $f = \sum_{j=1}^{\infty} g^{(j)}_{m_j}$ is such that $f \in \mathscr{L}^1_{\mathbf{C}}(\mathbf{R})$, $\mathscr{F}f \in \mathscr{L}^1_{\mathbf{C}}(\mathbf{R})$, $f(n) = 0$ for all $n \in \mathbf{Z}$, $\mathscr{F}f(0) = 1$, and $\mathscr{F}f(n) = 0$ for all $n \neq 0$ in $\mathbf{Z}$. Poisson's formula **(22.14.4.5)** does not apply to f: we have $\sum_{n \in \mathbf{Z}} |f(x+n)| < +\infty$ for all $x \in \mathbf{R}$, but the function $x \mapsto \sum_{n \in \mathbf{Z}} f(x+n)$ is not continuous on $\mathbf{R}$.

18. CONVOLUTION OF TEMPERED DISTRIBUTIONS AND THE PALEY–WIENER THEOREM

(22.18.1) *If* T *is a distribution on* $\mathbf{R}^n$ *with compact support, the Fourier transform of* T *is a tempered function on* $\mathbf{R}^n$, *given by the formula*

(22.18.1.1)
$$\mathscr{F}\mathrm{T}(\xi) = \langle \mathrm{T}, \exp(-2\pi i(\cdot \mid \xi)) \rangle$$

and extends to an entire *function on* $\mathbf{C}^n$ (*called the* Fourier–Laplace transform *of* T), *defined by the same formula for all* $\xi \in \mathbf{C}^n$.

We know **(17.12.4)** that T is a finite sum of distributions of the form $\mathrm{D}^{\alpha} f$, where α is a multi-index and $f \in \mathscr{K}(\mathbf{R}^n)$. By virtue of the formulas **(22.17.5.5)** and **(22.17.5.6)**, it is enough to prove the proposition for $\mathrm{T} = f$. We have then, for all $\xi \in \mathbf{R}^n$,

$$\mathscr{F}f(\xi) = \int \exp(-2\pi i(x \mid \xi)) f(x)\, dx.$$

If we define $(x \mid \zeta) = \sum_{k=1}^{n} x_k \zeta_k$ for all $\zeta = (\zeta_k) \in \mathbf{C}^n$, it is clear that this formula makes sense when ξ is replaced by any vector $\zeta \in \mathbf{C}^n$, and that the function $\zeta \mapsto \mathscr{F}f(\zeta)$ is analytic on $\mathbf{C}^n$, because as x runs through the (compact) support of f, and ζ runs through a compact neighborhood of a point $z_0 \in \mathbf{C}^n$, the function $\exp(-2\pi i(x \mid \zeta))$ remains bounded **(13.8.6(iii))**. Finally, $\mathscr{F}f$ is bounded on $\mathbf{R}^n$, hence tempered by virtue of **(22.17.5.6)**.

Since the functions $x \mapsto x^{\nu} f(x)$ also belong to $\mathscr{K}(\mathbf{R}^n)$ for each multi-index

ν, the derivatives of the entire function $\mathscr{F}\mathrm{T}$ are given, for each $\zeta \in \mathbf{C}^n$, by the formula

(22.18.1.2) $$D^\nu \mathscr{F}\mathrm{T}(\zeta) = (-2\pi i)^{|\nu|}\langle \mathrm{T}, p_\nu \exp(-2\pi i(\cdot \mid \zeta))\rangle$$

(where p_ν is the function $x \mapsto x^\nu$).

(22.18.2) *Let $f \in \mathscr{S}(\mathbf{R}^n)$ and $g \in \mathscr{S}(\mathbf{R}^n)$. Then the functions f and g are convolvable, and we have $f * g \in \mathscr{S}(\mathbf{R}^n)$ and $\mathscr{F}(f * g) = \mathscr{F}f \cdot \mathscr{F}g$. Moreover the bilinear mapping $(f, g) \mapsto f * g$ of $\mathscr{S}(\mathbf{R}^n) \times \mathscr{S}(\mathbf{R}^n)$ into $\mathscr{S}(\mathbf{R}^n)$ is continuous.*

The functions in $\mathscr{S}$ belong to $\mathrm{L}^1 \cap \mathrm{L}^2$, hence the fact that they are convolvable and the relation $\mathscr{F}(f * g) = \mathscr{F}f \cdot \mathscr{F}g$ are particular cases of (22.8.8). Since the product of two functions in $\mathscr{S}$ is also in $\mathscr{S}$, we see that $\mathscr{F}(f * g) \in \mathscr{S}$ and consequently (22.16.10) $f * g \in \mathscr{S}$. Finally, since $f * g = \bar{\mathscr{F}}(\mathscr{F}f \cdot \mathscr{F}g)$, the continuity of $(f, g) \mapsto f * g$ follows from the continuity of the mappings $(f, g) \mapsto fg$ (22.16.9) and $\mathscr{F}$ (22.16.10).

(22.18.3) *For each $f \in \mathscr{S}(\mathbf{R}^n)$ and $\mathrm{T} \in \mathscr{E}'(\mathbf{R}^n)$, $f * \mathrm{T}$ belongs to $\mathscr{S}(\mathbf{R}^n)$, and $\mathscr{F}(f * \mathrm{T}) = \mathscr{F}f \cdot \mathscr{F}\mathrm{T}$. Moreover, if a sequence (f_k) of functions in $\mathscr{S}(\mathbf{R}^n)$ converges to 0 in this space, and if a sequence (T_k) of distributions in $\mathscr{E}'(\mathbf{R}^n)$ remains bounded in $\mathscr{D}'(\mathbf{R}^n)$ and has its supports contained in a fixed compact set, then the sequence $(f_k * \mathrm{T}_k)$ converges to 0 in $\mathscr{S}(\mathbf{R}^n)$.*

Since $D^\nu(f * \mathrm{T}) = (D^\nu f) * \mathrm{T}$ (17.11.11.2), in order to show that $f * \mathrm{T} \in \mathscr{S}$ it is enough to show that $f * \mathrm{T}$ is rapidly decreasing. For each $x \in \mathbf{R}^n$ we have (17.12.1) $(f * \mathrm{T})(x) = \langle \mathrm{T}, \gamma(x)\check{f}\rangle$; since T is compactly supported, there exists a compact neighborhood K of Supp(T), a constant $a > 0$ and an integer $s \geqq 0$ such that

$$\langle \mathrm{T}, \gamma(x)\check{f}\rangle \leqq a \cdot \sup_{y \in \mathrm{K},\ |\nu| \leq s} |D^\nu f(x + y)|$$

for all $x \in \mathbf{R}^n$ (17.7.1).

We shall make use of the following elementary inequality:

(22.18.3.1) $$1 + r^2(x + y) \leqq 2(1 + r^2(x))(1 + r^2(y))$$

for all $x, y \in \mathbf{R}^n$. For we have $r(x + y) \leqq r(x) + r(y)$, hence

$$1 + r^2(x + y) \leqq 1 + r^2(x) + r^2(y) + 2r(x)r(y),$$

and the inequality follows now from the fact that $2r(x)r(y) \leqq r^2(x) + r^2(y)$.

Replacing (x, y) by $(x + y, -y)$ in this inequality, we see that

(22.18.3.2)
$$(1 + r^2(x))^m |D^\nu f(x + y)| \leqq 2(1 + r^2(y))^m |(1 + r^2(x + y))^m D^\nu f(x + y)|,$$

and since K is compact, there exists a constant $b_m > 0$ such that, for all $f \in \mathscr{S}(\mathbf{R}^n)$, we have

(22.18.3.3) $$(1 + r^2(x))^m |\langle T, \gamma(x)\check{f}\rangle| \leqq b_m \cdot q_{s,m}(f)$$

for all $x \in \mathbf{R}^n$, which establishes our assertion.

To show that $\mathscr{F}(f * T) = \mathscr{F}f \cdot \mathscr{F}T$, we note that T is the weak limit *in* $\mathscr{D}'(\mathbf{R}^n)$ of a sequence (g_k) of functions belonging to $\mathscr{D}(\mathbf{R}^n)$, with supports contained in a fixed compact set (17.12.3). It follows, first of all, that the sequence $(f * g_k)$ tends weakly to $f * T$ *in* $\mathscr{S}'(\mathbf{R}^n)$. Indeed, let $h \in \mathscr{S}(\mathbf{R}^n)$; we have $\langle f * g_k, h\rangle = \langle g_k, \check{f} * h\rangle$ by (14.10.9); since the supports of the g_k are contained in a fixed compact set, $\langle g_k, \check{f} * h\rangle$ tends to $\langle T, \check{f} * h\rangle$, and it remains to show that $\langle T, \check{f} * h\rangle = \langle f * T, h\rangle$. But h is the limit in $\mathscr{S}(\mathbf{R}^n)$ of a sequence (h_p) of functions belonging to $\mathscr{D}(\mathbf{R}^n)$, and we have $\langle T, \check{f} * h_p\rangle = \langle f * T, h_p\rangle$ by reason of (17.12.5.2); by virtue of (22.18.2), $\check{f} * h_p$ tends to $\check{f} * h$ in $\mathscr{S}(\mathbf{R}^n)$, and since T and $f * T$ are in $\mathscr{S}'(\mathbf{R}^n)$, this proves our assertion.

Because $\mathscr{F}$ is continuous on $\mathscr{S}'(\mathbf{R}^n)$, it follows that $\mathscr{F}(f * g_k)$ converges to $\mathscr{F}(f * T)$ in $\mathscr{S}'(\mathbf{R}^n)$; but by (22.18.2) we have $\mathscr{F}(f * g_k) = \mathscr{F}f \cdot \mathscr{F}g_k$; finally, since the supports of the g_k lie in a fixed compact set, the sequence (g_k) also converges to T in $\mathscr{S}'(\mathbf{R}^n)$, therefore $\mathscr{F}g_k$ tends to $\mathscr{F}T$ in $\mathscr{S}'(\mathbf{R}^n)$, and $\mathscr{F}f \cdot \mathscr{F}g_k$ tends to $\mathscr{F}f \cdot \mathscr{F}T$ (22.17.3).

It remains to prove the last assertion of the proposition. The assumptions imply the existence of a compact set K, an integer $s \geqq 0$ and a constant $a > 0$ such that, for all k and all x,

$$|\langle T_k, D^\alpha(\gamma(x)\check{f}_k)\rangle| \leqq a \cdot \sup_{y \in K, |\nu| \leq s} |D^{\alpha+\nu} f_k(x + y)|$$

by virtue of (17.8.5). The same argument as at the beginning of the proof, using (22.18.3.1), allows us to conclude that for each pair of integers $t \geqq 0$, $m \geqq 0$, there exists a constant $b_{t,m} > 0$ such that

$$q_{t,m}(f_k * T_k) \leqq b_{t,m} q_{s+t,m}(f_k)$$

for all k, and this inequality completes the proof.

(22.18.4) *If* $f \in \mathscr{D}(\mathbf{R}^n)$ *and* $S \in \mathscr{S}'(\mathbf{R}^n)$, *the function* $f * S$ *is tempered, and* $\mathscr{F}(f * S) = \mathscr{F}f \cdot \mathscr{F}S$.

Since $D^\nu(f * S) = (D^\nu f) * S$, in order to show that $f * S$ is a tempered function, it is enough to show that it is slowly increasing. By (17.12.1) we have $(f * S)(x) = \langle S, \gamma(x)\check{f}\rangle$, hence there exist integers s, m and a constant $a > 0$ such that $|(f * S)(x)| \leqq a \cdot q_{s,m}(\gamma(x)\check{f})$ for all functions $f \in \mathscr{D}(\mathbf{R}^n)$. But it follows immediately from (22.18.3.2), with x and y interchanged, that $q_{s,m}(\gamma(x)\check{f}) \leqq 2(1 + r^2(x))^m q_{s,m}(f)$, whence

(22.18.4.1) $$|(f * S)(x)| \leqq 2a(1 + r^2(x))^m q_{s,m}(f).$$

To prove that $\mathscr{F}(f * S) = \mathscr{F}f \cdot \mathscr{F}S$, observe this time that S is the weak limit *in* $\mathscr{S}'(\mathbf{R}^n)$ of a sequence (g_k) of functions in $\mathscr{D}(\mathbf{R}^n)$ (22.17.3). Let us first show that the sequence $(f * g_k)$ converges weakly to $f * S$ in $\mathscr{S}'(\mathbf{R}^n)$: if $h \in \mathscr{S}(\mathbf{R}^n)$, we have $\langle f * g_k, h\rangle = \langle g_k, \check{f} * h\rangle$ by (14.10.9), and $\langle g_k, \check{f} * h\rangle$ converges to $\langle S, \check{f} * h\rangle$. Next it is necessary to show that $\langle S, \check{f} * h\rangle = \langle f * S, h\rangle$; now h is the limit in $\mathscr{S}(\mathbf{R}^n)$ of a sequence (h_p) of functions in $\mathscr{D}(\mathbf{R}^n)$ (22.16.7), and we have $\langle S, \check{f} * h_p\rangle = \langle f * S, h_p\rangle$ by (17.12.5.2); since $\check{f} * h_p$ tends to $\check{f} * h$ in $\mathscr{S}(\mathbf{R}^n)$ (22.18.2), this proves our assertion.

Since $\mathscr{F}$ is continuous on $\mathscr{S}'(\mathbf{R}^n)$, $\mathscr{F}(f * g_k)$ tends to $\mathscr{F}(f * S)$, and we have $\mathscr{F}(f * g_k) = \mathscr{F}f \cdot \mathscr{F}g_k$; on the other hand $\mathscr{F}g_k$ tends to $\mathscr{F}S$ in $\mathscr{S}'(\mathbf{R}^n)$, and $\mathscr{F}f \cdot \mathscr{F}g_k$ tends to $\mathscr{F}f \cdot \mathscr{F}S$ (22.17.3), which completes the proof.

(22.18.5) *If* $T \in \mathscr{E}'(\mathbf{R}^n)$ *and* $S \in \mathscr{S}'(\mathbf{R}^n)$, *then* $T * S \in \mathscr{S}'(\mathbf{R}^n)$ *and* $\mathscr{F}(T * S) = \mathscr{F}T \cdot \mathscr{F}S$ (the product is defined because $\mathscr{F}T$ is a tempered function by (22.18.1)).

For each function $g \in \mathscr{D}(\mathbf{R}^n)$, we have $\langle T * S, g\rangle = \langle T, \check{S} * g\rangle$ by (17.12.5.2). On the other hand, there exists an integer $p \geqq 0$, a constant $b > 0$ and a compact subset K of $\mathbf{R}^n$ such that

$$|\langle T, h\rangle| \leqq b \cdot \sup_{x \in K, |\nu| \leq p} |D^\nu h(x)|$$

for all $h \in \mathscr{E}(\mathbf{R}^n)$ (17.7.1). Now $D^\nu(\check{S} * g) = \check{S} * D^\nu(g)$; if we apply (22.18.4.1), with S replaced by $\check{S}$, and f by each of the derivatives $D^\nu g$ such that $|\nu| \leqq p$, we see that there exist two integers s, m and a constant c such that

$$|\langle T * S, g\rangle| \leqq c \cdot q_{s+p,m}(g)$$

for each function $g \in \mathscr{D}(\mathbf{R}^n)$, which proves that $T * S$ is tempered.

To prove that $\mathscr{F}(T * S) = \mathscr{F}T \cdot \mathscr{F}S$, we remark that S is the weak limit in $\mathscr{S}'(\mathbf{R}^n)$ of a sequence (f_k) of functions in $\mathscr{D}(\mathbf{R}^n)$ (22.17.3). We shall first show that the sequence $(T * f_k)$ converges weakly to $T * S$ in $\mathscr{S}'(\mathbf{R}^n)$. For each function $h \in \mathscr{S}(\mathbf{R}^n)$, we have $\langle T * f_k, h\rangle = \langle f_k, \check{T} * h\rangle$, and since

$\check{T} * h \in \mathscr{S}(\mathbf{R}^n)$ by **(22.18.3)**, it follows that $\langle T * f_k, h \rangle$ tends to $\langle S, \check{T} * h \rangle$. It remains therefore to show that $\langle S, \check{T} * h \rangle = \langle T * S, h \rangle$; now h is the limit in $\mathscr{S}(\mathbf{R}^n)$ of a sequence (h_p) of functions in $\mathscr{D}(\mathbf{R}^n)$ **(22.16.7)**, and we have $\langle S, \check{T} * h_p \rangle = \langle T * S, h_p \rangle$ by **(17.12.5.2)**; but $\check{T} * h_p$ tends to $\check{T} * h$ in $\mathscr{S}(\mathbf{R}^n)$ by **(22.18.3)**, therefore $\langle S, \check{T} * h_p \rangle$ tends to $\langle S, \check{T} * h \rangle$, and $\langle T * S, h_p \rangle$ tends to $\langle T * S, h \rangle$ because $T * S \in \mathscr{S}'(\mathbf{R}^n)$.

This being so, we have $\mathscr{F}(T * f_k) = \mathscr{F}T \cdot \mathscr{F}f_k$ by **(22.18.3)**, and $\mathscr{F}(T * f_k)$ tends to $\mathscr{F}(T * S)$; on the other hand, $\mathscr{F}f_k$ tends to $\mathscr{F}S$ in $\mathscr{S}'(\mathbf{R}^n)$, and $\mathscr{F}T \cdot \mathscr{F}f_k$ tends to $\mathscr{F}T \cdot \mathscr{F}S$ **(22.17.3)**, which completes the proof.

(22.18.6) We shall now see that it is possible to *characterize* the Fourier transforms of distributions with compact support and C^∞-functions with compact support, by properties which do not depend on the distributions or functions of which they are the transforms. For each point $z = (z_k)_{1 \le k \le n} \in \mathbf{C}^n$ we define $|z| = \left(\sum_{k=1}^{n} |z_k|^2 \right)^{1/2}$, and we denote by $\mathscr{I}z$ the vector $(\mathscr{I}z_k)_{1 \le k \le n} \in \mathbf{R}^n$.

(22.18.7) (Paley–Wiener theorem) (i) *In order that an entire function* $\zeta \mapsto u(\zeta)$ *on* $\mathbf{C}^n$ *should be the Fourier–Laplace transform of a distribution on* $\mathbf{R}^n$ *with support contained in the closed ball* $B'(0; A)$, *it is necessary and sufficient that* there should exist an integer $N \geqq 0$ *and a constant* $C \geqq 0$ *such that*

(22.18.7.1)
$$|u(\zeta)| \leqq C(1 + |\zeta|)^N \exp(2\pi A |\mathscr{I}\zeta|)$$

for all $\zeta \in \mathbf{C}^n$.

(ii) *In order that an entire function* $\zeta \mapsto u(\zeta)$ *on* $\mathbf{C}^n$ *should be the Fourier–Laplace transform of a* C^∞*-function, with support contained in the closed ball* $B'(0; A)$, *it is necessary and sufficient that* for each integer $N \geqq 0$ *there should exist a constant* $C_N \geqq 0$ *such that*

(22.18.7.2)
$$|u(\zeta)| \leqq C_N(1 + |\zeta|)^{-N} \exp(2\pi A |\mathscr{I}\zeta|)$$

for all $\zeta \in \mathbf{C}^n$.

(I) *Necessity of* (i) *and* (ii). For a function $f \in \mathscr{D}(\mathbf{R}^n)$, the right-hand side of the formula **(22.16.10.1)** makes sense for any *complex* vector $\zeta \in \mathbf{C}^n$, as follows from the proof of **(22.18.1)**, and the rules for differentiating under the integral sign **(13.8.6)** show that for the entire function $u = \mathscr{F}f$ (extended to $\mathbf{C}^n$) we have

(22.18.7.3)
$$\zeta^\beta D^\alpha u(\zeta) = (2\pi i)^{|\alpha| - |\beta|} \int_{\mathbf{R}^n} \exp(-2\pi i(x | \zeta)) D^\beta((-x)^\alpha f(x))\, dx.$$

If $\operatorname{Supp}(f) \subset B'(0; A)$, we may replace $\mathbf{R}^n$ by $B'(0; A)$ in the integral. Now for $x \in B'(0; A)$ and $\zeta \in \mathbf{C}^n$, we have $\mathscr{R}(-2\pi i(x|\zeta)) \leqq 2\pi A|\mathscr{I}\zeta|$ by Cauchy–Schwarz; hence we have only to apply **(22.18.7.3)** for $\alpha = 0$ and $|\beta| \leqq N$ to obtain **(22.18.7.2)**.

We shall now prove that condition (i) is necessary. If $T \in \mathscr{E}'(\mathbf{R}^n)$, there exists an integer N and a constant $a \geqq 0$ such that

$$|\langle T, g\rangle| \leqq a \cdot \sup_{|\nu| \leq N,\ x \in \mathbf{R}^n} |D^\nu g(x)|$$

for all functions $g \in \mathscr{D}(\mathbf{R}^n)$, by virtue of **(17.7.1)**. Let h be a C^∞-function on $\mathbf{R}$ which takes its values in $[0, 1]$, is equal to 1 on $]-\infty, \frac{1}{2}]$ and is equal to 0 on $[1, +\infty[$ **(16.4.3)**. Also let

$$g_\zeta(x) = \exp(-2\pi i(x|\zeta))h(|\zeta|(|x| - A))$$

for all $\zeta \in \mathbf{C}^n$, so that $g_\zeta \in \mathscr{D}(\mathbf{R}^n)$ and agrees with the function $x \mapsto \exp(-2\pi i(x|\zeta))$ on a neighborhood of $B'(0; A)$ (depending on ζ). If u is the Fourier–Laplace transform of T, we have

$$|u(\zeta)| = |\langle T, g_\zeta\rangle| \leqq a \cdot \sup_{|\nu| \leq N,\ x \in \mathbf{R}^n} |D^\nu g_\zeta(x)|.$$

But by the definition of h, for each x in the support of g_ζ we have $|x| \leqq A + |\zeta|^{-1}$, and therefore

$$|\exp(-2\pi i(x|\zeta))| \leqq \exp(2\pi A(|\mathscr{I}\zeta| + 1)).$$

The inequality **(22.18.7.1)**, for a suitable value of C, now follows from Leibniz's formula.

(II) *Sufficiency of* (i) *and* (ii). Let u be an entire function on $\mathbf{C}^n$, satisfying the conditions of (ii). Then the restriction of u to $\mathbf{R}^n$ is a rapidly decreasing function **(22.16.4)**, and it follows therefore from **(22.16.2)** that its Fourier cotransform

$$f(x) = \int_{\mathbf{R}^n} \exp(2\pi i(x|\xi))u(\xi)\, d\xi$$

is defined and is a C^∞-function. We shall show that the inequalities **(22.18.7.2)** enable us to write, for *all* vectors $\eta \in \mathbf{R}^n$,

(22.18.7.4) $$f(x) = \int_{\mathbf{R}^n} \exp(2\pi i(x|\xi + i\eta))u(\xi + i\eta)\, d\xi;$$

this integral makes sense because, by virtue of **(22.18.7.2)**, the function $\xi \mapsto u(\xi + i\eta)$ is again rapidly decreasing. To prove **(22.18.7.4)**, we argue by induction on n. Write $\zeta = (\zeta_1, \ldots, \zeta_{n-1}, \zeta_n) \in \mathbf{C}^n$ in the form $\zeta = (\zeta', \zeta_n)$, with

$\zeta' = (\zeta_1, \ldots, \zeta_{n-1}) \in \mathbf{C}^{n-1}$. By virtue of the Lebesgue–Fubini theorem, it will be enough to show that the function

(22.18.7.5) $$v(\zeta') = \int_{-\infty}^{+\infty} \exp(2\pi i x_n \xi_n) u(\zeta', \xi_n)\, d\xi_n$$

is also given by the formula

(22.18.7.6) $$v(\zeta') = \int_{-\infty}^{+\infty} \exp(2\pi i x_n(\xi_n + i\eta_n)) u(\zeta', \xi_n + i\eta_n)\, d\xi_n$$

and satisfies the condition (ii) in $\mathbf{C}^{n-1}$. This last point is an immediate consequence of (22.18.7.5): for $\zeta = (\zeta', \zeta_n)$, it is clear that

$$(1 + |\zeta|)^2 \geqq (1 + |\zeta'|)(1 + |\xi_n|),$$

and therefore for each integer $N \geqq 1$

$$|v(\zeta')| \leqq C_{2N}(1 + |\zeta'|)^{-2N} \exp(2\pi A |\mathscr{I}\zeta'|) \int_{-\infty}^{+\infty} \frac{d\xi_n}{(1 + |\xi_n|)^2}.$$

To show that the right-hand sides of (22.18.7.5) and (22.18.7.6) are equal, we apply Cauchy's theorem to the entire function $z \mapsto \exp(2\pi i x_n z) u(\zeta', z)$ and the circuit whose image is the rectangle with vertices $\pm R$, $\pm R + i\eta_n$ (9.6.3). Then it is enough to show that the two integrals

$$\int_0^{\eta_n} \exp(2\pi i x_n(\pm R + it)) u(\zeta', \pm R + it)\, dt$$

tend to 0 with $1/R$. But it follows directly from (22.18.7.2) that they are bounded above in absolute value by $C(1 + R)^{-N}$ for some $N > 0$ and a constant C depending only on N and η_n; hence the result.

This being so, by majorizing the integral on the right-hand side of (22.18.7.4) by means of (22.18.7.2) (taking $N = n + 1$), we obtain

$$|f(x)| \leqq C_N \exp(2\pi(A|\eta| - (x|\eta))),$$

the integral being convergent (16.24.9.6). Since this relation is valid for all $\eta \in \mathbf{R}^n$, we may substitute $\eta = tx$, where $t > 0$; letting t tend to $+\infty$, it follows that $f(x) = 0$ if $|x| > A$.

We come now to the proof of sufficiency of (i). Since by virtue of (22.18.7.1) the restriction of u to $\mathbf{R}^n$ is a slowly increasing function, its Fourier cotransform $T = \overline{\mathscr{F}} u$ is a tempered distribution, and we have $u | \mathbf{R}^n = \mathscr{F} T$ (22.17.2). For each $\varepsilon > 0$, let g_ε be a nonnegative function of class C^∞ on $\mathbf{R}^n$, with support contained in the ball $B'(0; \varepsilon)$ and such that

$\int g_\varepsilon(x)\,dx = 1$. If v_ε is the Fourier–Laplace transform of g_ε, then for each integer $N \geqq 0$ there exists a constant C_N such that

$$|v_\varepsilon(\zeta)| \leqq C_N(1 + |\zeta|)^{-N} \exp(2\pi\varepsilon|\mathscr{I}\zeta|).$$

On the other hand, we have $\mathscr{F}(T * g_\varepsilon) = \mathscr{F}T \cdot \mathscr{F}g_\varepsilon$ **(22.18.4)**, which is therefore the restriction to $\mathbf{R}^n$ of the entire function uv_ε, and the hypothesis **(22.18.7.1)** implies that

$$|u(\zeta)v_\varepsilon(\zeta)| \leqq CC_{N'}(1 + |\zeta|)^{-(N'-N)} \exp(2\pi(A + \varepsilon)|\mathscr{I}\zeta|)$$

for all integers $N' > N$. By virtue of the sufficiency of condition (ii), it follows that $\operatorname{Supp}(T * g_\varepsilon)$ is contained in the ball $B'(0; A + \varepsilon)$ for all $\varepsilon > 0$. We shall show that this implies that the support of T is contained in $B'(0; A)$. Let U be an open set in $\mathbf{R}^n$ disjoint from $B'(0; A)$, and let $f \in \mathscr{D}(\mathbf{R}^n)$ be a function with support contained in U; for sufficiently small ε, $U \cap B'(0; A + \varepsilon) = \varnothing$, and consequently $\langle T * g_\varepsilon, f\rangle = 0$. But since the sequence $(T * g_{1/k})$ converges weakly to T in $\mathscr{D}'(\mathbf{R}^n)$ **(17.11.9)**, we have $\langle T, f\rangle = 0$, and the proof is complete.

PROBLEMS

1. For a function $f \in \mathscr{E}(\mathbf{R}^n)$, show that the following properties are equivalent:
   (α) f is tempered;
   (β) for each function $g \in \mathscr{S}(\mathbf{R}^n)$, we have $fg \in \mathscr{S}(\mathbf{R}^n)$;
   (γ) for each distribution $T \in \mathscr{S}'(\mathbf{R}^n)$, we have $f \cdot T \in \mathscr{S}'(\mathbf{R}^n)$.

   (To show that (β) implies (α), argue by contradiction, using Section 22.16, Problem 3(b). To show that (γ) implies (α), observe that $|f|$ is then a tempered distribution, and use Section 22.17, Problem 1.)

2. (a) For each function $f \in \mathscr{S}(\mathbf{R}^n)$ and each distribution $T \in \mathscr{S}'(\mathbf{R}^n)$, the function $x \mapsto \langle T, \gamma(x)\check{f}\rangle$ is tempered, and we denote it by $T * f$ (or $f * T$). When $f \in \mathscr{D}(\mathbf{R}^n)$ or $T \in \mathscr{E}'(\mathbf{R}^n)$ (so that T and f are strictly convolvable), the function so defined coincides with the convolution product of T and f defined in **(17.1.1)**. If g is another function in $\mathscr{S}(\mathbf{R}^n)$, we have $T * (f * g) = (T * f) * g$ and $\langle T * f, g\rangle = \langle T, \check{f} * g\rangle$.

   (b) If (f_k) is a sequence of functions in $\mathscr{S}(\mathbf{R}^n)$, and (T_k) a sequence of distributions in $\mathscr{S}'(\mathbf{R}^n)$, the sequence $(T_k * f_k)$ tends weakly to 0 in $\mathscr{S}'(\mathbf{R}^n)$ in each of the following two cases: (1) the sequence (f_k) is bounded in $\mathscr{S}(\mathbf{R}^n)$ and the sequence (T_k) tends weakly to 0 in $\mathscr{S}'(\mathbf{R}^n)$; (2) the sequence (f_k) tends to 0 in $\mathscr{S}(\mathbf{R}^n)$ and the sequence (T_k) is weakly bounded in $\mathscr{S}'(\mathbf{R}^n)$. (Use Section 22.16, Problem 4.)

   (c) Let f be a tempered function, and let h be a function belonging to $\mathscr{D}(\mathbf{R}^n)$ which is equal to 1 on some neighborhood of 0. Put $f_k(x) = f(x)h(x/k)$, so that $f_k \in \mathscr{D}(\mathbf{R}^n)$. Show that, for each function $g \in \mathscr{S}(\mathbf{R}^n)$, the sequence $(f_k g)$ converges to fg in $\mathscr{S}(\mathbf{R}^n)$. Deduce that the tempered function $(\bar{\mathscr{F}}f) * g$ belongs to $\mathscr{S}(\mathbf{R}^n)$. (Observe that the sequence $((\bar{\mathscr{F}}f_k) * g)$ is a Cauchy sequence in $\mathscr{S}(\mathbf{R}^n)$.) The mapping $g \mapsto (\bar{\mathscr{F}}f) * g$ of $\mathscr{S}(\mathbf{R}^n)$ into itself is continuous.

(d) With the same notation and hypotheses, for each distribution $T \in \mathcal{S}'(\mathbf{R}^n)$ the mapping $g \mapsto \langle T, (\mathcal{F}\check{f}) * g\rangle$ is a continuous linear form on $\mathcal{S}(\mathbf{R}^n)$, hence is a tempered distribution which we denote by $T * (\mathcal{F}f)$ or $(\mathcal{F}f) * T$. We have $\mathcal{F}(f \cdot T) = (\mathcal{F}f) * (\mathcal{F}T)$. When $f \in \mathcal{S}(\mathbf{R}^n)$, this definition agrees with (a) above. The distributions of the form $\mathcal{F}f$, where f is a tempered function, are called *declining* distributions. Every distribution S with compact support is declining, and the above definition of $T * S$ then agrees with that in (17.11.1).

If R and S are two declining distributions, then $R * S$ is declining, and $(T * R) * S = T * (R * S)$ for all tempered distributions T. Every declining function is a declining distribution. If $f \in \mathcal{S}(\mathbf{R}^n)$, and T is a tempered distribution, then $f \cdot T$ is a declining distribution.

3. (a) Show by an example that if μ is a tempered measure (when regarded as a distribution) on $\mathbf{R}^n$, and $f \in \mathcal{L}_{\mathbf{C}}^{\infty}(\mathbf{R}^n)$, the measure $f \cdot \mu$ need not be tempered.
(b) Let μ be a slowly increasing measure on $\mathbf{R}^n$, and let f be a locally integrable function on $\mathbf{R}^n$ such that $(1 + r^2)^{-m} f$ is integrable for sufficiently large m. Then $f \cdot \mu$ is a slowly increasing measure on $\mathbf{R}^n$. If $h \in \mathcal{D}(\mathbf{R}^n)$ is equal to 1 on some neighborhood of 0, and if $f_k(x) = f(x)h(x/k)$, then the sequence of distributions $(\mathcal{F}f_k) * (\mathcal{F}\mu)$ converges weakly in $\mathcal{S}'(\mathbf{R}^n)$ to $(f \cdot \mu)$, and we denote this limit by $(\mathcal{F}f) * (\mathcal{F}\mu)$. If $g \in \mathcal{L}_{\mathbf{C}}^{\infty}(\mathbf{R}^n)$, then with the definition above we have $((\mathcal{F}f) * (\mathcal{F}g)) * (\mathcal{F}\mu) = (\mathcal{F}f) * ((\mathcal{F}g) * (\mathcal{F}\mu))$. If f is a tempered function, this definition of $(\mathcal{F}f) * (\mathcal{F}\mu)$ agrees with that of Problem 2(d).

4. (a) Let T be a distribution on $\mathbf{R}^n$. Show that the set Γ (or $\Gamma(T)$) of points $\eta \in \mathbf{R}^n$ such that $\exp(2\pi(\cdot \mid \eta)) \cdot T$ is a tempered distribution, is a convex set (possibly empty). (Show that if η_j $(1 \leqq j \leqq r)$ are points of $\mathbf{R}^n$, the function

$$x \mapsto \alpha(x, \eta) = \frac{\exp(2\pi(x \mid \eta))}{\sum_{j=1}^{r} \exp(2\pi(x \mid \eta_j))}$$

is bounded on $\mathbf{R}^n$, together with all its derivatives, provided that η lies in the convex hull of the set of the η_j.)
(b) Suppose that the interior $\mathring{\Gamma}$ of Γ is not empty. Let Δ be a closed cube with center η_0 (i.e., a closed ball with respect to the norm $\sup_j |x_j|$ on $\mathbf{R}^n$) contained in the open set $\mathring{\Gamma}$. Show that there exists $\varepsilon > 0$ such that, for each $\eta \in \Delta$, the distribution

$$\exp(2\pi(\varepsilon(1 + r^2(x))^{1/2} + (x \mid \eta))) \cdot T$$

is tempered. (Observe that Δ is contained in the interior of the convex hull of the vertices of a slightly larger cube: if η_j $(1 \leqq j \leqq r)$ are these vertices (so that $r = 2^n$), construct as in (a) the function α, and show that if we put $\beta(x, \eta) = \alpha(x, \eta) \exp(2\pi\varepsilon(1 + r^2(x))^{1/2})$, the function $x \mapsto \beta(x, \eta)$ is bounded, together with all its derivatives, when $\eta \in \Delta$, provided that ε is sufficiently small.)
(c) Deduce from (b) that, for each $\eta \in \mathring{\Gamma}$, the distribution $\exp(2\pi(\cdot \mid \eta)) \cdot T$ is declining (Problem 2). (Use the fact that the function $\exp(-2\pi\varepsilon(1 + r^2)^{1/2})$ is declining, together with Problem 2.) For each $\eta \in \mathring{\Gamma}$, the distribution $\mathcal{F}(\exp(2\pi(\cdot \mid \eta)) \cdot T)$ is therefore a *tempered function*. Further, if we put

$$\mathcal{F}\mathcal{L}T(\zeta) = (\mathcal{F}(\exp(2\pi(\cdot \mid \eta)) \cdot T))(\xi)$$

for each $\zeta = \xi + i\eta$ belonging to the *tube* $B = \mathbf{R}^n \oplus i\mathring{\Gamma} \subset \mathbf{C}^n$, the function $\mathcal{F}\mathcal{L}T$ is holomorphic in $\mathbf{R}^n \oplus i\mathring{\Gamma}$, and for each compact set $K \subset \mathring{\Gamma}$ there exist $C > 0$ and $k > 0$ such that $|\mathscr{F}\mathscr{L}T(\zeta)| \leqq C(1 + r^2(\xi))^k$ for all $\zeta = \xi + i\eta \in \mathbf{R}^n \oplus iK$. (For each $\xi_0 \in \mathbf{R}^n$, let $\zeta_0 = \xi_0 + i\eta_0$. Show that when $|\zeta - \zeta_0|$ is sufficiently small we have

$$\mathscr{F}\mathscr{L}T(\zeta) = \langle \exp(2\pi(\varepsilon(1 + r^2)^{1/2} + (\cdot\,|\zeta_0))) \cdot T, \exp(-2\pi(\varepsilon(1 + r^2)^{1/2} + i(\cdot\,|\zeta - \zeta_0)))\rangle.$$

To show that $\mathscr{F}\mathscr{L}T$ is holomorphic in a neighborhood of ζ_0, argue as in **(17.10.1)**; for the majoration of $|\mathscr{F}\mathscr{L}T(\zeta)|$, argue as in **(22.18.4)**.)

The function $\mathscr{F}\mathscr{L}T$ is called the *Fourier–Laplace transform* of T. When T has a compact support, we have $\Gamma = \mathbf{R}^n$, and the preceding definition coincides with that of **(22.18.1)**. When $0 \in \mathring{\Gamma}$ (so that T is declining), the Fourier transform $\mathscr{F}T$ (which is a tempered function) is the restriction of $\mathscr{F}\mathscr{L}T$ to $\mathbf{R}^n$.

For each $\zeta_0 \in \mathbf{C}^n$, $\mathscr{F}\mathscr{L}(\exp(2\pi i(\cdot\,|\zeta_0)) \cdot T)$ is defined on the tube $B + \zeta_0$, and is equal to $\mathscr{F}\mathscr{L}T(\zeta - \zeta_0)$.

(d) Conversely, let $\mathring{\Gamma}$ be a nonempty convex open set in $\mathbf{R}^n$, and Φ a holomoorphic function defined on the tube $\mathbf{R}^n \oplus i\mathring{\Gamma}$. Suppose that, for each compact $K \subset \mathring{\Gamma}$, there exist $C > 0$ and $k \geqq 0$ such that $|\Phi(\zeta)| \leqq C(1 + r^2(\xi))^k$ for all $\zeta = \xi + i\eta \in \mathbf{R}^n \oplus iK$. Then there exists one and only one distribution T such that $\exp(2\pi(\cdot\,|\eta)) \cdot T$ is a tempered distribution for each $\eta \in \mathring{\Gamma}$, and such that $\Phi = \mathscr{F}\mathscr{L}T$. (With the aid of **(9.9.3)**, show first that all the partial derivatives of Φ admit analogous majorations; consequently, for each $\eta \in \mathring{\Gamma}$, the function $\xi \mapsto \Phi(\xi + i\eta)$ is tempered, hence is of the form $\mathscr{F}T_\eta$, where T_η is declining. Then prove that $\exp(-2\pi(\cdot\,|\eta)) \cdot T_\eta$ is a distribution independent of $\eta \in \mathring{\Gamma}$, by using the Cauchy conditions satisfied by Φ.) Furthermore, T is of order $\leqq k + 2$.

(e) Let h be a function belonging to $\mathscr{D}(\mathbf{R}^n)$ which is equal to 1 in some neighborhood of 0, and let $h_k(x) = h(x/k)$ for each integer $k \geqq 1$. Show that $\mathscr{F}\mathscr{L}(h_k \cdot T)$ tends uniformly to $\mathscr{F}\mathscr{L}T$ on each compact set contained in the tube $\mathbf{R}^n \oplus i\mathring{\Gamma}$, so that we may write

$$\mathscr{F}\mathscr{L}T(\zeta) = \lim_{k \to \infty} \langle T, h_k \exp(-2\pi i(\cdot\,|\zeta))\rangle.$$

The function $p \mapsto \mathscr{L}T(p) = \mathscr{F}\mathscr{L}T((1/2\pi i)p)$, which is holomorphic on $-2\pi\mathring{\Gamma} \oplus i\mathbf{R}^n$, is called the *Laplace transform* of T, and is denoted by $\mathscr{L}T$.

If T is a distribution on the open interval $]0, +\infty[$ in $\mathbf{R}$, and if S is its image under the diffeomorphism $u: t \mapsto \log t$ of this open set onto $\mathbf{R}$, the analytic function $s \mapsto \mathscr{L}S(1 - s)$ is called the *Mellin transform* of T, denoted by $\mathscr{M}T$.

When T is a locally integrable function f, the Laplace transform is defined by

$$\mathscr{L}f(p) = \int e^{-(x|p)} f(x)\, dx$$

and (when $n = 1$) the Mellin transform is defined by

$$\mathscr{M}f(s) = \int_0^\infty x^{s-1} f(x)\, dx,$$

where p and s are such that the integrals (with respect to Lebesgue measure) are defined. When $f(x) = e^{-x}$, $\mathscr{M}f$ is the gamma function $s \mapsto \Gamma(s)$ (defined for $\mathscr{R}s > 0$). When $f(x) = F(a, b, c; -x)$, where F is the hypergeometric function (Section **17.11**, Problem 2(b)), $\mathscr{M}f$ is the function

$$s \mapsto \frac{\Gamma(c)}{\Gamma(a)\Gamma(b)} \cdot \frac{\Gamma(s)\Gamma(a - s)\Gamma(b - s)}{\Gamma(c - s)}$$

(defined for $0 < \mathscr{R}s < \inf(\mathscr{R}a, \mathscr{R}b)$).

5. Let S, T be two distributions on $\mathbf{R}^n$ such that $\Gamma(S) \cap \Gamma(T)$ (Problem 4) has a nonempty interior. Show that there exists a unique distribution U such that $\mathring{\Gamma}(U) \supset \mathring{\Gamma}(S) \cap \mathring{\Gamma}(T)$ and such that

$$\exp(2\pi(\cdot\,|\,\eta)) \cdot U = (\exp(2\pi(\cdot\,|\,\eta)) \cdot S) * (\exp(2\pi(\cdot\,|\,\eta)) \cdot T)$$

for all $\eta \in \mathring{\Gamma}(S) \cap \mathring{\Gamma}(T)$ (cf. Problems 2(d) and 4(e)). If S or T has compact support, then $U = S * T$; in the general case we put $U = S * T$ by definition, and we have $\mathscr{F}\mathscr{L}(S * T) = (\mathscr{F}\mathscr{L}S)(\mathscr{F}\mathscr{L}T)$.

6. (a) Suppose that T is a tempered distribution whose support is contained in a *closed convex cone* C with vertex 0 (i.e., a closed set C such that the relations $x \in C$, $y \in C$, $\lambda \geqq 0$, $\mu \geqq 0$, imply $\lambda x + \mu y \in C$), not the whole of $\mathbf{R}^n$. The set C^* of points $\eta \in \mathbf{R}^n$ such that $(x\,|\,\eta) \leqq 0$ for all $x \in C$ is then a closed convex cone, not consisting of the origin alone, and we have $(C^*)^* = C$. (Use Section 12.15, Problem 4.) Show that the set $\Gamma(T)$ contains C^*.
(b) Suppose that the interior of C^* is nonempty. If a function $f \in \mathscr{S}(\mathbf{R}^n)$ has its support contained in C, its Fourier transform extends to a continuous function u defined on $\mathbf{R}^n \oplus iC^*$, holomorphic in the interior of $\mathbf{R}^n \oplus iC^*$ and such that *for each* integer $N > 0$ there exists a constant $A_N > 0$ for which $|u(\zeta)| \leqq A_N(1 + |\zeta|)^{-N}$ for all $\zeta \in \mathbf{R}^n \oplus iC^*$. Conversely, if u is holomorphic and satisfies this condition in the interior of $\mathbf{R}^n \oplus iC^*$, there exists a function $f \in \mathscr{E}(\mathbf{R}^n)$, with support contained in C, such that the function $x \mapsto \exp(2\pi(x\,|\,\eta))f(x)$ belongs to $\mathscr{S}(\mathbf{R}^n)$ for all $\eta \in \mathring{C}^*$, and such that u is equal, in the interior of $\mathbf{R}^n \oplus iC^*$, to the Fourier–Laplace transform of f. (Argue as in the proof of (22.18.7)(ii)).)
(c) Under the same hypotheses, if T is a tempered distribution with support contained in C, its Fourier–Laplace transform u is holomorphic in the interior of $\mathbf{R}^n \oplus iC^*$, and such that *there exists* an integer $N > 0$ and a constant $A > 0$ for which $|u(\zeta)| \leqq A(1 + |\zeta|)^N$ for all ζ in the interior of $\mathbf{R}^n \oplus iC^*$. Conversely, if u is holomorphic and satisfies this condition in the interior of $\mathbf{R}^n \oplus iC^*$, there exists a distribution T on $\mathbf{R}^n$, with support contained in C, such that $\exp(2\pi(\cdot\,|\,\eta)) \cdot T$ is tempered for all $\eta \in \mathring{C}^*$, and such that u is equal, in the interior of $\mathbf{R}^n \oplus iC^*$, to the Fourier–Laplace transform of T. (Argue as in the proof of (22.18.7(i)).)
(d) Show that if T is tempered, and $\mathrm{Supp}(T) \subset C$, the function $\xi \mapsto \mathscr{F}\mathscr{L}T(\xi + i\eta)$, considered as a tempered distribution, tends to $\mathscr{F}T$ *in the space* $\mathscr{S}'(\mathbf{R}^n)$ as η tends to 0 whilst remaining in the interior of C^* (cf. Problem 7(d)).
(e) Generalize the preceding results to the situation where the interior of C^* is empty. Show that by means of a linear transformation of $\mathbf{R}^n$, we may assume that C^* is contained in a vector subspace spanned by $p < n$ vectors of the canonical basis of $\mathbf{R}^n$, and that the interior of C^* with respect to this subspace is not empty.

7. With the same notation as in Problem 6, suppose that both C and C^* have nonempty interiors. The *Cauchy kernel* (relative to C) is defined to be the Fourier–Laplace transform $K(\zeta) = \mathscr{F}\mathscr{L}\varphi_C(\zeta)$, where the characteristic function φ_C is considered as a tempered distribution.
(a) For each η interior to C^*, put $K_\eta(\xi) = K(\xi + i\eta)$, which is a tempered function on $\mathbf{R}^n$. Show that K_η and all its derivatives belong to all the spaces $\mathscr{L}^p_{\mathbf{C}}(\mathbf{R}^n)$ for $2 \leqq p \leqq +\infty$. (Observe that the hypothesis on C implies that, for each η interior to C^*, there exists a number $a_\eta > 0$ such that $(x\,|\,\eta) \leqq -a_\eta\|x\|$ for all $x \in C$, by using the compactness of $\mathbf{S}_{n-1}$.)
(b) For each distribution T which is the sum of a tempered measure with support contained in C, and a tempered distribution with support contained in the interior $\mathring{C}$, we

may define the convolution $K_\eta * (\mathscr{F}T)$ to be the function $\xi \mapsto \langle T, \mathscr{F}(\varepsilon_\xi * K_\eta)\rangle$, and then we have

$$(K_\eta * \mathscr{F}T)(\xi) = \mathscr{F}\mathscr{L}T(\xi + i\eta) \qquad (\text{"Cauchy's formula"})$$

for $\xi + i\eta$ interior to $\mathbf{R}^n \oplus iC^*$. This definition coincides with the usual one when $\mathscr{F}T$ is a function belonging to $\mathscr{L}^1_C(\mathbf{R}^n)$ or $\mathscr{L}^2_C(\mathbf{R}^n)$.

(c) We have $\int |K(\xi + i\eta)|^2 \, d\xi = K(2i\eta)$ for η in the interior of C^*. The *Poisson kernel* (relative to C) is defined to be the function

$$(\xi, \eta) \mapsto P(\xi, \eta) = |K(\xi + i\eta)|^2/K(2i\eta)$$

defined in the interior of $\mathbf{R}^n \times C^*$. For each η interior to C^*, the function $\xi \mapsto P(\xi, \eta)$ is tempered, and belongs to all the spaces $\mathscr{L}^p_C(\mathbf{R}^n)$ for $1 \leqq p \leqq +\infty$. Show that for each distribution T satisfying the conditions of (b) we have

$$(P(\cdot, \eta) * \mathscr{F}T)(\xi) = \mathscr{F}\mathscr{L}T(\xi + i\eta) \qquad (\text{"Poisson's formula"}).$$

(Use (b) to express the product $K(\xi + \xi' + i\eta)\mathscr{F}\mathscr{L}T(\xi + i\eta)$ as the value of a convolution, by means of Problem 5, and choose ξ' appropriately.)

(d) Show that for each neighborhood V of 0 in $\mathbf{R}^n$, the integral $\int_{\complement V} P(\xi, \eta) \, d\xi$ tends to 0 as η tends to 0 whilst remaining in the interior of C^*. (Apply (c) by taking for T a function $f \geqq 0$ in $\mathscr{K}(\mathbf{R}^n)$, with support contained in C and such that $\int f(x) \, dx = 1$.) Deduce that, for each function $g \in \mathscr{L}^1_C(\mathbf{R}^n)$ (resp. each locally integrable function g such that $\mathscr{F}g \in \mathscr{L}^1_C(\mathbf{R}^n)$), with support contained in C, the integral

$$\int |\mathscr{F}\mathscr{L}g(\xi + i\eta) - \mathscr{F}g(\xi)|^2 \, d\xi$$

(resp. the integral

$$\int |\mathscr{F}\mathscr{L}g(\xi + i\eta) - \mathscr{F}g(\xi)| \, d\xi)$$

tends to 0 as η tends to 0 whilst remaining in the interior of C^* (cf. **(14.11.1)**). What can be said when in addition $\mathscr{F}g$ is continuous and bounded in $\mathbf{R}^n$?

(e) When $n = 1$ and C is the interval $[0, +\infty[$, we have $K(\zeta) = 1/(2\pi i\zeta)$ and $P(\xi, \eta) = -\eta/(\pi(\xi^2 + \eta^2))$ for $\eta < 0$.

8. (a) With the same condition on C as in Problem 7, let T be a distribution which is the sum of a tempered distribution with support contained in the interior of C, and of a tempered measure μ with support contained in C and such that $|\mu|(\{0\}) = 0$. Show that $K_\eta * \overline{\mathscr{F}T} = 0$ for η in the interior of C^*. (Observe that $C \cap (-C) = \{0\}$.)

(b) Put $\mathscr{F}T = U + iV$, where U and V are real tempered distributions. Show that $\mathscr{F}\mathscr{L}T(\xi + i\eta) = 2K_\eta * U$ for η in the interior of C^*. The *Hilbert distribution* (relative to C) is defined to be the tempered distribution

$$H = -i\mathscr{F}(\varphi_C - \varphi_{-C});$$

show that $V = H * U$ (we say that V is the *Hilbert transform* of U). We have $H * H = -\mathscr{F}(\varphi_C + \varphi_{-C})$. The Hilbert transformation, restricted to the space $\mathscr{L}^2_C(C \cup (-C))$, is unitary and of square $-I$. If $n = 1$ and $C = [0, +\infty[$, then $H = (1/\pi)\text{P.V.}(1/x)$ (cf. Section 22.17, Problem 4).

9. (a) Let f be a locally integrable function defined on $\mathbf{R}$. The set of $\eta \in \mathbf{R}$ such that $\exp(2\pi(\cdot|\eta))f \in \mathscr{L}^2_{\mathbf{C}}(\mathbf{R})$ is an interval $\mathrm{I} \subset \mathbf{R}$ (possibly empty). (Same method as in Problem 4.) The interval I is contained in the interval $\Gamma(f)$, and the Fourier–Laplace transform $u = \mathscr{F}\mathscr{L}f$ is holomorphic on $\mathbf{R} \oplus i\mathring{\mathrm{I}} = \mathrm{B}$; moreover, the integral

$$\int_{-\infty}^{+\infty} |u(\xi + i\eta)|^2 \, d\xi = \int_{-\infty}^{+\infty} |f(x)|^2 e^{4\eta x} \, dx$$

is a continuous function of η on $\mathring{\mathrm{I}}$, hence is bounded on each compact interval contained in $\mathring{\mathrm{I}}$.

(b) Conversely, let u be a holomorphic function on B, such that for each $\eta \in \mathring{\mathrm{I}}$ the function $\xi \mapsto u(\xi + i\eta)$ belongs to $\mathscr{L}^2_{\mathbf{C}}(\mathbf{R})$ and such that the function $\eta \mapsto \int |u(\xi + i\eta)|^2 \, d\xi$ is bounded on every compact interval contained in $\mathring{\mathrm{I}}$. Let $\mathrm{J} = [\eta_1, \eta_2]$ be a compact interval, containing more than one point, and contained in $\mathring{\mathrm{I}}$. Show that there exist $r > 0$ and $\mathrm{C} > 0$ such that

$$|u(\zeta)| \leq \mathrm{C}\left(\int_{\eta_1}^{\eta_2} dt \int_{\xi - r}^{\xi + r} |u(s + it)|^2 \, ds\right)^{1/2}$$

for all $\zeta = \xi + i\eta \in \mathbf{R} + i\mathrm{J}$ (use Section 9.3, Problem 6, and the Cauchy–Schwarz inequality). Deduce first that u is bounded on $\mathbf{R} + i\mathrm{J}$, and then that for each $\eta \in \mathrm{J}$ the function $\xi \mapsto u(\xi + i\eta)$ tends to 0 as ξ tends to $\pm\infty$, uniformly in $\eta \in \mathrm{J}$. (Use the dominated convergence theorem.)

(c) For each $\eta \in \mathring{\mathrm{I}}$, let f_η be the Fourier cotransform of the function $\xi \mapsto u(\xi + i\eta)$, which belongs to $\mathscr{L}^2_{\mathbf{C}}(\mathbf{R})$. Show that

$$f_{\eta_1} \exp(-2\pi\eta_1 x) = f_{\eta_2} \exp(-2\pi\eta_2 x)$$

for any two points $\eta_1, \eta_2 \in \mathring{\mathrm{I}}$, (Apply Cauchy's theorem to the function $\zeta \mapsto u(\zeta)\exp(2\pi i x\zeta)$, integrated around a suitable rectangle.) Hence there exists a function $f \in \mathscr{L}^2_{\mathbf{C}}(\mathbf{R})$ such that $u = \mathscr{F}\mathscr{L}f$.

(d) Let u be a holomorphic function on the lower half-plane $\mathscr{I}\zeta < 0$. In order that $u = \mathscr{F}\mathscr{L}f$, where $f \in \mathscr{L}^2_{\mathbf{C}}(\mathbf{R})$ and $f(x) = 0$ for all $x < 0$, it is necessary and sufficient that for each $\eta < 0$ the function $\xi \mapsto u(\xi + i\eta)$ should belong to $\mathscr{L}^2_{\mathbf{C}}(\mathbf{R})$ and that the function $\eta \mapsto \int |u(\xi + i\eta)|^2 \, d\xi$ should be bounded for $-\infty < \eta < 0$. (Use (b) and Problem 6(c).)

(e) Let $f \in \mathscr{L}^2_{\mathbf{C}}(\mathbf{R})$; for each $\eta < 0$, put

$$u(\xi + i\eta) = (\mathrm{P}(\cdot, \eta) * f)(\xi) = \int f(x)\mathrm{P}(\xi - x, \eta) \, dx.$$

If u is holomorphic on $\mathscr{I}\zeta < 0$, then u satisfies the conditions in (d).

(f) Generalize these results to $\mathbf{R}^n$, $n \geq 2$.

10. Give an example of an entire function f on $\mathbf{C}$ whose restriction to $\mathbf{R}$ is bounded (and hence is a tempered distribution) and which is such that $|f(it)| \geq \varphi(t)$ for all sufficiently large real $t > 0$, where φ is an *arbitrary* positive, increasing, real-valued function. (Take $f(z) = e^{ig(z)}$, where g is real on the real axis, and use Section 9.16, Problem 4.)

11. (a) Let S be the set of complex numbers $z = re^{i\theta}$ such that $0 \leqq \theta \leqq \pi/\alpha$, where $\alpha > \frac{1}{2}$. Let f be a holomorphic function defined on an open neighborhood of S, and suppose that

(1) $|f(z)| \leqq A \exp(|z|^{\beta})$ for all $z \in S$, where $A > 0$ and $0 \leqq \beta < \alpha$;

(2) $|f(z)| \leqq M$ for $z = r \geqq 0$ and for $z = re^{i\pi/\alpha}$, $r \geqq 0$. Then $|f(z)| \leqq M$ for all $z \in S$. (If $\beta < \gamma < \alpha$, consider the function $g(z) = \exp(z^{\gamma})$; reduce S to a relatively compact set by means of a conformal transformation, and apply the Phragmen–Lindelöf principle (Section 9.5, Problem 16).)

(b) With the notation of **(22.18.6)**, let u be an entire function on $\mathbf{C}^n$ such that for each $\varepsilon > 0$ there exists a constant $B_\varepsilon > 0$ for which $|u(z)| \leqq B_\varepsilon \exp((A + \varepsilon)|z|)$, where A is a given positive real number. Suppose also that $|u(z)| \leqq 1$ for all $z \in \mathbf{R}^n$. Show that $|u(z)| \leqq \exp(A|\mathscr{I}z|)$ for all $z \in \mathbf{C}^n$. (Reduce to the case $n = 1$, and consider the function $v_\varepsilon(z) = u(z) \exp(i(A + \varepsilon)z)$; deduce from (a), taking $\alpha = 2$, that $v_\varepsilon(z)$ is bounded for $\mathscr{I}z \geqq 0$; then apply (a) again, taking $\alpha = 1$, to show that $|v_\varepsilon(z)| \leqq 1$ for $\mathscr{I}z \geqq 0$.)

12. In order that an entire function $\zeta \mapsto u(\zeta)$ on $\mathbf{C}^n$ should be the Fourier–Laplace transform of a distribution on $\mathbf{R}^n$ with support contained in the ball $B'(0; A)$, it is necessary and sufficient that for each $\varepsilon > 0$ there should exist a constant $B_\varepsilon > 0$ such that $|u(\zeta)| \leqq B_\varepsilon \exp(2\pi(A + \varepsilon)|\zeta|)$ for all $\zeta \in \mathbf{C}^n$, and also that the restriction of u to $\mathbf{R}^n$ should be a tempered distribution (in which case it follows from **(22.18.1)** that this restriction is in fact a tempered *function*). (Consider successively the following three cases: (1) the restriction of u to $\mathbf{R}^n$ is bounded: in that case apply Problem 11(b); (2) the restriction of u to $\mathbf{R}^n$ is a tempered function: in that case multiply u by the Fourier transform of a suitable function in $\mathscr{D}(\mathbf{R}^n)$, and so reduce to the first case; (3) the general case: consider the convolution of $u|\mathbf{R}^n$ with a function of $\mathscr{D}(\mathbf{R}^n)$ whose Fourier transform does not vanish (Section 22.17, Problem 5).)

13. (a) Every tempered distribution T on $\mathbf{R}$ may be written as $T = T_1 - T_2$, where T_1 and T_2 are tempered, the support of T_1 is bounded below, and the support of T_2 is bounded above. The Fourier–Laplace transform $\mathscr{F}\mathscr{L}T_1$ (resp. $\mathscr{F}\mathscr{L}T_2$) is then defined and holomorphic on the half-plane $\mathscr{I}\zeta < 0$ (resp. $\mathscr{I}\zeta > 0$). Show that there exists an integer $N \geqq 0$ such that, for each compact interval $J \subset \mathbf{R}$, there is a constant A_J with the property that $|\mathscr{F}\mathscr{L}T_1(\xi + i\eta)| \leqq A_J \cdot |\eta|^{-N}$ for $\xi \in J$ and $\eta < 0$ (resp. $|\mathscr{F}\mathscr{L}T_2(\xi + i\eta)| \leqq A_J \cdot |\eta|^{-N}$ for $\xi \in J$ and $\eta > 0$). (Reduce to the case where T is a tempered function, by replacing T by $T * g$, where $g \in \mathscr{D}(\mathbf{R})$ is such that $\mathscr{F}g$ never vanishes (Section 22.17, Problem 5).)

(b) Let I be a bounded open interval in $\mathbf{R}$ and let $D \subset \mathbf{C}$ be the open disc on I as diameter; let (u_n) be a sequence of functions holomorphic on D, and such that there exist $A > 0$ and $N > 0$ for which $|u_n(\xi + i\eta)| \leqq A \cdot |\eta|^{-N}$ for all $\zeta = \xi + i\eta$ in D and all integers $n \geqq 1$. Suppose also that the sequence $(u_n(\zeta))$ converges at each point $\zeta \in D$ *not belonging to* I. Under these conditions, prove that the sequence $(u_n(\zeta))$ converges for *all* $\zeta \in D$, and that its limit is holomorphic in D (*Carleman's principle*). (Consider a disc D_1 concentric with D and of smaller radius; if α and β are the ends of the diameter of D_1 contained in the real line $\mathbf{R}$, consider the functions $(\zeta - \alpha)^N(\zeta - \beta)^N u_n(\zeta)$ on the boundary of D, and use **(9.13.1)** and **(9.13.2)**.)

(c) A point $\xi_0 \in \mathbf{R}$ is said to be a *point of prolongation* for T if there exists a disc Δ with center ξ_0 in $\mathbf{C}$ and a holomorphic function v on Δ which coincides with $\mathscr{F}\mathscr{L}T_1$ (resp. $\mathscr{F}\mathscr{L}T_2$) in the intersection of Δ with the half-plane $\mathscr{I}\zeta < 0$ (resp. $\mathscr{I}\zeta > 0$). Show that the set of points of prolongation is the complement of the support of the distribution $\mathscr{F}T$. (To show that a point of prolongation does not belong to the support of $\mathscr{F}T$, apply Problem 6(d) to T_1 and T_2. To show that a point ξ_0 not in the support of $\mathscr{F}T$ is a point of prolongation, consider a regularizing sequence (g_k) (17.1.2); if Δ is a sufficiently small disc with center ξ_0, there exists a holomorphic function u_k on Δ which is equal to $\mathscr{F}\mathscr{L}(g_k * T_1)$ (resp. $\mathscr{F}\mathscr{L}(g_k * T_2)$) at the points of Δ such that $\mathscr{I}\zeta < 0$ (resp. $\mathscr{I}\zeta > 0$) (Section 9.9, Problem 2); then use (a) and (b).)

14. Let μ be a slowly increasing *positive* measure on $\mathbf{R}$. Then the set of $z \in \mathbf{C}$ such that the function $x \mapsto e^{-zx}$ is μ-integrable is a vertical strip $\alpha < \mathscr{R}z < \beta$ (possibly empty). Show that α and β are singular points for the Laplace transform $\mathscr{L}\mu(z) = \int e^{-zx}\,d\mu(x)$. (Reduce to the case where the support of μ is contained in $[0, +\infty[$ (Problem 13) and $\alpha = 0$. Argue by contradiction: suppose that, for the function $u(z) = \mathscr{L}\mu(z)$, the Taylor series

$$\sum_{k=0}^{\infty} \frac{1}{k!} u^{(k)}(\rho)(z-\rho)^k,$$

for $\rho > 0$ sufficiently small, is absolutely convergent in a circle with center ρ and radius $> \rho$. Deduce that the series with positive terms

$$\sum_{k=0}^{\infty} \frac{(\rho+\varepsilon)^k}{k!} \int_0^{\infty} e^{-\rho x} x^k\,d\mu(x)$$

converges for sufficiently small $\varepsilon > 0$, and conclude that this would contradict the definition of the interval $]\alpha, \beta[$.)

15. (a) Let μ be a positive bounded measure on $\mathbf{R}$, so that $\mathscr{F}\mu$ is continuous and bounded on $\mathbf{R}$. Show that if $\mathscr{F}\mu$ is analytic at the point $\xi = 0$, the Fourier–Laplace transform $\mathscr{F}\mathscr{L}\mu$ is defined and holomorphic on a horizontal strip $-\alpha < \mathscr{I}\zeta < \beta$, with $\alpha > 0$ and $\beta > 0$ (α, β finite or not). (Write $\mu = \mu_1 + \mu_2$, where the support of μ_1 (resp. μ_2) is contained in $[0, +\infty[$ (resp. $]-\infty, 0]$). Show that if $\mathscr{F}\mu$ is analytic at the point $\xi = 0$, there exists a holomorphic function defined on a disc with center 0 which coincides with $\mathscr{F}\mu_1$ on the intersection of this disc with $\mathbf{R}$, by using Section 9.9, Problem 2, then make use of Problem 14.)

(b) Deduce from (a) that for $\mathscr{F}\mu$ to be analytic at the point 0 it is necessary and sufficient that there should exist a real number $r > 0$ such that $e^{-rt}\mu(\complement[-t, t])$ tends to 0 as t tends to $+\infty$.

(c) If $\mathscr{F}\mathscr{L}\mu$ is holomorphic in the strip $-\alpha < \mathscr{I}\zeta < \beta$, show that

$$|\mathscr{F}\mathscr{L}\mu(\xi + i\eta)| \leqq \mathscr{F}\mathscr{L}\mu(i\eta)$$

for $-\alpha < \eta < \beta$; deduce that $\mathscr{F}\mathscr{L}\mu(i\eta) \neq 0$ for $-\alpha < \eta < \beta$, if $\mu \neq 0$.

(d) Show that the function $\eta \mapsto \mathscr{F}\mathscr{L}\mu(i\eta)$ and the function $\eta \mapsto \log \mathscr{F}\mathscr{L}\mu(i\eta)$ are convex on the interval $]-\alpha, \beta[$.

(e) Suppose that $\mu = \mu_1 * \mu_2$, where μ_1 and μ_2 are positive bounded measures. Show that if $\mathscr{F}\mathscr{L}\mu$ is holomorphic for $-\alpha < \mathscr{I}\zeta < \beta$, then so also are $\mathscr{F}\mathscr{L}\mu_1$ and $\mathscr{F}\mathscr{L}\mu_2$ (use (13.21.11)).

16. Let α be a real number >0, and let f be the function which is 0 for $x < 0$, and equal to $\exp(-x^{1+\alpha})$ for $x \geqq 0$. Show that $\mathscr{F}\mathscr{L}f$ is an entire function and that there exists a constant $c > 0$ such that $|\mathscr{F}\mathscr{L}f(i\eta)| \geqq \exp(c|\eta|^{(\alpha+1)/\alpha})$ for η sufficiently large. (Use Laplace's method.†)

† See my book, *Infinitesimal Calculus*, Boston (Houghton–Mifflin), 1971.

17. Let f be a function defined on $\mathbf{R}$, which is zero for $x < 0$, and equal to an entire function $\sum_{n=0}^{\infty} a_n x^n$ for $x \geqq 0$. In order that the Laplace transform $\mathscr{L}f$ should be defined on the exterior of a disc $|s| > \mathrm{R}$ in $\mathbf{C}$, and be such that $\mathscr{L}f(1/s)$ is holomorphic for $|s| \leqq 1/\mathrm{R}$, it is necessary and sufficient that there should exist $\mathrm{A} > 0$ and $\mathrm{R} > 0$ such that $|a_n| \leqq \mathrm{AR}^n/n!$ for all n.

18. (a) Let μ be a measure on $\mathbf{R}$ with support contained in the interval $[-1, 1]$, so that its Fourier–Laplace transform f satisfies the inequality

$$|f(z)| \leqq \mathrm{C} \exp(2\pi |\mathscr{I} z|)$$

for some constant $\mathrm{C} > 0$. Show that if γ_k is a circuit whose image is the square with center 0 and side-length $4k$, the integral

$$\int_{\gamma_k} \frac{f(z)\, dz}{z^2 \cos 2\pi z}$$

tends to 0 as the integer k tends to $+\infty$. Deduce from the residue theorem (**9.16.1**) that

$$f'(0) = \frac{8}{\pi} \sum_{m \in \mathbf{Z}} \frac{(-1)^m}{(2m+1)^2} f\left(\frac{2m+1}{4}\right)$$

and, more generally, that for all $x \in \mathbf{R}$ we have

(1)
$$f'(x) = \frac{8}{\pi} \sum_{m \in \mathbf{Z}} \frac{(-1)^m}{(2m+1)^2} f\left(x + \frac{2m+1}{4}\right),$$

which can also be written in the form $f' = f * \nu$, where ν is the measure defined by the mass $8 \cdot (-1)^{m+1}/(2m+1)^2\pi$ at each of the points $\frac{1}{4}(2m+1)$, $m \in \mathbf{Z}$.
(b) Deduce from (1) that $\|f'\| \leqq 2\pi \|f\|$ (norms in $\mathscr{B}_{\mathbf{C}}(\mathbf{R})$). Moreover, there is equality only if $f(z) = a \sin 2\pi(z - h)$ for $a \in \mathbf{C}$ and $h \in \mathbf{R}$. (Reduce to the case where $(1/2\pi) f'(0) = 1 = \|f\|$, and deduce that all the derivatives of f are equal to those of $\sin 2\pi z$ at the points $n/4$, $n \in \mathbf{Z}$: argue by induction on the order of the derivative.)
(c) Let T be a distribution on $\mathbf{R}^n$ with support contained in the ball $\mathrm{B}'(0; \mathrm{A})$. Suppose also that $\mathscr{F}\mathrm{T}$ is a *bounded* function. Show that (with norms in $\mathscr{B}_{\mathbf{C}}(\mathbf{R}^n)$)

(2)
$$\|\mathrm{D}(\mathscr{F}\mathrm{T})\| \leqq 2\pi \mathrm{A} \cdot \|\mathscr{F}\mathrm{T}\|$$

(*Bernstein's inequality*). (By regularization and orthogonal transformation, reduce to the case where T is a function in $\mathscr{D}(\mathbf{R}^n)$, and the maximum of the Euclidean norm $\|\mathrm{D}(\mathscr{F}\mathrm{T})(\xi)\|$ is attained at the point $\xi = 0$ and is equal to $\mathrm{D}_1(\mathscr{F}\mathrm{T})(0)$; then use the result of (b).)

19. Let f be a holomorphic function defined on a disc $|z| < \mathrm{R}$, such that $f'(0) \neq 0$. For each $r \in \mathbf{R}$, let $r_1, r_2, \ldots, r_n$ be the absolute values of the zeros of f (each counted according to its multiplicity) such that $|z| \leqq r$. Show that the function $\theta \mapsto \log|f(re^{i\theta})|$ is integrable on $[0, 2\pi]$ and that

(*)
$$\log \frac{r^n |f(0)|}{r_1 r_2 \cdots r_n} = \frac{1}{2\pi} \int_0^{2\pi} \log|f(re^{i\theta})|\, d\theta$$

(*Jensen's formula*). (Consider the two sides of $(*)$ as functions of r. Show that the function $t \mapsto \int_0^{2\pi} \log|1 - te^{i\theta}|\, d\theta$ is continuous at $t = 1$, and deduce that the right-hand side of $(*)$ is a continuous function of r. Show also that if r is not the absolute value of a zero of f, the right-hand side of $(*)$ has a derivative at the point r, equal to

$$\mathscr{R}\left(\frac{1}{2\pi}\int_0^{2\pi} \frac{f'(re^{i\theta})}{f(re^{i\theta})} e^{i\theta}\, d\theta\right) = \frac{n}{r}$$

(cf. (9.17.1)).

20. (a) Let f be an entire function on $\mathbf{C}$ which satisfies the inequality

$$|f(z)| \leqq a \exp(\mathrm{A}|z|^\rho)$$

for all $z \in \mathbf{C}$, where A, a, and ρ are three positive constants. If (r_n) is the sequence of absolute values of the zeros of f (each counted according to its multiplicity), arranged in increasing order of magnitude, show that there exists a constant $b > 0$ such that the $r_n \neq 0$ satisfy the inequality $|r_n| \geqq b \cdot n^{1/\rho}$. (Apply Jensen's inequality (Problem 19) to prove that when $f(0) \neq 0$ we have $n \log 2 \leqq \mathrm{A}(2r_n)^\rho - \log|f(0)| + \log a$.) Consequently, for each $\alpha > \rho$, the series $\sum_n' r_n^{-\alpha}$ is convergent, where $\sum'$ means that the r_n equal to 0 are excluded from the summation.

(b) Let (a_n) be an infinite sequence of nonzero complex numbers, such that $|a_n| \leqq |a_{n+1}|$ and such that the series $\sum_n |a_n|^{-\alpha}$ converges. For each $\beta > \alpha$ and each integer n, let D_n be the closed disc with center a_n and radius $|a_n|^{-\beta}$, and let D be the union of the D_n. Show that for each $\varepsilon > 0$ there exists $\mathrm{R} > 0$ such that, for each $z \notin \mathrm{D}$ satisfying $|z| \geqq \mathrm{R}$, the infinite product of primary factors (Section 9.12, Problem 1) $\prod_{n=1}^{\infty} \mathrm{E}(z/a_n, p)$, where p is an integer $\geqq \alpha - 1$, satisfies the inequality

$$\left|\prod_{n=1}^{\infty} \mathrm{E}(z/a_n, p)\right| \geqq \exp(-|z|^{\alpha+\varepsilon}).$$

(Use Section 9.12, Problem 1(b), and note that if $|z| \notin \mathrm{D}_n$ and $|z| \geqq \frac{1}{2}|a_n|$, then $|1 - (z/a_n)| \geqq (2|z|)^{-1-\beta}$.)

In particular, there exists an increasing sequence (R_n) of positive real numbers, tending to $+\infty$, such that

$$\left|\prod_{n=1}^{\infty} \mathrm{E}(z/a_n, p)\right| \geqq \exp(-\mathrm{R}_n^{\alpha+\varepsilon})$$

when $|z| = \mathrm{R}_n$.

(c) Deduce from (a) and (b) that if f is an entire function satisfying an inequality $|f(z)| \leqq a \exp(b|z|^p)$ for some integer $p > 0$ and two constants $a > 0$, $b > 0$, then we may write

$$f(z) = z^m e^{\mathrm{P}(z)} \prod_n \mathrm{E}(z/a_n, p)$$

where m is an integer $\geqq 0$, P is a polynomial of degree $\leqq p$, and (a_n) is the (finite or infinite) sequence of zeros $\neq 0$ of f (each counted according to its multiplicity), the series $\sum_n |a_n|^{-p-\varepsilon}$ being convergent for all $\varepsilon > 0$ (*Hadamard's factorization theorem*).

(d) In particular, for each distribution $T \in \mathscr{E}'(\mathbf{R})$, we may write

$$\mathscr{F}T(z) = z^m e^{az+b} \prod_n \left(1 - \frac{z}{a_n}\right) e^{z/a_n} \tag{1}$$

where (a_n) is the sequence of zeros $\neq 0$ of $\mathscr{F}T$, each counted according to its multiplicity, a and b are complex numbers, and m is an integer $\geqq 0$, the series $\sum_n |a_n|^{-1-\varepsilon}$ being convergent. Give an example of an entire function equal to the right-hand side of (1), with $a_n = n$, but which is not the Fourier transform of a distribution belonging to $\mathscr{E}'(\mathbf{R})$.

(e) Let (α_n) be a sequence of positive real numbers such that $\sum_n \alpha_n < +\infty$. Let g_n be the function defined by $g_n(x) = 1/2\alpha_n$ for $|x| \leqq \alpha_n$, and $g_n(x) = 0$ for $|x| > \alpha_n$. Then the infinite product $\mathop{\ast}\limits_{n=1}^{\infty} g_n$ (each g_n being identified with a measure on $\mathbf{R}$) converges, and its Fourier–Laplace transform is the entire function

$$f(z) = \prod_n \frac{\sin 2\pi\alpha_n z}{2\pi\alpha_n z}$$

(Section 22.17, Problem 24). Show that $G_n = g_1 * g_2 * \cdots * g_n$ is of class C^{n-1}; $G = \mathop{\ast}\limits_{n=1}^{\infty} g_n$ is a function of class C^∞, with support contained in $\left[-\sum_n \alpha_n, \sum_n \alpha_n\right]$; and for each index k, the sequence $(D^k G_{n+k})_{n \geq 1}$ converges uniformly to $D^k G$.

(f) Deduce from (e) that the multiplicities of the real zeros of $\mathscr{F}T$, where $T \in \mathscr{E}'(\mathbf{R})$, may form an unbounded sequence. Show that for each strictly increasing sequence (λ_n) of positive real numbers, such that $\sum_n \lambda_n^{-1} < +\infty$, there exists a distribution $T \in \mathscr{E}'(\mathbf{R})$ such that the λ_n are zeros of $\mathscr{F}T$ (which in general will have other real zeros).

21. (a) Let $T \in \mathscr{E}'(\mathbf{R})$ be such that $\mathscr{F}T(0) = 0$, or equivalently $T * 1 = 0$. Then $S = T * Y = -T * (1 - Y)$ (where Y is Heaviside's function) is the unique distribution with compact support such that $DS = T$, and hence we have $\mathscr{F}T = 2\pi i x \cdot \mathscr{F}S$.

(b) Let $f \in \mathscr{D}(\mathbf{R})$ and suppose that $\mathscr{F}f$ (extended to an entire function on $\mathbf{C}$) is given by formula (1) of Problem 20. Show that if we put

$$F_0(x) = 2\pi i \int_{-\infty}^{x} f(t)\, dt, \qquad F_n(x) = 2\pi i \int_{-\infty}^{x} e^{2\pi i a_n(x-t)} f(t)\, dt,$$

then we have

$$-2\pi i x f(x) = af(x) + mF_0(x) + \sum_n (F_n(x) + a_n^{-1} f(x)),$$

the series on the right-hand side being normally convergent.

22. (a) A distribution $T \in \mathscr{D}'(\mathbf{R})$ is said to be *mean-periodic* if there exists a *nonzero* distribution $S \in \mathscr{E}'(\mathbf{R})$ with compact support, such that $S * T = 0$. An equivalent condition is that there exists a nonzero function $f \in \mathscr{D}(\mathbf{R})$ such that $f * T = 0$, or again that $\langle \gamma(s)T, \check{f} \rangle = 0$ for all $s \in \mathbf{R}$. Show that another equivalent condition is that the weakly closed subspace of $\mathscr{D}'(\mathbf{R})$ generated by the $\gamma(s)T$ is not the whole of $\mathscr{D}'(\mathbf{R})$ (cf. Section 12.15, Problem 13).

(b) A locally integrable periodic function is mean-periodic. The function $f(x) = e^x$ is mean-periodic, because $f * (\varepsilon_a - e^{-a}\varepsilon_0) = 0$ for all $a \neq 0$. A tempered distribution T cannot be mean-periodic unless the support of $\mathscr{F}$T is a discrete denumerable subset of **R**. There exists no mean-periodic distribution $\neq 0$ which has compact support or belongs to $L^1_C(\mathbf{R})$ or $L^2_C(\mathbf{R})$.

(c) Every distribution $T \in \mathscr{D}'(\mathbf{R})$ may be written in the form $T_1 - T_2$, where the support of T_1 (resp. T_2) is bounded below (resp. above). If $f * T = 0$ for some nonzero $f \in \mathscr{D}(\mathbf{R})$, the distribution $S = f * T_1 = f * T_2$ has compact support, hence $\mathscr{F}S/\mathscr{F}f$ extends to a *meromorphic* function on **C**. If $T = T'_1 - T'_2$ is another decomposition of T with the same properties, and if $S' = f * T'_1 = f * T'_2$, then $\mathscr{F}S'/\mathscr{F}f$ is the sum of $\mathscr{F}S/\mathscr{F}f$ and an *entire* function. If $g \in \mathscr{D}(\mathbf{R})$ is another nonzero function such that $g * T = 0$, and if $U = g * T_1 = g * T_2$, then we have $\mathscr{F}U/\mathscr{F}g = \mathscr{F}S/\mathscr{F}f$. We have thus defined, to within the addition of an entire function (the Fourier–Laplace transform of a distribution with compact support), a meromorphic function on **C**. Any one of these functions is called the *Carleman transform* of the mean-periodic distribution T, and is denoted by $\mathscr{C}$T (by abuse of notation). The singular parts (9.15) of all the Carleman transforms of T are the same.

(d) The monomials x^k $(k \geq 0)$ are mean-periodic functions (observe that $D^{k+1}x^k = 0$), and the function $k!/(2\pi iz)^{k+1}$ is a Carleman transform of x^k. If T is mean-periodic, so also is $e^{2\pi i\lambda x} \cdot T$ for each $\lambda \in \mathbf{C}$, and the function $z \mapsto \mathscr{C}T(z - \lambda)$ is a Carleman transform of this distribution. For each distribution S with compact support, $S * T$ is mean-periodic, and $(\mathscr{C}T)(\mathscr{F}S)$ is a Carleman transform of $S * T$.

(e) Derivatives (of any order) and primitives (Section 17.5, Problem 3) of a mean-periodic distribution are mean-periodic. (If $DS = T$ and if $f * T = 0$, then $Df * S = c$, where c is a constant.)

(f) Every mean-periodic *tempered* distribution may be obtained as follows: start with a distribution $T \in \mathscr{E}'(\mathbf{R})$ and the expression of $\mathscr{F}$T by formula (1) of Problem 20. Let (α_k) be the (finite or infinite) sequence of *distinct real zeros* of $\mathscr{F}$T, and let ν_k be the order of multiplicity of α_k; then

$$(*) \qquad \mathscr{F}S = \sum_k \sum_{j<\nu_k} b_{kj} D^j \varepsilon_{\alpha_k},$$

where the coefficients b_{kj} are such that the series $\sum_k \sum_{j<\nu_k} b_{kj} D^j g(\alpha_k)$ converges for all functions $g \in \mathscr{S}(\mathbf{R})$; this implies in particular that there exists an integer $m > 0$ such that $b_{kj} = 0$ for $j > m$ (Section 22.17, Problem 2). Conversely, if $b_{kj} = 0$ for $j > m$, and if there exists an integer $N > 0$ such that $\sum_k \sum_{j<\nu_k} |b_{kj}|(1 + |\alpha_k|)^{-N} < +\infty$, then the formula (*) defines the Fourier transform of a mean-periodic tempered distribution S. The set of poles of a Carleman transform $\mathscr{C}$S is then the set of α_k such that $b_{kj} \neq 0$ for at least one value of $j < \nu_k$. If we write $S = S_1 - S_2$, where the support of S_1 (resp. S_2) is bounded below (resp. above), then the corresponding Carleman transform $\mathscr{C}$S coincides with the Fourier–Laplace transform $\mathscr{F}\mathscr{L}S_1$ for $\mathscr{I}z < 0$ and with $\mathscr{F}\mathscr{L}S_2$ for $\mathscr{I}z > 0$, the set of poles of $\mathscr{C}$S being the support of $\mathscr{F}$S (cf. Problem 13).

23. (a) Let T be a mean-periodic distribution on **R**, and let $f \in \mathscr{D}(\mathbf{R})$ be a nonzero function such that $f * T = 0$ and $\mathscr{F}f(0) = 0$. Put $g = f * Y = -f * (1 - Y)$, so that $\mathscr{F}f(x) = 2\pi ix \cdot \mathscr{F}g(x)$. Show that if 0 is not a pole of a Carleman transform $\mathscr{C}$T, then $g * T = 0$.

(b) Show that if $\mathscr{C}$T has no poles, then $T = 0$. (Let $f \in \mathscr{D}(\mathbf{R})$ be nonzero and such that $f * T = 0$. By using (a) and Problem 21(b), show that $(Pf) * T = 0$ for each polynomial P(x), and deduce first that $g * T = 0$ for all $g \in \mathscr{D}(\mathbf{R})$ whose support is contained in the

interior of the support of f, then that $g * \mathrm{T} = 0$ for all $g \in \mathscr{D}(\mathbf{R})$, from which it follows that $\mathrm{T} = 0$.)
(c) Show that if $\lambda \in \mathbf{C}$ is a pole of $\mathscr{C}\mathrm{T}$ of order n, there exists $f \in \mathscr{D}(\mathbf{R})$ such that $f \neq 0$, $f * \mathrm{T} = 0$ and such that $\mathscr{F}f$ has a zero of order n at the point λ. (Reduce to the case $\lambda = 0$ (Problem 22(d)), then to the case $n = 0$ by considering $\mathrm{D}^n\mathrm{T} = (\mathrm{D}^n\varepsilon_0) * \mathrm{T}$; then use (a).)

24. (a) For each nonzero function $f \in \mathscr{D}(\mathbf{R})$, let V_f (resp. W_f) denote the subspace of $\mathscr{E}(\mathbf{R})$ (resp. $\mathscr{D}'(\mathbf{R})$) consisting of the functions $g \in \mathscr{E}(\mathbf{R})$ (resp. the distributions $\mathrm{T} \in \mathscr{D}(\mathbf{R})$) such that $f * g = 0$ (resp. $f * \mathrm{T} = 0$); V_f is closed in the Fréchet space topology of $\mathscr{E}(\mathbf{R})$, and W_f is closed in the weak topology of $\mathscr{D}'(\mathbf{R})$. We have $\mathrm{V}_f = \mathrm{W}_f \cap \mathscr{E}(\mathbf{R})$, and W_f is the weak closure of V_f; more precisely, for each distribution $\mathrm{T} \in \mathrm{W}_f$, there exists a sequence (g_n) of functions in V_f which converges to T for the weak topology (cf. (17.12.3)).
(b) For each mean-periodic function $g \in \mathscr{E}(\mathbf{R})$ (resp. each mean-periodic distribution $\mathrm{T} \in \mathscr{D}'(\mathbf{R})$), let L_g be the closed subspace of the Fréchet space $\mathscr{E}(\mathbf{R})$ generated by the $\gamma(s)g$, $s \in \mathbf{R}$ (resp. let $\mathrm{M_T}$ be the weakly closed subspace of $\mathscr{D}'(\mathbf{R})$ generated by the $\gamma(s)\mathrm{T}$, $s \in \mathbf{R}$). The subspace L_g (resp. $\mathrm{M_T}$) is the intersection of the V_f (resp. W_f) such that $f \in \mathscr{D}(\mathbf{R})$ and $f * g = 0$ (resp. $f * \mathrm{T} = 0$). (Cf. Section 12.15, Problems 4(f) and 13(b).)
(c) Let T be a mean-periodic distribution on $\mathbf{R}$. Show that if $\lambda \in \mathbf{C}$ is a pole of order $\geqq n + 1$ of $\mathscr{C}\mathrm{T}$, then $\mathscr{F}f$ has a zero of order $\geqq n + 1$ at the point λ, for each $f \in \mathscr{D}(\mathbf{R})$ such that $f * \mathrm{T} = 0$ (i.e., $\mathrm{T} \in \mathrm{W}_f$). Deduce that for $0 \leqq k \leqq n$ the mean-periodic function $x \mapsto x^k e^{2\pi i \lambda x}$ belongs to $\mathrm{M_T} \cap \mathscr{E}(\mathbf{R})$.
(d) Conversely, if a polynomial P of degree n is such that the function $x \mapsto \mathrm{P}(x)e^{2\pi i \lambda x}$ belongs to $\mathrm{M_T}$, then $\mathscr{F}f$ has a zero of order $\geqq n + 1$ at the point λ, for each $f \in \mathscr{D}(\mathbf{R})$ such that $\mathrm{T} \in \mathrm{W}_f$; deduce that $\mathscr{C}\mathrm{T}$ has a pole of order $\geqq n + 1$ at the point λ. (Use Problem 23(c).)
(e) If $g \in \mathscr{E}(\mathbf{R})$ is mean-periodic, show that g is the limit in the Fréchet space $\mathscr{E}(\mathbf{R})$ of a sequence of linear combinations of functions belonging to L_g and of the form $x \mapsto \mathrm{P}(x)\exp(2\pi i \lambda x)$, where λ is a pole of $\mathscr{C}g$ of order ν, and P is a polynomial of degree $\leqq \nu$; these functions are called the *exponential-polynomials* associated with $\mathscr{C}g$ (cf. Section 12.15, Problem 4(f)). Deduce that if $\mathrm{T} \in \mathrm{W}_f$, T is the limit (in the weak topology of $\mathscr{D}'(\mathbf{R})$) of a sequence of linear combinations of exponential-polynomials belonging to W_f.

25. Let f be a nonzero function belonging to $\mathscr{D}(\mathbf{R})$, and let Λ be the set of *distinct* zeros of the entire function $\mathscr{F}f$. For each $\lambda \in \Lambda$, let $n(\lambda)$ denote the multiplicity of λ as a zero of $\mathscr{F}f$; recall that the series $\sum_\lambda n(\lambda)/|\lambda|^{1+\alpha}$ converges for each $\alpha > 0$ (Problem 20).
(a) For each locally integrable function $g \in \mathrm{W}_f$ (i.e., such that $g * f = 0$) and each $a \in \mathbf{R}$, define $g_a(x)$ to be equal to $g(x)$ for $x \geqq a$, and $g_a(x) = 0$ for $x < a$; then

$$\operatorname{Supp}(g_a * f) \subset a + \operatorname{Supp}(f).$$

For $\lambda \in \Lambda$, if we take $g(x) = x^j e^{2\pi i \lambda x}$ for $0 \leqq j \leqq n(\lambda) - 1$, show that

$$(1) \qquad \mathscr{F}(g_a * f)(\zeta) = \left(\frac{i}{2\pi}\right)^{j+1} \mathscr{F}f(\zeta) \frac{d^j}{d\zeta^j}\left(\frac{e^{-2\pi i(\zeta - \lambda)a}}{\lambda - \zeta}\right).$$

(b) For locally integrable functions $g, h \in \mathrm{W}_f$, put

$$\Phi_a(g, h) = \langle g_a, \check{h}_a * \check{f} \rangle = \langle h_a, \check{g}_a * \check{f} \rangle = \langle g_a * h_a, \check{f} \rangle.$$

Let E_λ denote the vector subspace of dimension $n(\lambda)$ in V_f spanned by the exponential-polynomials $x^j e^{2\pi i \lambda x}$ for $0 \leqq j \leqq n(\lambda) - 1$, where λ is a zero of $\mathscr{F}f$. Show that if

λ, μ are distinct zeros of $\mathscr{F}f$, we have $\Phi_a(g, h) = 0$ for all $g \in \mathrm{E}_\lambda$ and $h \in \mathrm{E}_\mu$. (The calculation of the integral

$$\int_a^{x-a} e^{\xi t} e^{\eta(x-t)} t^j (x-t)^k \, dt$$

can be reduced to the case $j = k = 0$ by differentiation with respect to ξ and η.)

(c) Show that there exists no function $g \neq 0$ in E_λ such that $\Phi_0(g, h) = 0$ for all $h \in \mathrm{E}_\lambda$. Deduce that, for each $\lambda \in \Lambda$, the space V_f is the topological direct sum **(12.13)** of E_λ and the closure of the sum of the E_μ, $\mu \neq \lambda$ (use Problem 24(e)). On the other hand, for $\lambda \in \Lambda$ and $0 \leqq j \leqq n(\lambda) - 1$, there exists a unique polynomial $\mathrm{P}_{\lambda j}$ of degree $< n(\lambda)$ such that

$$(2) \qquad \Phi_0(x^j e^{2\pi i \lambda x}, \mathrm{P}_{\lambda j}(x) e^{2\pi i \lambda x}) = 1,$$
$$\Phi_0(x^k e^{2\pi i \lambda x}, \mathrm{P}_{\lambda j}(x) e^{2\pi i \lambda x}) = 0 \qquad \text{for} \quad 0 \leqq k \leqq n(\lambda) - 1 \quad \text{and} \quad k \neq j.$$

Show also that

$$\Phi_0(x^j e^{2\pi i \lambda x}, x^k e^{2\pi i \lambda x}) = \Phi_a(x^j e^{2\pi i \lambda x}, x^k e^{2\pi i \lambda x})$$

for $0 \leqq j, k \leqq n(\lambda) - 1$ and all $a \in \mathbf{R}$, so that the relations (2) remain valid with Φ_0 replaced by Φ_a.

(d) For each locally integrable function $g \in \mathrm{W}_f$, define

$$(3) \qquad c_{\lambda j}(g) = \Phi_a(g, \mathrm{P}_{\lambda j} \chi_\lambda)$$

for all $\lambda \in \Lambda$ and $0 \leqq j \leqq n(\lambda) - 1$, where $\chi_\lambda(x) = e^{2\pi i \lambda x}$. The $c_{\lambda j}$ are called the *Fourier coefficients* of g relative to f. Show that if $c_{\lambda j} = 0$ for all $\lambda \in \Lambda$ and $0 \leqq j \leqq n(\lambda) - 1$, then $g(x) = 0$ almost everywhere.

If I is an interval in $\mathbf{R}$ containing the support of f, and if the restriction of g to some interval $a + \mathrm{I}$ is negligible, show that g is negligible.

(e) With the $c_{\lambda j}(g)$ as defined by (3), show that for each $\lambda \in \Lambda$ the sum

$$s_\lambda(g)(x) = \sum_{j=0}^{n(\lambda)-1} c_{\lambda j}(g) x^j e^{2\pi i \lambda x}$$

is equal to $2\pi i \cdot \mathrm{Res}_\lambda(\mathscr{C}g(\zeta) e^{2\pi i \zeta x})$, where $\mathscr{C}g$ is any Carleman transform of g (use formula (1)). Let (H_n) be an increasing sequence of finite subsets of Λ, with union Λ. In order that the sequence of sums $\sum_{\lambda \in \mathrm{H}_n} s_\lambda(g)$ should converge to g with respect to the weak topology of $\mathscr{D}'(\mathbf{R})$, it is necessary and sufficient that for each function $u \in \mathscr{D}(\mathbf{R})$ the sequence of sums $\sum_{\lambda \in \mathrm{H}_n} \mathrm{Res}_\lambda(\mathscr{C}g(\zeta) \mathscr{F}u(\zeta))$ should converge.

Conversely, if a family of complex numbers $(a_{\lambda j})$ (where $\lambda \in \Lambda$, $0 \leqq j \leqq n(\lambda) - 1$) is such that the sequence of functions $x \mapsto \sum_{\lambda \in \mathrm{H}_n} \sum_j a_{\lambda j} x^j e^{2\pi i \lambda x}$ converges in $\mathscr{D}'(\mathbf{R})$ to a locally integrable function g, then $g \in \mathrm{W}_f$ and $c_{\lambda j}(g) = a_{\lambda j}$ for all pairs (λ, j). In particular, this is the case if the sequence above converges uniformly on all compact subsets of $\mathbf{R}$.

(f) If we put $h_\lambda(x) = 2\pi i \cdot \mathrm{Res}_\lambda((\mathscr{F}f(\zeta))^{-1} e^{2\pi i \zeta x})$, then we have

$$s_\lambda(g)(x) = (h_\lambda * (g_a * f))(x)$$

for all $a \in \mathbf{R}$. Deduce that if the support of f is contained in the interval $[b, c]$, there exists a constant $C(\lambda)$, depending only on f and $\lambda \in \Lambda$, such that for all $x \in \mathbf{R}$

$$|s_\lambda(g)(x)| \leqq C(\lambda) \int_{x+b-c}^{x-b+c} |g(t)| \, dt.$$

(Observe that $\operatorname{Supp}(g_a * f) \subset a + \operatorname{Supp}(g)$ and that we may take $a = x$.)
(g) Deduce from (f) that if there exist two constants $A > 0$ and $\alpha \in \mathbf{R}$ such that $|g(x)| \leqq A \exp(2\pi\alpha x)$ whenever $x > 0$, then $s_\lambda(g) = 0$ for all $\lambda \in \Lambda$ such that $\mathscr{I}\lambda < -\alpha$, and if $\mathscr{I}\lambda = -\alpha$ we have $c_{\lambda j}(g) = 0$ for $j > 0$. In particular, if $|g(x)| \leqq A \exp(-x^{1+\beta})$ for all $x > 0$, where A, β are two positive real constants, then $s_\lambda(g) = 0$ for all $\lambda \in \Lambda$, so that g is negligible. If $A > 0$ and $\alpha > 0$ are such that $|g(x)| \leqq A \exp(2\pi\alpha|x|)$ for all $x \in \mathbf{R}$, then we have $s_\lambda(g) = 0$ for $\mathscr{I}\lambda > \alpha$ or $\mathscr{I}\lambda < -\alpha$.
(h) If the derivative Dg of a locally integrable mean-periodic function is itself (as a distribution) a locally integrable function, then $s_\lambda(Dg) = D(s_\lambda(g))$ for all $\lambda \in \Lambda$.

26. With the notation of Problem 25, suppose that $g \in W_f$ *belongs to* $\mathscr{L}^\infty_{\mathbf{C}}(\mathbf{R})$.
(a) Show that the poles of $\mathscr{C}g$ are real and simple, and that $|c_{\lambda 0}(g)| \leqq N_\infty(g)$. (Use Problem 22(f) to majorize $\mathscr{C}g$ in a neighborhood of one of its poles.) Deduce that $g = D^2 g_2$, where g_2 is an almost periodic continuous function on $\mathbf{R}$, the Fourier transform of a bounded measure carried by the discrete set $\Lambda_0 \subset \Lambda$ of poles of $\mathscr{C}g$ (observe that $\sum_{\lambda \in \Lambda} (1 + |\lambda|)^{-2} < +\infty$). The Fourier transform $\mathscr{F}g$ is then the (in general unbounded) atomic measure defined by the masses $c_{\lambda 0}(g)$ at the points $\lambda \in \Lambda_0$.
(b) Put $\Psi_\varepsilon(t) = (1/\varepsilon)(1 - (|t|/\varepsilon))^+$ and $g_\varepsilon = g * \Psi_\varepsilon$, which is a continuous function belonging to W_f, such that $\|g_\varepsilon\| \leqq N_\infty(g)$; its Fourier coefficients are

$$c_{\lambda 0}(g_\varepsilon) = c_{\lambda 0}(g) \frac{\sin^2 \pi\lambda\varepsilon}{\pi^2\lambda^2\varepsilon^2}.$$

Show that g_ε is almost periodic and that

$$\sum_{\lambda \in \Lambda_0} \left| c_{\lambda 0}(g) \frac{\sin^2 \pi\lambda\varepsilon}{\pi^2\lambda^2\varepsilon^2} \right|^2 = \lim_{T \to \infty} \frac{1}{2T} \int_{-T}^{T} |g_\varepsilon(t)|^2 \, dt.$$

Deduce that $\sum_{\lambda \in \Lambda_0} |c_{\lambda 0}(g)|^2 < +\infty$.
(c) Deduce that $g = c_{00}(g) + Dg_1$ (as a distribution), where g_1 is an almost periodic continuous function on $\mathbf{R}$, whose Fourier transform is the bounded measure defined by the masses $c_{\lambda 0}(g)/2\pi i\lambda$ at the points $\lambda \neq 0$ of Λ_0. Show that g is correlated (Section 22.17, Problem 10) to each function in $\mathscr{P}(\mathbf{R})$. If we put $\chi_\xi(t) = e^{-2\pi i\xi t}$, we have $M(g\chi_\xi) = 0$ if $\xi \notin \Lambda_0$, and $M(g\chi_\lambda) = c_{\lambda 0}(g)$ for all $\lambda \in \Lambda_0$.
(d) Show that if g is uniformly continuous, then it is almost periodic. (Consider the function $g * \Psi_\varepsilon * \Psi_\varepsilon$, which belongs to W_f and is twice continuously differentiable on $\mathbf{R}$, and observe that it converges uniformly to g as $\varepsilon \to 0$.)

27. Let $\sigma_1 < \sigma_2$ be real numbers, B the vertical strip $\sigma_1 \leqq \mathscr{R}s \leqq \sigma_2$ in $\mathbf{C}$, and let F be the closed subspace of the Banach space $\mathscr{C}^\infty_{\mathbf{C}}(B)$ consisting of the functions which are continuous and bounded on B and holomorphic in the interior of B. The additive group $\mathbf{R}$ acts continuously on F by the rule $(x, f) \mapsto \gamma(ix)f$, where $\gamma(ix)f$ is the function $s \mapsto f(s - ix)$. A function $f \in F$ is said to be *almost periodic* in B if f is almost periodic (Section 22.10, Problem 8) with respect to this action of $\mathbf{R}$. An equivalent condition is that for each $\varepsilon > 0$

there exists a number $T > 0$ such that, in each interval of length T in **R**, there exists a number x for which $|f(s) - f(s - ix)| \leqq \varepsilon$ for all $s \in B$ (Section 22.17, Problem 8).
(a) Show that if we put $f_\sigma(t) = f(\sigma + it)e^{-\lambda\sigma}$ for $\sigma_1 \leqq \sigma \leqq \sigma_2$, the number $M(f_\sigma \chi_\lambda)$, where $\chi_\lambda(t) = e^{-i\lambda t}$, is independent of σ (for λ real). (Use Section 9.9, Problem 1.) This number is therefore zero except for a finite or infinite sequence (λ_n) of values of $\lambda \in \mathbf{R}$, and if we put $M(f_\sigma \chi_{\lambda_n}) = a_n$, we have

$$\lim_{T \to \infty} \frac{1}{2T} \int_{-T}^{T} |f(\sigma + it)|^2 \, dt = \sum_n |a_n|^2 e^{2\lambda_n \sigma}.$$

(b) Deduce from (a) that on B the function f is the uniform limit of a sequence of linear combinations of a finite number of the functions $e^{\lambda_n s}$. (Use Section 22.10, Problem 9(d).)

28. With the notation of Problem 27, let $f \in F$. Suppose that there exists $\sigma \in]\sigma_1, \sigma_2[$ such that the function $t \mapsto f(\sigma_0 + it)$ is almost periodic. Show that if $\sigma_1 < \sigma'_1 < \sigma_0 < \sigma'_2 < \sigma_2$, the function f is almost periodic in the strip $\sigma'_1 \leqq \mathscr{R}s \leqq \sigma'_2$. (Consider the function

$$s \mapsto f(s) - f(s - ix)$$

for some $x \in \mathbf{R}$, and apply the three-line theorem (Section 13.17, Problem 6) in the two strips $\sigma_1 \leqq \mathscr{R}s \leqq \sigma_0$ and $\sigma_0 \leqq \mathscr{R}s \leqq \sigma_2$.)

29. Let T be a pseudomeasure on **R** (Section 22.17, Problem 14) whose Fourier transform is almost periodic. Show that if T is a bounded measure, its Laplace transform is almost periodic in every strip $-\sigma \leqq \mathscr{R}s \leqq 0$, where $\sigma > 0$ is arbitrary. If the support of T is compact, its Laplace transform is an entire function which is almost periodic in every vertical strip $B \subset \mathbf{C}$.

30. For each positive real number α and each distribution $T \in \mathscr{E}'(\mathbf{R})$ with compact support, the sum $\boldsymbol{\sigma}(\alpha)T = \sum_{k \in \mathbf{Z}} \gamma(k\alpha)T$ is defined in $\mathscr{D}'(\mathbf{R})$.
(a) For a distribution T on **R** the following conditions are equivalent:
(1) There exists a distribution $U \in \mathscr{E}'(\mathbf{R})$ such that $T = U - \gamma(\alpha)U$.
(2) $T \in \mathscr{E}'(\mathbf{R})$ and $\boldsymbol{\sigma}(\alpha)T = 0$.
(3) $T \in \mathscr{E}'(\mathbf{R})$ and $\mathscr{F}T(k/\alpha) = 0$ for all $k \in \mathbf{Z}$.
(To show that (2) implies (1), consider the sum $\sum_{k \geq 0} \gamma(k\alpha)T$, and show that its support is compact.)
(b) Let $\beta > 0$ be such that there exists an integer $m \geqq 2$ and a positive real number A with the property that

$$|1 - \exp(-2\pi i n\alpha/\beta)|^{-1} \leqq A \cdot |n|^{m-1}$$

for all integers $n \neq 0$ (cf. (d) below). Put

$$a_n = \frac{(\beta/n)^{m+1}}{1 - \exp(-2\pi i n\alpha/\beta)}$$

for $n \in \mathbf{Z}$, $n \neq 0$, and

$$h(z) = \sum_{n \in \mathbf{Z},\, n \neq 0} a_n/(\beta z - n),$$

a series which is absolutely and uniformly convergent on the complement of the union of the discs $|\beta z - n| < \beta\varepsilon$ for all $\varepsilon > 0$ (where $n \in \mathbf{Z}$, $n \neq 0$). Deduce that the function

$$\Phi(z) = z^{m+2}(1 - e^{-2\pi i\beta z})h(z)$$

is an entire function such that

$$\Phi(n/\beta) = \frac{2\pi in/\beta}{1 - \exp(-2\pi in\alpha/\beta)}$$

for $n \in \mathbf{Z}$, $n \neq 0$, and that there exist $C > 0$ and $b > 0$ such that

$$|\Phi(z)| \leqq C|z|^{m+2} \exp(b|\mathscr{I}z|)$$

for all $z \in \mathbf{C}$, so that $\Phi = \mathscr{F}S$ where S is a distribution with compact support. Deduce that there exists a distribution $T \in \mathscr{E}'(\mathbf{R})$ such that

$$(*) \qquad \varepsilon_0 = S * (Y - \gamma(\alpha)Y) + T * (Y - \gamma(\beta)Y)$$

where Y is Heaviside's function. (Use (a).)

(c) Let u be a linear form (which need *not* be continuous) on one of the vector spaces of distributions $\mathscr{D}$, $\mathscr{S}$, $\mathscr{E}$, $\mathscr{D}'$, $\mathscr{S}'$, $\mathscr{E}'$, such that $u(\gamma(s)T) = u(T)$ for all $s \in \mathbf{R}$ and all distributions T in the space in question. Show that $u(DT) = 0$ for all distributions T in the space in question, and deduce the expression of u (*Meisters' theorem*). (Observe that, for each distribution U, it follows from $(*)$ that

$$DU = (S * U) - \gamma(\alpha)(S * U) + (T * U) - \gamma(\beta)(T * U).)$$

(d) Let ξ be a real root of an irreducible polynomial $P \in \mathbf{Q}[X]$ of degree $m \geqq 2$, so that $P'(\xi) \neq 0$. Show that for each rational number p/q in its lowest terms we have

$$\left|\xi - \frac{p}{q}\right| \leqq \frac{C}{q^m}$$

where C is a constant independent of p/q (use the mean value theorem). Deduce that there exists a constant $B > 0$ such that

$$|1 - e^{2n\pi i\xi}|^{-1} \leqq B \cdot |n|^{m-1}$$

for all integers $n \neq 0$.

19. PERIODIC DISTRIBUTIONS AND FOURIER SERIES

(22.19.1) Recall ((22.14.3) and (22.13.1)) that the dual group of the torus $\mathbf{T}^n = \mathbf{R}^n/\mathbf{Z}^n$ may be canonically identified with the discrete group $\mathbf{Z}^n$. If we denote by p the canonical mapping $\mathbf{R}^n \to \mathbf{T}^n$, this identification assigns to the number $\langle p(x), v\rangle$ (where $x = (x_1, \ldots, x_n) \in \mathbf{R}^n$ and $v = (v_1, \ldots, v_n) \in \mathbf{Z}^n$) the value $\exp(2\pi i(x|v))$, which we shall also write as $\exp(2\pi i(\dot{x}|v))$, where $\dot{x} = p(x)$. Also, *in this section*, we define $|v| = \sup_k |v_k|$, so that $|v| \leqq \sum_{k=1}^{n} |v_k| \leqq n|v|$. The normalized Haar measure $m_{\mathbf{T}^n}$ on $\mathbf{T}^n$ (with total mass 1) is associated to Lebesgue measure on $\mathbf{R}^n$ and to the normalized Haar measure on $\mathbf{Z}^n$ (for which each point has mass 1). Let I be the interval $[0, 1] \subset \mathbf{R}$; the frontier $\mathrm{I}^n - \mathring{\mathrm{I}}^n$ of the cube I^n is negligible with respect to Lebesgue measure, hence (14.4.5) $p(\mathrm{I}^n - \mathring{\mathrm{I}}^n)$ is negligible in $\mathbf{T}^n$; on the other hand, the restriction of p to $\mathring{\mathrm{I}}^n$ is a diffeomorphism of this open subset of $\mathbf{R}^n$ onto the complement of $p(\mathrm{I}^n - \mathring{\mathrm{I}}^n)$ in $\mathbf{T}^n$, and the image under p of Lebesgue measure on $\mathring{\mathrm{I}}^n$ is the measure induced on $p(\mathring{\mathrm{I}}^n)$ by the measure $m_{\mathbf{T}^n}$. It follows that the space $\mathrm{L}^1_{\mathbf{C}}(\mathbf{T}^n)$ (resp. $\mathrm{L}^2_{\mathbf{C}}(\mathbf{T}^n)$, $\mathrm{L}^\infty_{\mathbf{C}}(\mathbf{T}^n)$) may be identified canonically with $\mathrm{L}^1_{\mathbf{C}}(\mathrm{I}^n)$ (resp. $\mathrm{L}^2_{\mathbf{C}}(\mathrm{I}^n)$, $\mathrm{L}^\infty_{\mathbf{C}}(\mathrm{I}^n)$). The Fourier transform of a function $f \in \mathscr{L}^1_{\mathbf{C}}(\mathbf{T}^n)$ is the mapping $\nu \mapsto c_\nu$ of $\mathbf{Z}^n$ into $\mathbf{C}$ defined by

(22.19.1.1)
$$c_\nu = \int_{\mathrm{I}^n} \exp(-2\pi i(x \mid \nu)) f(p(x))\, dx;$$

the complex numbers c_ν are called the *Fourier coefficients* of f. They tend to 0 at infinity (22.10.13).

In the other direction, the Fourier transform of a function $\nu \mapsto c_\nu$ belonging to $\mathrm{L}^1_{\mathbf{C}}(\mathbf{Z}^n)$, that is to say an *absolutely summable family* $(c_\nu)_{\nu \in \mathbf{Z}^n}$ of complex numbers (5.3), is the continuous function on $\mathbf{T}^n$

(22.19.1.2)
$$\dot{x} \mapsto \sum_{\nu \in \mathbf{Z}^n} c_\nu \exp(-2\pi i(\dot{x} \mid \nu)).$$

If the function f belongs to $\mathscr{P}^1(\mathbf{T}^n)$, and if (c_ν) is the family of its Fourier coefficients, then this family is absolutely summable, and Fourier's reciprocity formula (22.10.13) takes the form

(22.19.1.3)
$$f(\dot{x}) = \sum_{\nu \in \mathbf{Z}^n} c_\nu \exp(2\pi i(\dot{x} \mid \nu)).$$

The series on the right-hand side is called the *Fourier series* of f.

The analogues of the propositions (22.16.2) and (22.16.3) are the following:

(22.19.2) (i) *If the family* $(c_\nu)_{\nu \in \mathbf{Z}^n}$ *is such that* $(|\nu|^m c_\nu)_{\nu \in \mathbf{Z}^n}$ *is absolutely summable, the function*

$$\dot{x} \mapsto \sum_{\nu \in \mathbf{Z}^n} c_\nu \exp(-2\pi i(\dot{x} \mid \nu))$$

is of class C^m *on* $\mathbf{T}^n$.

(ii) *If f is a function of class C^m on $\mathbf{T}^n$, and if $(c_\nu)_{\nu \in \mathbf{Z}^n}$ is the family of its Fourier coefficients, then the family $(|\nu|^m c_\nu)_{\nu \in \mathbf{Z}^n}$ tends to 0 at infinity.*

(i) Since $\mathbf{R}^n$ is a covering of $\mathbf{T}^n$, it is enough to show that the periodic function $g: x \mapsto \sum_{\nu \in \mathbf{Z}^n} c_\nu \exp(-2\pi i(x|\nu))$ is of class C^m on $\mathbf{R}^n$. But the assumptions made imply, by virtue of **(8.6.3)**, that for each α such that $|\alpha| \leqq m$ the derivative $D^\alpha g$ exists and that

(22.19.2.1) $$D^\alpha g(x) = (-2\pi i)^{|\alpha|} \sum_{\nu \in \mathbf{Z}^n} \nu^\alpha c_\nu \exp(-2\pi i(x|\nu)).$$

(ii) The hypotheses imply that $f = g \circ p$, where g is a periodic function of class C^m on $\mathbf{R}^n$. Using the formula **(22.19.1.1)**, the periodicity of g and of its derivatives of order α for $|\alpha| \leqq m$, and the Lebesgue–Fubini theorem, we obtain on integrating by parts α_k times with respect to x_k, for $1 \leqq k \leqq n$,

(22.19.2.2) $$(2\pi i)^{|\alpha|}\nu^\alpha c_\nu = \int_{\mathbf{I}^n} \exp(-2\pi i(x|\nu)) D^\alpha g(x)\, dx,$$

from which the result follows, because the Fourier coefficients of $D^\alpha g$ tend to 0 at infinity.

(22.19.3) A family $(c_\nu)_{\nu \in \mathbf{Z}^n}$ is said to be *rapidly decreasing* if for *each* integer $m \geqq 0$ the family $(|\nu|^m c_\nu)_{\nu \in \mathbf{Z}^n}$ tends to 0 at infinity (or, equivalently, is bounded). Since the number of multi-indices $\nu \in \mathbf{Z}^n$ such that $|\nu| = k$ is equal to $(2k+1)^n - (2k-1)^n \sim 2n(2k)^{n-1}$, it follows that if the family $(|\nu|^m c_\nu)_{\nu \in \mathbf{Z}^n}$ is bounded, the family $(|\nu|^{m-n-1} c_\nu)$ is absolutely summable. Let $\mathscr{S}(\mathbf{Z}^n)$ be the set of rapidly decreasing families (c_ν), and endow this vector space with the topology of a locally convex metrizable space by means of the sequence of seminorms

(22.19.3.1) $$q_m((c_\nu)) = \sup_{\nu \in \mathbf{Z}^n} (1 + |\nu|)^m |c_\nu|.$$

The space $\mathbf{C}^{(\mathbf{Z}^n)}$ of families $(c_\nu)_{\nu \in \mathbf{Z}^n}$ with *finite support* is *dense* in $\mathscr{S}(\mathbf{Z}^n)$, because for each integer $m \geqq 0$, each family $(c_\nu)_{\nu \in \mathbf{Z}^n}$ in $\mathscr{S}(\mathbf{Z}^n)$ and each $\varepsilon > 0$, there exists an integer $N > 0$ such that $(1 + |\nu|)^m |c_\nu| \leqq \varepsilon$ for $|\nu| > N$; if we put $c'_\nu = c_\nu$ for $|\nu| \leqq N$, and $c'_\nu = 0$ for $|\nu| > N$, then the family (c'_ν) has a finite support, and $q_m((c_\nu - c'_\nu)) \leqq \varepsilon$.

(22.19.4) *The mapping which takes each family $(c_\nu) \in \mathscr{S}(\mathbf{Z}^n)$ to the function $\dot{x} \mapsto \sum_{\nu \in \mathbf{Z}^n} c_\nu \exp(-2\pi i(\dot{x}|\nu))$ on $\mathbf{T}^n$ is an isomorphism of the space $\mathscr{S}(\mathbf{Z}^n)$ onto the Fréchet space $\mathscr{E}(\mathbf{T}^n)$.*

This follows immediately from the definitions and the formulas (22.19.2.1) and (22.19.2.2), which provide the inequalities

$$\sup_{x \in \mathbf{I}^n,\ |\alpha| \leq m} |\mathrm{D}^\alpha g(x)| \leqq b_m q_m((c_\nu))$$

and

$$q_m((c_\nu)) \leqq b'_m \sup_{x \in \mathbf{I}^n,\ |\alpha| \leq m} |\mathrm{D}^\alpha g(x)|$$

where b_m and b'_m are constants depending only on m and n.

(22.19.5) A family $(a_\nu)_{\nu \in \mathbf{Z}^n}$ of complex numbers is said to be *slowly increasing* if there exists an integer $m \geqq 0$ such that $|a_\nu| \leqq (1 + |\nu|)^m$ for sufficiently large $|\nu|$. Clearly these families form a complex vector space. We shall show that this space may be canonically identified with the *dual* $\mathscr{S}'(\mathbf{Z}^n)$ of the Fréchet space $\mathscr{S}(\mathbf{Z}^n)$: a slowly increasing family (a_ν) being identified with the linear form $(c_\nu) \mapsto \sum_\nu a_\nu c_\nu$ on $\mathscr{S}(\mathbf{Z}^n)$. Let (a_ν) be a slowly increasing family, so that there exists an integer $m \geqq 0$ and a constant $b > 0$ such that $|a_\nu| \leqq b(1 + |\nu|)^m$ for all $\nu \in \mathbf{Z}^n$; since the family $(|\nu|^{-n-1})$ is summable, we deduce from the fact that $q_{m+n+1}((c_\nu))$ is finite for each family $(c_\nu) \in \mathscr{S}(\mathbf{Z}^n)$ that the family $(a_\nu c_\nu)$ is absolutely summable, and that

$$\sum_\nu |a_\nu c_\nu| \leqq b\Big(\sum_\nu |\nu|^{-n-1}\Big) q_{m+n+1}((c_\nu)),$$

which proves that $(c_\nu) \mapsto \sum a_\nu c_\nu$ is a continuous linear form on $\mathscr{S}(\mathbf{Z}^n)$. Conversely, let u be such a form, so that there exists an integer $m \geqq 0$ and a constant $b > 0$ such that $|u((c_\nu))| \leqq b \cdot q_m((c_\nu))$. For each multi-index $\nu \in \mathbf{Z}^n$, put $e_\nu = (\delta_{\mu\nu})_{\mu \in \mathbf{Z}^n}$, and let $a_\nu = u(e_\nu)$; then we have

$$|a_\nu| \leqq b(1 + |\nu|)^m$$

for all $\nu \in \mathbf{Z}^n$, in other words the family (a_ν) is slowly increasing, and it is clear that u coincides with the continuous linear form $(c_\nu) \mapsto \sum_\nu a_\nu c_\nu$ on $\mathbf{C}^{(\mathbf{Z}^n)}$, hence everywhere because this subspace is dense in $\mathscr{S}(\mathbf{Z}^n)$.

(22.19.6) By virtue of (22.19.4), the Fourier transformation $\mathscr{F}$ is an isomorphism of $\mathscr{S}(\mathbf{Z}^n)$ onto $\mathscr{E}(\mathbf{T}^n)$; its *transpose* (12.15.3) is therefore an *isomorphism* of the space $\mathscr{E}'(\mathbf{T}^n)$ of distributions on $\mathbf{T}^n$ (endowed with the weak topology) onto the weak dual $\mathscr{S}'(\mathbf{Z}^n)$ of $\mathscr{S}(\mathbf{Z}^n)$. We denote this transpose also by $\mathscr{F}$: it is called the *Fourier transformation* on $\mathscr{E}'(\mathbf{T}^n)$. If we identify $\mathscr{S}'(\mathbf{Z}^n)$ with the space of slowly increasing families, the Fourier transform

$\mathscr{F}$T of a distribution $T \in \mathscr{E}'(\mathbf{T}^n)$ is therefore a slowly increasing family (a_ν) such that, for each function $f \in \mathscr{E}(\mathbf{T}^n)$,

(22.19.6.1) $$\langle T, f \rangle = \sum_\nu a_\nu c_{-\nu},$$

if $f(\dot{x}) = \sum_\nu c_\nu \exp(2\pi i(\dot{x} \mid \nu))$ is the expansion of f as a Fourier series.

The a_ν are called the *Fourier coefficients* of the distribution T.

(22.19.7) *The Fourier coefficients of a distribution* $T \in \mathscr{E}'(\mathbf{T}^n)$ *are given by the formula*

(22.19.7.1) $$a_\nu = \langle T, \exp(-2\pi i(\cdot \mid \nu)) \rangle$$

and we have

(22.19.7.2) $$T = \sum_{\nu \in \mathbf{Z}^n} a_\nu \exp(2\pi i(\cdot \mid \nu)),$$

the family on the right-hand side being weakly summable in $\mathscr{E}'(\mathbf{T}^n)$ (which means that for each function $f \in \mathscr{E}(\mathbf{T}^n)$, the family $(a_\nu \langle \exp(2\pi i(\cdot \mid \nu)), f \rangle)$ is absolutely summable, with sum $\langle T, f \rangle$).

The formula (22.19.7.1) is a particular case of (22.19.6.1), with f replaced by the character $\dot{x} \mapsto \exp(-2\pi i(\dot{x} \mid \nu))$, and the formula (22.19.7.2) is just another way of writing (22.19.6.1).

(22.19.8) For each distribution $T \in \mathscr{E}'(\mathbf{T}^n)$, its *inverse image* $T' = {}^t p(T)$ under the canonical mapping $p: \mathbf{R}^n \to \mathbf{T}^n$ (17.4.4) is a *periodic* distribution with period group $\mathbf{Z}^n$, that is to say $\gamma(\nu)T' = T'$ for all $\nu \in \mathbf{Z}^n$; and conversely all the periodic distributions on $\mathbf{R}^n$ with period group $\mathbf{Z}^n$ are obtained in this way. Let (a_ν) be the family of Fourier coefficients of T, so that T is given by the formula (22.19.7.2), on the right-hand side of which $\exp(2\pi i(\cdot \mid \nu))$ is the *character* $\dot{x} \mapsto \exp(2\pi i(\dot{x} \mid \nu))$ *of* $\mathbf{T}^n$. Then we have the formula

(22.19.8.1) $$T' = \sum_{\nu \in \mathbf{Z}^n} a_\nu \exp(2\pi i(\cdot \mid \nu))$$

where this time $\exp(2\pi i(\cdot \mid \nu))$ is the *character* $x \mapsto \exp(2\pi i(x \mid \nu))$ *of* $\mathbf{R}^n$, and the family on the right-hand side is *weakly summable* in $\mathscr{S}'(\mathbf{R}^n)$. Let us first prove this last assertion: for each function $g \in \mathscr{S}(\mathbf{R}^n)$, we have by definition

$\langle g, \exp(2\pi i(\cdot \mid \nu))\rangle = \mathscr{F}g(-\nu)$; since by definition the family $(\mathscr{F}g(-\nu))$ is rapidly decreasing, the family $(a_\nu\langle g, \exp(2\pi i(\cdot \mid \nu))\rangle)$ is absolutely summable (22.19.5), which proves our assertion. By virtue of (22.17.8), the right-hand side of (22.19.8.1) is therefore a tempered distribution. To see that it is equal to T′, it is enough to show that the two sides of (22.19.8.1) have the same restriction to an open set $U \subset \mathbf{R}^n$ such that $p \mid U$ is a diffeomorphism of U onto the open set $p(U) \subset \mathbf{T}^n$ (17.4.3); but this follows immediately in this case from the relation (22.19.7.2) and the definition of T′ (17.4.4).

Since the Fourier transformation is continuous on $\mathscr{S}(\mathbf{R}^n)$, the formula (22.19.8.1) gives the value of $\mathscr{F}T'$: for $\mathscr{F}(\exp(2\pi i(\cdot \mid \nu)))$ is the Dirac measure ε_ν (22.8.6.1), and hence we have

(22.19.8.2) $$\mathscr{F}T' = \sum_{\nu \in \mathbf{Z}^n} a_\nu \varepsilon_\nu,$$

the *discrete measure on* $\mathbf{R}^n$ defined by the masses a_ν at the points $\nu \in \mathbf{Z}^n$.

PROBLEMS

1. (a) Every distribution $T \in \mathscr{E}'(\mathbf{T}^n)$ is the sum of a finite number of translates $\gamma(s_j)T_j$, where each T_j has support contained in $p(\mathring{I}^n)$. If T_j is canonically identified with a distribution on $\mathring{I}^n$ and hence with a distribution $T_j' \in \mathscr{E}'(\mathbf{R}^n)$, then the Fourier coefficient c_ν of T_j is equal to $\mathscr{F}T_j'(\nu)$.
(b) Show that if $T \in \mathscr{E}'(\mathbf{T}^n)$ has order $\leqq m$, its Fourier coefficients c_ν satisfy
$$|c_\nu| \leqq a|\nu|^m$$
for some constant $a > 0$. (Use (a) and Section 17.7, Problem 1.)

2. (a) Let f be a continuous periodic function on $\mathbf{R}$, with period 1. For each $h > 0$, put $\omega(h) = \sup_{t \in \mathbf{R}} |f(t+h) - f(t)|$. Show that the Fourier coefficients c_n of f satisfy the inequalities $|c_n| \leqq \frac{1}{2}\omega(1/2n)$. (Observe that $c_n = -\int_0^1 f(t)\exp(-2\pi i n(t + (1/2n)))\,dt$.)
(b) Consider the function
$$f(t) = \sum_{n=1}^{\infty} \frac{\cos(3^n t)}{3^{n\alpha}}, \qquad \text{where} \quad 0 < \alpha \leqq 1.$$
Show that there exist two constants $a > 0$, $b > 0$ such that $a \cdot h^\alpha \leqq \omega(h) \leqq b \cdot h^\alpha$, and hence that the inequality in (a) cannot be improved (up to a constant factor). (Evaluate $\omega(h)$ for $\pi/3^n \leqq h \leqq \pi/3^{n-1}$.)

3. (a) Let
$$D_n(x) = \frac{\sin(2n+1)\pi x}{\sin \pi x} = \sum_{k=-n}^{n} e^{2\pi i k x}$$
for $x \in \mathbf{R}$.

For each function $f \in \mathscr{L}^1_{\mathbf{C}}(\mathbf{T})$, identified with a function in $\mathscr{L}^1_{\mathbf{C}}(\mathrm{I})$, with Fourier coefficients c_n, let

$$\mathrm{S}_n(f)(x) = \sum_{k=-n}^{n} c_k e^{2\pi i k x}$$

for $x \in \mathbf{R}$: the $\mathrm{S}_n(f)$ are the "partial sums" of the Fourier series of f, and we have $\mathrm{S}_n(f) = \mathrm{D}_n * f$. Let $x \in \mathring{\mathrm{I}} =]0, 1[$; in order that $\mathrm{S}_n(f)(x)$ should tend to a finite limit as $n \to +\infty$, it is necessary and sufficient that the integral

$$\int_0^a \frac{\sin nu}{u} g(u)\, du,$$

where $g(t) = \frac{1}{2}(f(x+t) + f(x-t))$, and a is any positive real number, should tend to a finite limit as $n \to +\infty$. This is the case in particular if, in some neighborhood of x, the function f is equal almost everywhere to a function f_0 which is continuous and differentiable at x, and then $\mathrm{S}_n(f)(x)$ converges to $f_0(x)$ (cf Section **22.17**, Problem 18). What happens at $x = 0$ or $x = 1$?

(b) Show that if f is continuous on $\mathbf{T}$, and therefore may be identified with a periodic function on $\mathbf{R}$, and if $\mathrm{D}f$ (the derivative of f considered as a distribution) belongs to $\mathscr{L}^1_{\mathbf{C}}(\mathrm{I})$, then the sequence $(\mathrm{S}_n(f))$ converges uniformly to f on I. (Remark that we may write

$$f(x) = \int_0^1 f(t)\, dt + \int_0^1 (t - \tfrac{1}{2})\mathrm{D}f(t + x)\, dt$$

and replace $t - \frac{1}{2}$ by its Fourier series expansion, using Section **22.17**, Problem 16(a).)

(c) Let μ be a measure on $\mathbf{T}$, identified with a measure on $\mathbf{R}$ carried by the interval $[0, 1[$. Show that if $0 \leq a < b < 1$ and if $\mu(\{a\}) = \mu(\{b\}) = 0$, then

$$\mu([a, b]) = \lim_{\mathrm{N} \to \infty} \sum_{-\mathrm{N}}^{\mathrm{N}} \frac{c_n}{2\pi i n}(e^{2\pi i n b} - e^{2\pi i n a})$$

where (c_n) is the sequence of Fourier coefficients of μ.

(d) Let μ, ν be two measures on $\mathbf{T}$ and (c_n), (d_n) their respective Fourier coefficients. Show that

$$\lim_{\mathrm{N} \to \infty} \frac{1}{2\mathrm{N} + 1} \sum_{-\mathrm{N}}^{\mathrm{N}} c_n \bar{d}_n = \sum_x \mu(\{x\})\overline{\nu(\{x\})},$$

the sum on the right being over the (at most denumerable) set of points $x \in \mathbf{T}$ such that $\mu(\{x\}) \neq 0$ and $\nu(\{x\}) \neq 0$. (Same method as in Section **22.17**, Problem 11; observe that D_n converges uniformly to 0 on the complement of a neighborhood of 0.)

In particular, we have

$$\sum_x |\mu(\{x\})|^2 = \lim_{\mathrm{N} \to \infty} \frac{1}{2\mathrm{N} + 1} \sum_{-\mathrm{N}}^{\mathrm{N}} |c_n|^2.$$

4. (a) Let (Φ_λ) be a family of continuous functions on $\mathbf{T}$, depending on a parameter λ belonging to a subset L of $\mathbf{R}$, such that (when Φ_λ is identified with a periodic function on $\mathbf{R}$)

$$\Phi_\lambda(x) = \sum_{k=-\infty}^{+\infty} \alpha_k(\lambda) e^{2\pi i k x},$$

where the series $\sum_{k=-\infty}^{+\infty} |\alpha_k(\lambda)|$ is convergent. Suppose also that: (1) $\int_0^1 \Phi_\lambda(x)\,dx = 1$ for all $\lambda \in \mathrm{L}$; (2) $\int_0^1 |\Phi_\lambda(x)|\,dx \leqq \mathrm{M}$ for all $\lambda \in \mathrm{L}$, where M is a constant; (3) as λ tends to λ_0 in the closure of L in $\bar{\mathbf{R}}$, the integral $\int_0^{1-\delta} |\Phi_\lambda(x)|\,dx$ tends to 0, for all δ such that $0 < \delta < \frac{1}{2}$. Show that, for $1 \leqq p < +\infty$ and all functions $f \in \mathscr{L}_{\mathbf{C}}^p(\mathrm{I})$, if (c_n) is the sequence of the Fourier coefficients of f, the continuous function

$$(*) \qquad \Phi_\lambda * f : x \mapsto \sum_{k \in \mathbf{Z}} \alpha_k(\lambda) c_k e^{2\pi i k x}$$

converges in pth power mean to f as $\lambda \to \lambda_0$ (cf. (**14.11.1**)). When $p = +\infty$, the function $(*)$ converges weakly to f as $\lambda \to \lambda_0$, the functions in $\mathscr{L}_{\mathbf{C}}^\infty(\mathrm{I})$ being considered as linear forms on $\mathrm{L}_{\mathbf{C}}^1(\mathrm{I})$).

This result applies in particular in the following two cases:

(1) $\mathrm{L} = \mathbf{N}$, $\lambda_0 = +\infty$, $\alpha_k(n) = (1 - (|k|/(n+1)))^+$, so that the corresponding function Φ_λ is here

$$\mathrm{K}_n(x) = \sum_{k=-n}^{n} \left(1 - \frac{|k|}{n+1}\right) e^{2\pi i k x} = \frac{1}{n+1}\left(\frac{\sin(n+1)\pi x}{\sin \pi x}\right)^2$$
$$= \frac{1}{n+1}(\mathrm{D}_0(x) + \mathrm{D}_1(x) + \cdots + \mathrm{D}_n(x))$$

(*Féjer's kernel*); the function $(*)$ is here, in the notation of Problem 3.

$$\sigma_n(f) = \frac{1}{n+1}(\mathrm{S}_0(f) + \mathrm{S}_1(f) + \cdots + \mathrm{S}_n(f)),$$

and the corresponding method of summation is "Cesàro summation."

(2) $\mathrm{L} =]0, 1[$, $\lambda_0 = 1$, $\alpha_k(r) = r^{|k|}$ for $0 < r < 1$, so that the corresponding function Φ_λ is here

$$\mathrm{P}_0(r, x) = \sum_{k=-\infty}^{+\infty} r^{|k|} e^{2\pi i k x} = 1 + 2\sum_{k=1}^{\infty} r^k \cos 2\pi k x = \mathscr{R}\left(\frac{1 + re^{2\pi i x}}{1 - re^{2\pi i x}}\right) = \frac{1 - r^2}{1 - 2r\cos 2\pi x + r^2}$$

(*Poisson's kernel* (relative to the disc $|z| < 1$)); the function $(*)$ is now the series

$$x \mapsto \sum_{n=-\infty}^{+\infty} c_n r^{|n|} e^{2\pi i n x}$$

and the corresponding method of summation is called "Abel summation."

Furthermore, in this last case, for each function $f \in \mathscr{L}_{\mathbf{C}}^1(\mathbf{R})$, the function $x \mapsto \sum_{n=-\infty}^{+\infty} c_n r^{|n|} e^{2\pi i n x}$ converges almost everywhere to f as r tends to 1 (Section **14.11**, Problem 17(c)).

(b) Suppose in addition that $\Phi_\lambda(x) = \Phi_\lambda(-x)$ and that in each interval $[\delta, 1-\delta]$ (with $\delta > 0$) the functions Φ_λ tend uniformly to 0 as $\lambda \to \lambda_0$. Show that if f is continuous at a point x_0, the value of $(*)$ at x_0 tends to $f(x_0)$ as $\lambda \to \lambda_0$. These conditions are satisfied by the two examples in (a).

If f is periodic and continuous (resp. of class C^k, for some integer k), the sequence of functions $(*)$ converges uniformly to f on I (resp. the hth derivative of $(*)$ converges uniformly to $f^{(h)}$ on I, for $0 \leqq h \leqq k$).

(c) Under the hypotheses of (b), in order that a sequence $(c_n)_{n \in \mathbf{Z}}$ should be the sequence of Fourier coefficients of a positive measure on $\mathbf{T}$, it is necessary and sufficient that as $\lambda \to \lambda_0$ the sequence of norms in $L^1_{\mathbf{C}}(I)$ of the functions $\sum_k \alpha_k(\lambda) c_k e^{2\pi i k x}$ should be bounded. (Argue as in Section 22.17, Problem 19(c).) Hence derive another proof of the fact that the distribution which is the sum of the series $\sum_{n=0}^{\infty} e^{2\pi i n x}$ is not a measure (cf. Section 22.11, Problem 9(f)).

(d) In order that the sequence $(c_n)_{n \in \mathbf{Z}}$ should be the sequence of Fourier coefficients of a positive measure on $\mathbf{T}$, it is necessary and sufficient that the trigonometric polynomials

$$\sum_{k=-n}^{n} \left(1 - \frac{|k|}{n+1}\right) c_k e^{2\pi i k x}$$

should be $\geqq 0$ for all n (use (c)).

(e) If $f \in \mathscr{L}^1_{\mathbf{R}}(I)$ is *real* and continuous at a point $x_0 \in I$, show that there exists a subsequence $(S_{n_k}(f))$ of partial sums of the Fourier series of f such that $(S_{n_k}(f)(x_0))$ converges to $f(x_0)$. (Argue by contradiction: if the result were false, the sequence $(S_n(f)(x_0) - f(x_0))$ would be of constant sign from a certain stage onwards, and of absolute value $\geqq \alpha > 0$ for some α independent of n, thereby contradicting the result of (b).)

(f) Show that if the sequence n_j tends to $+\infty$ sufficiently fast, the Fourier series of the function

$$f(x) = \sum_{j=1}^{\infty} 2^{-j} K_{n_j}(x)$$

does not converge in mean to f (compare with (a)).

5. (a) Show that there exist two constants $a > 0$, $b > 0$, which depend only on n and the integer $k > 0$, such that

$$a|\nu|^{2k} \leqq \sum_{\alpha_1 + \cdots + \alpha_n = k} \nu^{2\alpha} \leqq b|\nu|^{2k}$$

for all $\nu \in \mathbf{Z}^n$.

(b) Let $f \in \mathscr{E}^{(k)}(\mathbf{T}^n)$. If (c_ν) is the family of its Fourier coefficients, show that the family (c_ν) is absolutely summable for $k > \frac{1}{2}n$, and more precisely that there exists a constant C, depending only on n and k, such that

$$(1) \qquad \sum_{\nu \in \mathbf{Z}^n} |c_\nu| \leqq C \cdot \left(\sum_{\alpha_1 + \cdots + \alpha_n = k} (N_2(D^\alpha f))^2 \right)^{1/2},$$

the norm N_2 being that of the space $L^2_{\mathbf{C}}(I^n)$. (Observe that $D^\alpha f \in L^2_{\mathbf{C}}(I^n)$ and use the Plancherel theorem and the Cauchy–Schwarz inequality, together with (a).)

(c) Deduce from (b) that if a distribution T on $\mathbf{T}^n$ is such that its family (c_ν) of Fourier coefficients satisfies an inequality of the form $|c_\nu| \leqq A \cdot |\nu|^m$, where A is a constant and m is an integer $\geqq 0$, then T is of order $\leqq m + 1 + [n/2]$ if m is even, and of order $\leqq m + 2 + [n/2]$ if m is odd. When $n = 1$, T is always of order $\leqq m + 1$. (Consider the family (d_ν) defined (for $\nu \neq 0$) by

$$c_\nu = \begin{cases} d_\nu \left(\sum_{j=1}^{n} \nu_j^2 \right)^{m/2} & (m \text{ even}) \\ d_\nu \left(\sum_{j=1}^{n} \nu_j^2 \right)^{(m+1)/2} & (m \text{ odd}) \end{cases}$$

and $c_\nu = \nu \cdot d_\nu$ if $n = 1$; then the family $(d_\nu \exp(2\pi i(\cdot \mid \nu)))$ is weakly summable in the space $\mathscr{E}'^{(k)}(\mathbf{T}^n)$ for all integers $k > n/2$.)

(d) Deduce from (b) that if $k > n/2$, for each function $f \in \mathscr{D}^{(k)}(\mathbf{R}^n)$, the Fourier transform $\mathscr{F}f$ belongs to $\mathscr{L}^1_{\mathbf{C}}(\mathbf{R}^n)$. (Reduce to the case where the support of f is contained in $\mathbf{I}^n$; decompose $\mathbf{R}^n$ into the union of the cubes $\mathbf{I}^n + \nu$ for $\nu \in \mathbf{Z}^n$, apply the inequality (1) to the functions $\gamma(s)f$, $s \in \mathbf{I}^n$, and use (13.21.9).)

(e) Deduce from (d) that if a measurable function f on $\mathbf{R}^n$ is such that

$$|f(x)| \leqq \mathrm{A}(1 + r(x))^m$$

for some constant $\mathrm{A} > 0$ and an integer $m \geqq 0$, then its Fourier transform is a distribution of order $\leqq m + 1 + [\frac{1}{2}n]$ if m is even or $n = 1$, and of order $\leqq m + 2 + [\frac{1}{2}n]$ if m is odd (argue as in (c)).

6. With the notation of Problem 2, suppose that $\omega(h) \leqq a \cdot h^\alpha$ for some $\alpha > 0$. Show that there exists a constant $b > 0$ such that

$$\sum_{2^m \leq |n| < 2^{m+1}} |c_n| \leqq b \cdot 2^{m(\frac{1}{2} - \alpha)}$$

for all integers $m > 0$. (Consider the Fourier coefficients of the function $x \mapsto f(x - 1/(3 \cdot 2^m)) - f(x)$, and by using the Plancherel theorem obtain a majoration of $\sum_{2^m \leq n < 2^{m+1}} |c_n|^2$.) Conclude that if $\alpha > \frac{1}{2}$, the Fourier series of f is absolutely convergent (*Bernstein's theorem*).

7. Let f be a twice continuously differentiable real function defined on an interval $[a, b] \subset \mathbf{R}$.

(a) Suppose that on $[a, b]$ the derivative f' is monotone and that there exists $\lambda > 0$ such that either $f'(x) \geqq \lambda$ or $f'(x) \leqq -\lambda$ for all $x \in [a, b]$. Show that

$$\left| \int_a^b e^{2\pi i f(x)}\, dx \right| \leqq \frac{2}{\pi\lambda}.$$

$$\left(\text{Write } e^{2\pi i f(x)} = \frac{1}{2\pi i f'(x)} \frac{d}{dx} (e^{2\pi i f(x)}). \right)$$

(b) Show that if there exists $\rho > 0$ such that either $f''(x) \geqq \rho$ or $f''(x) \leqq -\rho$ for $x \in [a, b]$, then

$$\left| \int_a^b e^{2\pi i f(x)}\, dx \right| \leqq 4\rho^{-1/2}.$$

(Reduce to the case where f' is of constant sign on $[a, b]$, by decomposing $[a, b]$ into two intervals; then observe that for each $t \in]a, b[$ we have $\left| \int_a^t e^{2\pi i f(x)}\, dx \right| \leqq t - a$, and $|f'(x)| \geqq (t - a)\rho$ for $x \in [t, b]$; now use (a), and choose t suitably.)

(c) Suppose that f' is monotone and that $|f'(x)| \leqq 1 - \alpha$ for all $x \in [a, b]$, where $0 < \alpha < 1$. Show that for each integer $n \geqq 1$ we have

$$\left| \int_a^b f'(x) e^{2\pi i (f(x) \pm nx)}\, dx \right| \leqq \frac{2(1 - \alpha)}{n - 1 + \alpha}. \tag{1}$$

(Same method as in (a).) Deduce that there exists a constant M depending only on α, and not on a or b, such that

$$\left| \int_a^b e^{2\pi i f(x)}\, dx - \sum_{a<n\leq b} e^{2\pi i f(n)} \right| \leqq \mathrm{M}.$$

(Use the Euler–Maclaurin formula†

$$\mathrm{F}(m) + \mathrm{F}(m+1) + \cdots + \mathrm{F}(n) = \int_m^n \mathrm{F}(t)\, dt + \tfrac{1}{2}(\mathrm{F}(m) + \mathrm{F}(n)) - \sum_{k=1}^{\infty} \frac{1}{k\pi} \int_m^n \mathrm{F}'(t) \sin 2k\pi t\, dt$$

and the majoration (1).)

8. Let α, β be two real numbers such that $0 < \alpha < 1$ and $\beta > 1 - \frac{1}{2}\alpha$ (so that $\beta > \frac{1}{2}$). Show that the series

$$(*) \qquad \sum_{n=1}^{\infty} n^{-\beta} e^{in^\alpha} e^{2\pi i n x}$$

is uniformly convergent in **R** and hence is the Fourier series of a periodic continuous function. (With the help of Problem 7, show that there exists a constant $\mathrm{A} > 0$ such that

$$\left| \sum_{n=1}^{\mathrm{N}} e^{in^\alpha} e^{2\pi i n x} \right| \leqq \mathrm{A}\mathrm{N}^{1-\frac{1}{2}\alpha}$$

for all $x \in \mathbf{R}$ and all integers $\mathrm{N} \geqq 1$. Then use Abel's partial summation formula.) Hence there exist continuous functions on **T** whose Fourier coefficients c_n satisfy $|c_n| = n^{-\beta}$, for each $\beta > \frac{1}{2}$ (cf. Section **13.21**, Problem 14).

9. (a) Let (r_n) be a sequence of real numbers $\geqq 0$, and let (θ_n) be a sequence of real numbers. In order that the series whose general term is $r_n \cos(nx - \theta_n)$ should converge at the points of an integrable set of measure > 0 in **R**, it is necessary that $r_n \to 0$ as $n \to \infty$ (*Cantor–Lebesgue theorem*). (Show that the hypothesis implies that the series is *uniformly* convergent on an integrable set E of measure $\lambda(\mathrm{E}) > 0$. Then argue by contradiction: if the result were false, there would exist a strictly increasing sequence of integers (n_k) such that $\cos(n_k x - \theta_{n_k})$ tends uniformly to 0 for $x \in \mathrm{E}$, and show that this is absurd by considering the integral $\int_{\mathrm{E}} \cos^2(n_k x - \theta_{n_k})\, dx$ and using the Riemann–Lebesgue theorem.)

(b) Show that if the series with general term $r_n \cos(nx - \theta_n)$ is absolutely convergent at all the points of an integrable set of positive measure, then the series with general term r_n is convergent (and consequently the series with general term $r_n \cos(nx - \theta_n)$ is normally convergent in **R**). (*Denjoy–Lusin theorem:* same method.)

Likewise, if the series with general term $r_n^2 \cos^2(nx - \theta_n)$ converges at all the points of an integrable set of measure > 0, then $\sum_n r_n^2 < +\infty$.

10. Let T be a periodic distribution on **R**, and let $(c_n)_{n \in \mathbf{Z}}$ be the sequence of its Fourier coefficients. The series

$$\sum_{n=1}^{\infty} (c_n e^{2\pi i n x} + c_{-n} e^{-2\pi i n x})$$

† See my book, *Infinitesimal Calculus*, Boston (Houghton–Mifflin), 1971, Chap. IX.

cannot converge at the points of an integrable set of measure > 0 unless $\lim_{n \to +\infty} c_n = \lim_{n \to -\infty} c_n = 0$ (Problem 9). We shall assume that these conditions are satisfied in this problem; they imply that T is a distribution of order $\leqq 1$ (Problem 5).

(a) The set $E \subset I = [0, 1]$ where the series with general term $c_n e^{2\pi inx} + c_{-n} e^{-2\pi inx}$ converges, is integrable. If T is a measure, show that the sum of this series is an integrable function on E (use Egoroff's theorem). If the sum is not integrable, T must therefore be a distribution of order 1.

(b) Give an example of a nonzero measure on **T** such that the corresponding Fourier series $c_0 + \sum_{n=1}^{\infty} (c_n e^{2\pi inx} + c_{-n} e^{-2\pi inx})$ converges almost everywhere to 0 (cf. Section 22.17, Problem 29(h)).

(c) Suppose that $c_n = c_{-n}$ for all $n \in \mathbf{Z}$, so that the Fourier series of T is of the form $\frac{1}{2}a_0 + \sum_{n=1}^{\infty} a_n \cos 2\pi nx$, where the sequence (a_n) tends to 0 with $1/n$. Put $\Delta a_n = a_{n+1} - a_n$ and $\Delta^2 a_n = \Delta a_{n+1} - \Delta a_n = a_n - 2a_{n+1} + a_{n+2}$. In order that the Fourier series of T should converge for $0 < x < 1$, it is necessary and sufficient that the series $\sum_{n=1}^{\infty} \Delta a_n \cdot D_n(x)$ should converge for $0 < x < 1$ (Problem 3); this is always the case when $\sum_{n=1}^{\infty} |\Delta a_n| < +\infty$, and then the function $f(x) = \frac{1}{2}a_0 + \sum_{n=1}^{\infty} a_n \cos 2\pi nx$ is continuous on $]0, 1[$. In particular, this is the case when the sequence (a_n) is a decreasing sequence of nonnegative numbers.

If in addition $n \cdot \Delta a_n$ tends to 0, the Fourier series of T will converge for $0 < x < 1$ if and only if the series

$$\sum_{n=1}^{\infty} (n+1)\, \Delta^2 a_n \cdot K_n(x)$$

(Problem 4) converges; this is always the case when $\sum_{n=1}^{\infty} (n+1)|\Delta^2 a_n|$ converges, and in this case the function

$$f(x) = \tfrac{1}{2}a_0 + \sum_{n=1}^{\infty} a_n \cos 2\pi nx$$

is integrable on I and is (identified with) the distribution T. (Use the fact that $K_n \geqq 0$ and $\int_0^1 K_n(x)\,dx = 1$.)

When the a_n are $\geqq 0$ and $\Delta^2 a_n \geqq 0$ for all n, the sequence (a_n) is said to be *convex*. In this case the conditions $\lim_{n \to \infty} n \cdot \Delta a_n = 0$ and $\sum_{n=1}^{\infty} (n+1)\, \Delta^2 a_n < +\infty$ are always satisfied.

(d) Suppose that $c_n = -c_{-n}$, so that the Fourier series of T is of the form $\sum_{n=1}^{\infty} b_n \sin 2\pi nx$, where the sequence (b_n) tends to 0 with $1/n$. For the Fourier series of T to converge for $0 \leqq x \leqq 1$, it is necessary and sufficient that the series $\sum_{n=1}^{\infty} \Delta b_n \cdot E_n(x)$ should converge for $0 \leqq x \leqq 1$, where

$$E_n(x) = \sum_{k=1}^{n} \sin 2\pi kx = \tfrac{1}{2}\sin 2\pi nx + \frac{\sin^2 \pi nx}{\tan \pi x}$$

This is always the case when $\sum_{n=1}^{\infty} |\Delta b_n| < +\infty$, and in particular when the sequence (b_n) is a decreasing sequence of nonnegative numbers; the function $f(x) = \sum_{n=1}^{\infty} b_n \sin 2\pi nx$ is then continuous on $]0, 1[$. When $\sum_{n=1}^{\infty} |\Delta b_n| < +\infty$, the convergence of the series $\sum_{n=1}^{\infty} b_n \sin 2\pi nx$ on $]0, 1[$ is also equivalent to the convergence of the series with general term

$$\Delta b_n \cdot \frac{\sin^2 \pi nx}{\tan \pi x}.$$

If in addition the series $\sum_{n=1}^{\infty} |\Delta b_n| \cdot \log n$ converges, the function f is integrable on I and is (identified with) the distribution T. (Begin by showing that

$$\int_0^{1/2} \frac{\sin^2 n\pi x}{\tan \pi x}\, dx \sim \frac{\log n}{\pi}$$

by dividing the interval $[0, \frac{1}{2}]$ at the points k/n, $1 \leqq k \leqq [\frac{1}{2}n]$.) If the sequence (b_n) is nonnegative and decreasing, then f is integrable on I if and only if $\sum_{n=1}^{\infty} b_n/n < +\infty$. If this condition is not satisfied, then T is a distribution of order 1, whose restriction to $]0, 1[$ is identified with f.

(c) Show that as x tends to 0 through values > 0, we have

$$\sum_{n=1}^{\infty} \frac{\cos 2\pi nx}{n \log n} \sim \log \log \frac{1}{x},$$

$$\sum_{n=1}^{\infty} \frac{\log n}{n} \sin 2\pi nx \sim \frac{\pi}{2} \log \frac{1}{x}.$$

(For the first formula, split the sum into two parts; majorize the sum of the terms for which $n \geqq 1/x$, and compare the sum of the others with the sum $\sum_{n<1/x} 1/(n \log n)$. For the second formula, split the sum into three parts given by $n < \varepsilon/x$, $\varepsilon/x \leqq n \leqq 1/\varepsilon x$, and $n > 1/\varepsilon x$, where $\varepsilon > 0$ is fixed and arbitrarily small, and show that the principal part comes from the second of these three sums.)

Show likewise that

$$\sum_{n=1}^{\infty} \frac{\sin 2\pi nx}{\log n} \sim \frac{1}{2\pi x \log(1/x)}$$

(use (d)). Generalize.

11. Let $(c_n)_{n \in \mathbf{Z}}$ be a sequence of complex numbers such that $\lim_{n\to\infty} c_n = \lim_{n\to\infty} c_{-n} = 0$ and $\sum_{n \in \mathbf{Z}} |c_n|^2 < +\infty$. If (r_n) is the orthonormal system of Rademacher functions (Section **13.21**, Problem 10), define

$$S_n(x, t) = \sum_{k=-n}^{n} \left(1 - \frac{|k|}{n+1}\right)^{+} r_{|n|}(t) c_n e^{2\pi i n x}.$$

Show that there exists a negligible set $N \subset I$ such that for each $t \notin N$ there exists a negligible set $N_t \subset I$ with the property that $\sup_n |S_n(x, t)| = +\infty$ for all $x \notin N_t$. (Note that by virtue of Problem 9 there exists a negligible set $N' \subset I$ such that

$$\sum_{n \in \mathbf{Z}} |c_n e^{2\pi inx} + c_{-n} e^{-2\pi inx}|^2 = +\infty$$

for all $x \notin N'$. Then apply Section **13.21**, Problem 12, and consider the set of pairs $(x, t) \in I \times I$ such that $\sup_n |S_n(x, t)| = +\infty$.) Deduce that if T_t is the distribution of order $\leqq 1$ whose Fourier coefficients are $c_n r_{|n|}(t)$, where $t \notin N$, then there exists no nonempty open interval $J \subset I$ such that $T_t | J$ can be identified with a continuous function (use the local character of Cesàro summation, cf. Problem 4); it can also be shown that $T_t | J$ cannot be identified with an integrable function on J. Compare with Problem 10.

12. (a) If n is an integer $\geqq 3$, we may write $\log \log n = a + 2k\pi$ where k is an integer $\geqq 0$ and $0 \leqq a < 2\pi$. Show that for each integer h such that $0 \leqq h < \frac{1}{8}\pi \log n$, we have

$$|(n + h)(a - \log \log(n + h)) - 2k(n - h)\pi| \leqq \tfrac{1}{4}\pi.$$

(Consider the function $x(a - \log \log x)$ and its first two derivatives, for $n \leqq x \leqq n + h$.)
(b) Deduce from (a) that the partial sums of the series

$$\sum_{n=0}^{\infty} \frac{e^{-in \log \log n}}{(\log n)^{1/2}} e^{2\pi inx}$$

are unbounded, for all $x \in \mathbf{R}$. (Consider the sum of the terms of the series for which $N \leqq n \leqq N + \frac{1}{8}\pi \log N$.)†

13. Let $f \in \mathscr{L}^1_{\mathbf{C}}(I)$. With the notation of Problem 4, we have

$$\lim_{n \to \infty} N_1(f - K_n * f) = 0.$$

(a) Let (α_n) be a sequence of nonnegative real numbers such that (1) $\sum_n \alpha_n/n < +\infty$; (2) $\sum_n \alpha_n N_1(f - K_n * f) < +\infty$. Then the series $g = f + \sum_n \alpha_n(f - K_n * f)$ converges in $\mathscr{L}^1_{\mathbf{C}}(I)$; if (c_n) is the sequence of Fourier coefficients of f, then the Fourier coefficients of g are

$$d_n = c_n\left(1 + \sum_m \alpha_m\left(1 - \left(1 - \frac{m}{n+1}\right)^+\right)\right)$$

where the series on the right is convergent.
(b) Let $b_n = 1 + \sum_m \alpha_m(1 - (1 - (m/(n + 1)))^+)$. Show that the sequence (b_n) is increasing, and that if the sequence (α_n) also satisfies the condition (3) $\sum_n \alpha_n = +\infty$, then the sequence (b_n) tends to $+\infty$, and the sequence $(1/b_n)$ is convex (Problem 10). (To establish the existence of sequences (α_n) satisfying conditions (1), (2), and (3), consider for each integer

† There exist functions $f \in \mathscr{L}^1_{\mathbf{C}}(I)$ whose Fourier series does not converge at any point ([83], [88]).

$k \geqq 1$ the smallest integer n_k such that $N_1(f - K_{n_k} * f) \leqq 2^{-k}$, and distinguish two cases, according as $\sum_k 1/n_k$ is finite or not.)

(c) Deduce from (a) and (b) and Problem 10 that there exists a function $h \in \mathscr{L}_{\mathbf{C}}^1(\mathrm{I})$ such that $f = g * h$ almost everywhere.

(d) Prove likewise that if $f \in \mathscr{L}_{\mathbf{C}}^p(\mathrm{I})$, where $1 < p < +\infty$ (resp. if f is of class C^k and periodic), there exists a function $h \in \mathscr{L}_{\mathbf{C}}^1(\mathrm{I})$ and a function $g \in \mathscr{L}_{\mathbf{C}}^p(\mathrm{I})$ (resp. a periodic function g of class C^k) such that $f = g * h$.

14. The *Hilbert distribution* on $\mathbf{T}$ is the distribution H whose Fourier coefficients are $c_n = -i$ for $n > 0$, $c_n = i$ for $n < 0$, and $c_0 = 0$. For each distribution $S = U + iV$ on $\mathbf{T}$, where U and V are real and with Fourier coefficient $c_0 = 0$, we have $V = H * U$. If S is any distribution on $\mathbf{T}$, $H * S$ is called the *Hilbert transform* of S. For each function $f \in \mathscr{L}_{\mathbf{C}}^2(\mathbf{T})$, we have $H * f \in \mathscr{L}_{\mathbf{C}}^2(\mathbf{T})$ and $N_2(H * f) \leqq N_2(f)$. Show that if H is identified with a periodic distribution on $\mathbf{R}$, its restriction to $]-\frac{1}{2}, \frac{1}{2}[$ is P.V.(cot πx), in the notation of Section 17(9, Problem 1. (Use the expression of $E_n(x)$ given in Problem 10.) The Hilbert distribution H is therefore of order 1.

If the restriction of f to an open $A \subset \mathbf{T}$ is of class C^k, then the restriction of $H * f$ to A is a function of class C^{k-1} (17.12.1).

If the restriction of f to A is analytic, then the restriction of $H * f$ to A is analytic. (Use the method of Section 17.12, Problem 10.)

15. (a) Let S be a distribution on $\mathbf{T}^n$. Show that if there exists a nonmeager subset A of the Banach space $L_{\mathbf{C}}^1(\mathbf{T}^n)$ such that, for each $\tilde{f} \in A$, $S * f$ is a measure on $\mathbf{T}^n$, then S is a measure. (Use Problem 29 of Section 12.16 to show that $S * f$ is a measure for each $f \in \mathscr{L}_{\mathbf{C}}^1(\mathbf{T}^n)$, and that $\|S * f\| \leqq c \cdot N_1(f)$ for some constant $c > 0$; then consider the measures $S * g_k$, where (g_k) is a regularizing sequence (17.1).)

In particular, the set of $\tilde{f} \in L_{\mathbf{C}}^1(\mathbf{T})$ such that $H * f$ (Problem 14) is a measure on $\mathbf{T}$ is meager. Give an example of a function $f \in \mathscr{L}_{\mathbf{C}}^1(\mathbf{T})$ such that $H * f$ is a distribution of order 1 (Problem 10(d)).

(b) Show that if there exists a nonmeager subset A of the Banach space $\mathscr{C}_{\mathbf{C}}(\mathbf{T}^n)$ such that $S * f \in \mathscr{L}_{\mathbf{C}}^\infty(\mathbf{T}^n)$ for all $f \in A$, then S is a measure. (Let B be the set of functions $g \in \mathscr{E}(\mathbf{T}^n)$ such that $N_1(g) \leqq 1$; show that $|\langle S * g, f\rangle| \leqq N_\infty(S * \check{f})$ for $g \in B$ and $f \in A$. Then apply Section 12.16, Problem 30 to the $S * g$ considered as measures on $\mathbf{T}^n$; finally consider the $S * g_k$, where (g_k) is a regularizing sequence.) In particular, the set of functions $f \in \mathscr{C}_{\mathbf{C}}(\mathbf{T})$ such that $H * f$ is a bounded function is meager. Give an example of a function $f \in \mathscr{C}_{\mathbf{C}}(\mathbf{T})$ such that $H * f$ is unbounded (Problem 10(e)).

16. Identify the torus $\mathbf{T} = \mathbf{R}/\mathbf{Z}$ with the circle $\mathbf{U} \subset \mathbf{C}$ of complex numbers of absolute value 1, the point $e^{2\pi i\theta}$ of $\mathbf{U}$ being identified with the class of θ mod 1. Consider the following vector spaces:

(1) The complex vector space $\mathscr{E}'^+(\mathbf{U})$ of distributions on $\mathbf{U}$ whose Fourier coefficients of index < 0 are zero, and its subspace $\mathscr{E}_0'^+(\mathbf{U})$ of codimension 1 consisting of the distributions whose Fourier coefficients of index $\leqq 0$ are zero.

(2) The real vector space $\mathscr{E}'_{\mathbf{R}}(\mathbf{U})$ of *real* distributions on $\mathbf{U}$ and its subspace $\mathscr{E}'_{\mathbf{R},0}(\mathbf{U})$ consisting of real distributions whose Fourier coefficient of index 0 is zero.

(3) The complex vector space $\mathscr{A}\mathscr{L}(\mathbf{U})$ of holomorphic functions $f(z) = \sum_{n=0}^{\infty} c_n z^n$ on the disc $|z| < 1$, for which the sequence $(c_n)_{n \geq 0}$ is slowly increasing.

(a) The mapping $S \mapsto \mathscr{R}S$ is an **R**-linear isomorphism of $\mathscr{E}_0'^{+}(\mathbf{U})$ onto $\mathscr{E}'_{\mathbf{R},0}(\mathbf{U})$, and the inverse isomorphism is $U \mapsto U + iH * U = (\varepsilon_0 + iH) * U$, in the notation of Problem 15.

(b) The mapping which takes each function $f(z) = \sum_{n=0}^{\infty} c_n z^n$ in $\mathscr{A}\mathscr{L}(\mathbf{U})$ to the distribution $S = \sum_{n=0}^{\infty} c_n e^{2\pi i n\theta}$ on $\mathbf{U}$ is a **C**-linear isomorphism of $\mathscr{A}\mathscr{L}(\mathbf{U})$ onto $\mathscr{E}'^{+}(\mathbf{U})$. The inverse isomorphism takes a distribution S to a holomorphic function f such that the function $\theta \mapsto f(re^{2\pi i\theta})$ (for $0 \leqq r < 1$) may be written as

$$K_0(r, \cdot) * S = P_0(r, \cdot) * S,$$

where

$$K_0(r, \theta) = \frac{1}{1 - re^{2\pi i\theta}}$$

is the Cauchy kernel for the unit disc, and

$$P_0(r, \theta) = \frac{1 - r^2}{1 - 2r\cos(2\pi\theta) + r^2}$$

is the Poisson kernel for the unit disc. Furthermore, as $r \to 1$, the function $f(re^{2\pi i\theta})$ tends to the distribution S in $\mathscr{E}'^{+}(\mathbf{U})$ (cf. Problem 4).

(c) The algebra $B \subset \mathscr{C}_{\mathbf{C}}(\mathbf{U})$ defined in (15.3.9) is the intersection $\mathscr{C}_{\mathbf{C}}(\mathbf{U}) \cap \mathscr{E}'^{+}(\mathbf{U})$, which corresponds under the isomorphism in (b) to the set of functions in $\mathscr{A}\mathscr{L}(\mathbf{U})$ which can be extended to continuous functions on the closed unit disc $|z| \leqq 1$; moreover B is dense in $\mathscr{E}'^{+}(\mathbf{U})$ with respect to the weak topology.

(d) If μ is normalized Haar measure on **U**, we have $\mathscr{E}'^{+}(\mathbf{U}) \cap M_{\mathbf{C}}(\mathbf{U}) = H^1(\mu)$ (Section 15.3, Problem 16), and $\mathscr{E}'^{+}(\mathbf{U}) \cap L^p_{\mathbf{C}}(\mu) = H^p(\mu)$ for $1 \leqq p \leqq +\infty$. (Use Problem 4.)

(e) Show that for $1 < p < +\infty$ the mapping $\tilde{f} \mapsto (H * f)^{\sim}$ is a continuous linear mapping of the Banach space $L^p_{\mathbf{C}}(\mu)$ into itself (*M. Riesz' theorem*). (Consider first the case $p \geqq 2$, using Problem 14 for $p = 2$, and then (a) to show that $H * f \in \mathscr{L}^p_{\mathbf{C}}(\mu)$ for all $f \in \mathscr{L}^p_{\mathbf{C}}(\mu)$ when $p \geqq 2$; then use the closed graph theorem (12.16.11). To pass to the case $1 < p < 2$, observe that if f, g are trigonometric polynomials, we have

$$|\langle H * f, g\rangle| \leqq N_p(f) N_q(H * \check{g}),$$

where $p^{-1} + q^{-1} = 1$; then use Section 13.17, Problem 1.)

Deduce that the restriction to $H^p(\mu)$ of the mapping $S \mapsto \mathscr{R}S$ is a surjection of $H^p(\mu)$ onto $L^p_{\mathbf{R}}(\mu)$ for $1 < p \leqq +\infty$. When $H^p(\mu)$ is identified with a subspace of $\mathscr{A}\mathscr{L}(\mathbf{U})$, this mapping is bijective on the subspace of $H^p(\mu)$ consisting of holomorphic functions whose value at $z = 0$ is a real number.

On the other hand, the restriction of $S \mapsto \mathscr{R}S$ to $H^1(\mu)$ is not surjective (Problem 15(a)).

(f) The space $H^p(\mu)$, for $1 \leqq p < +\infty$, may be identified with the subspace of $L^p_{\mathbf{C}}(\mu)$ which is the closure of the set of trigonometric polynomials $\sum_{n=0}^{N} c_n e^{2\pi i n\theta}$ with Fourier coefficients $c_n = 0$ for $n < 0$. Deduce that the partial Fourier sums $S_n(f)$ converge in pth power mean to f, for all $f \in \mathscr{H}^p(\mu)$ (observe that this property is trivial for the trigonometric polynomials). Deduce from (e) that the same is true for functions $f \in \mathscr{L}^p_{\mathbf{C}}(\mu)$, where $1 < p < +\infty$ (cf. Problem 4(f)).

(g) For $1 \leqq p \leqq +\infty$, the space $H^p(\mu)$ may be identified with the subspace of $\mathscr{A}\mathscr{L}(\mathbf{U})$ consisting of holomorphic functions F such that the set of functions $\theta \mapsto F(re^{2\pi i\theta})$, $0 \leqq r < 1$, is bounded in $L^p_{\mathbf{C}}(\mu)$. (Use Problem 4(a).)

17. (a) With the notation of Problem 16, for each nonnegligible function $f \in \mathscr{H}^1(\mu)$, the function $\log|f|$ is μ-integrable (Section **15**.3, Problem 16). Let $\mathrm{F}(z)$ be the corresponding holomorphic function in $\mathrm{H}^1(\mu)$, so that $\theta \mapsto \mathrm{F}(re^{2\pi i\theta})$ is the function $\mathrm{K}_0(r, \cdot) * f$. If (ζ_n) is the finite or infinite sequence of zeros of f in the disc $|z| < 1$, each counted according to its multiplicity, show that the series $\sum_n (1 - |\zeta_n|)$ is convergent. (Use Jensen's formula, cf. Section **22**.18, Problem 19.)
(b) For $|z| < 1$ let

$$b(z, \zeta) = \frac{\bar{\zeta}(\zeta - z)}{|\zeta|(1 - z\bar{\zeta})} \quad (0 < |\zeta| < 1),$$

$$b(z, 0) = z.$$

Let $(\zeta_n)_{n \geq 1}$ be a finite or infinite sequence of complex numbers such that $|\zeta_n| < 1$ for all n, and $\sum_n (1 - |\zeta_n|) < +\infty$. Show that the product $\mathrm{B}(z) = \prod_n b(z, \zeta_n)$ converges absolutely and uniformly in every disc $|z| \leq r < 1$. A holomorphic function B of this form is called a *Blaschke product*; we have $|\mathrm{B}(z)| < 1$ whenever $|z| < 1$, so that B corresponds (Problem 16(b)) to a function in $\mathscr{H}^\infty(\mu)$, defined to within a μ-negligible function, and denoted by $\theta \mapsto \mathrm{B}(e^{2\pi i\theta})$ by abuse of notation; it is therefore the weak limit, and the limit almost everywhere, of the functions $\theta \mapsto \mathrm{B}(re^{2\pi i\theta})$ as $r \to 1$. Show that $|\mathrm{B}(e^{2\pi i\theta})| = 1$ almost everywhere in $\mathbf{U}$. (Argue by contradiction, by noting that if we put $\mathrm{B}_\mathrm{N}(z) = \prod_{n=1}^{\mathrm{N}} b(z, \zeta_n)$, then the class of the function $\theta \mapsto \mathrm{B}(e^{2\pi i\theta})$ is the weak limit in $\mathrm{L}_\mathbf{C}^\infty(\mu)$ of the sequence of functions $\theta \mapsto \mathrm{B}_\mathrm{N}(e^{2\pi i\theta})$.)

18. Let f be a holomorphic function on the disc $|z| < 1$. For each $p > 0$ and $0 \leq r < 1$, put

$$h_p(f, r) = \int_0^1 |f(re^{2\pi i\theta})|^p \, d\theta.$$

Show that the mapping $r \mapsto h_p(f, r)$ is increasing. (Consider first the case $p = 2$ (Section **9**.9, Problem 4). If f does not vanish for $|z| \leq r < 1$, there exists a holomorphic function g defined on a disc $|z| < r + \varepsilon$ such that $|g(z)| = |f(z)|^{p/2}$ in this disc (**16**.**28**.**9**), and $h_p(f, \rho) = h_2(g, \rho)$ for $0 \leq \rho \leq r$. In the general case, if $\zeta_1, \ldots, \zeta_n$ are the zeros of f (each counted according to its multiplicity) in the disc $|z| < r$, consider the function

$$f_1(z) = f(z)\left(\prod_{k=1}^{n} b(z/r, \zeta_k/r)\right)^{-1}$$

and note that $|f(z)| \leq |f_1(z)|$ for $|z| < r$ and $|f_1(z)| = |f(z)|$ for $|z| = r$.)

19. (a) With the hypotheses and notation of Problem 17(a), consider the Blaschke product $\mathrm{B}(z) = \prod_n b(z, \zeta_n)$, so that $\mathrm{G} = \mathrm{FB}^{-1}$ is holomorphic and nonzero in the disc $|z| < 1$. Show that G corresponds to a function $g \in \mathscr{H}^1(\mu)$ such that $|g| = |f|$ almost everywhere. (If we put $\mathrm{B}_\mathrm{N}(z) = \prod_{n=1}^{\mathrm{N}} b(z, \zeta_n)$, use Problem 18 to show that the functions $\theta \mapsto \mathrm{G}(re^{2\pi i\theta})$ from a bounded set in $\mathscr{L}_\mathbf{C}^1(\mu)$, and then use Problem 16(f).)
(b) Show that a function $f \in \mathscr{H}^1(\mu)$ is *exterior* (Section **15**.3, Problem 12) if and only if the corresponding holomorphic function F does not vanish in the disc $|z| < 1$. (Use

Section 15.3, Problem 13(a), by observing that if $F(z) \neq 0$ for $|z| < 1$, we may write $F = e^u$, where u is holomorphic and corresponds to a function $v \in \mathscr{H}^1(\mu)$. For this purpose, use the fact that $\log|f|$ is integrable.) Deduce that in the decomposition $F = GB$ of (a) above, G corresponds to the exterior component and B to the interior component of the canonical decomposition of f (Section 15.3, Problem 13(d)).

(c) If $F(z) = \sum_{n=0}^{\infty} c_n z^n$ is the holomorphic function corresponding to a function $f \in \mathscr{H}^1(\mu)$, show that the series $\sum_{n=0}^{\infty} c_n/n$ is absolutely convergent. (Observe that $\sum_{n=0}^{\infty} n^{-1}c_n e^{2\pi i n\theta}$ is the Fourier series of a continuous function, and consequently that the series $\sum_{n=0}^{\infty} n^{-1}c_n r^n$ tends to a finite limit as $r \to 1$; hence deduce the result in the particular case where all the coefficients c_n are $\geqq 0$. In the general case, write $F(z) = F_1(z)F_2(z)$, where $F_1(z) = \sum_{n=0}^{\infty} a_n z^n$ and $F_2(z) = \sum_{n=0}^{\infty} b_n z^n$ both belong to $H^2(\mu)$ (Section 15.3, Problem 16), and consider the series $\sum_n |a_n| z^n$ and $\sum_n |b_n| z^n$, which also belong to $H^2(\mu)$.)

20. (a) The mapping $z \mapsto (z - i)/(z + i)$ transforms the half-plane $\mathscr{I}z > 0$ into the open disc $|w| < 1$, and the real line $\mathbf{R} : \mathscr{I}z = 0$ into the complement of $\{1\}$ in the circle $|w| = 1$. The inverse mapping is $w \mapsto i(1 + w)/(1 - w)$. Deduce that if $F(w)$ is a function in $H^p(\mu)$, where $1 < p < +\infty$, then $G(z) = F((z - i)/(z + i))$ is a function of the form

$$z \mapsto c + i \int_{\mathbf{R}} \frac{g(t)\,dt}{z - t} \tag{1}$$

where c is a complex constant and g is a measurable real-valued function such that $t \mapsto |g(t)|^p/(1 + t^2)$ is integrable. Consider the converse (cf. Problem 16).
(b) Deduce from (a) that if $G(z)$ is a function of the form (1), the function $t \mapsto |\log|g(t)||/(1 + t^2)$ is integrable (cf. Section 15.3, Problem 16(c)).

21. (a) Let $t \mapsto U(t)$ be a continuous unitary representation of $\mathbf{R}$ on a separable Hilbert space E, and let $\mathbf{a} \in E$ be a unit vector. For each $t \in \mathbf{R}$, let $\mathbf{a}_t = U(t) \cdot \mathbf{a}$, and let E_t be the closed subspace of E generated by the $\mathbf{a}_s$ for $s \leqq t$. Let E_∞ denote the closure of the union of the E_t, $t \in \mathbf{R}$; let $E_{-\infty}$ denote the intersection of the E_t, $t \in \mathbf{R}$; and let E' be the orthogonal supplement of $E_{-\infty}$ in E_∞. Assume that $E_{-\infty} \neq E_\infty$. Show that there exist two closed subspaces M_1, M_2 of E which are orthogonal, have E_∞ as their (direct) sum, are stable under U, and are such that

$$E_0 = (M_1 \cap E_0) \oplus (M_2 \cap E_0);$$

also that there exists an isometry T of $L^2_{\mathbf{C}}(m_{\mathbf{R}})$ onto M_1, such that $T^{-1}U(t)T$ is the linear mapping $\tilde{g} \mapsto (\gamma(t)g)^\sim$; finally, $M_1 \cap E_0$ is the image under T of the subspace of classes of functions in $\mathscr{L}^2_{\mathbf{C}}(m_{\mathbf{R}})$ which are *zero for* $t \geqq 0$. (Apply Section 22.15, Problem 10, and consider the image under T_0 of the space of classes of functions in $\mathscr{L}^2_{\mathbf{F}}(m_{\mathbf{R}})$ of the form $s \mapsto g(s)\mathbf{e}$, where $\mathbf{e}$ is a unit vector in F.)
(b) There exists on the other hand an isometry S of E_∞ onto $L^2_{\mathbf{C}}(\mu)$, where μ is a bounded positive measure on $\mathbf{R}$, such that if we put $\chi_t(s) = \exp(2\pi i t s)$, then

$$S \cdot (U(t) \cdot \mathbf{x}) = \chi_t \cdot (S \cdot \mathbf{x})$$

for all $\mathbf{x} \in E_{\infty}$, and $S \cdot \mathbf{a} = 1$ (the constant function equal to 1), whence $S \cdot \mathbf{a}_t = \chi_t$ (cf. **22.15.1**). We have then $S \cdot (P_{M_j} \cdot \mathbf{x}) = \tilde{\varphi}_{A_j}(S \cdot \mathbf{x})$ for $j = 1, 2$, where A_1 and A_2 form a partition of $\mathbf{R}$ into universally measurable sets such that, if $\mathbf{b}_j = P_{M_j} \cdot \mathbf{a}$, we have $S \cdot b_j = \tilde{\varphi}_{A_j}$ $(j = 1, 2)$ **(15.10.6)**. Let $h \in \mathscr{L}^2_{\mathbf{C}}(m_{\mathbf{R}})$ be the function which is zero for $t \geqq 0$ and is such that $T \cdot \tilde{h} = \mathbf{b}_1$; show that

$$(\mathbf{b}_1 | U(t) \cdot \mathbf{b}_1) = (\tilde{\varphi}_{A_1} | \chi_t \tilde{\varphi}_{A_1}) = \int \varphi_{A_1}(\xi) e^{2\pi i \xi t} \, d\mu(\xi) = \int |\mathscr{F}h(\xi)|^2 e^{2\pi i \xi t} \, d\xi,$$

and deduce that $\varphi_{A_1} \cdot \mu$ is a measure with base $m_{\mathbf{R}}$, whose density relative to $m_{\mathbf{R}}$ is $|\mathscr{F}h|^2$.
(c) Show that the set of functions $\gamma(t)h$, $t \leqq s$, form a total set in the set of functions in $\mathscr{L}^2_{\mathbf{C}}(m_{\mathbf{R}})$ which are zero for $t > s$. Deduce that the set Z of $\xi \in \mathbf{R}$ such that $\mathscr{F}h(\xi) = 0$ is negligible with respect to Lebesgue measure λ. (Observe that φ_Z is orthogonal to all the functions $\gamma(t)h$.) Show that the Fourier–Laplace transform $\mathscr{F}\mathscr{L}h$, which is defined and holomorphic on the half-plane $\mathscr{I}\zeta > 0$ (Section **22.18**, Problem 6), does not vanish in this half-plane. (If $\mathscr{F}\mathscr{L}h(\zeta) = 0$, then $\exp(-2\pi i \zeta x)$ would be orthogonal to all the functions $\gamma(t)h$, $t \leqq 0$.) Moreover the function $\log|h(t)|/(1 + t^2)$ is integrable with respect to Lebesgue measure (use Problem 20).
(d) Deduce from (c) that $\mu = |\mathscr{F}h|^2 \cdot \lambda + \nu$, where $\nu = \varphi_{A_2} \cdot \mu$ is disjoint from λ. Also we have

$$(\mathbf{b}_2 | U(t) \cdot \mathbf{b}_2) = \int e^{2\pi i \xi t} \, d\nu(\xi);$$

deduce that M_2 is the closed subspace generated by the $U(t) \cdot \mathbf{b}_2$ for $t \leqq s$, where s is any real number. (Repeat the argument of (a), replacing E by M_2 and $\mathbf{a}$ by $\mathbf{b}_2$, and show that if M_2 were not generated by the $U(t) \cdot \mathbf{b}_2$, $t \leqq s$, then ν would not be disjoint from λ.) Conclude that $M_2 \subset E_s$ for all $s \in \mathbf{R}$, hence that $M_2 \subset E_{-\infty}$, and finally that $M_1 = E'$, $M_2 = E_{-\infty}$.
(e) For $\theta > 0$, the projection $\mathbf{c}_{t,\theta}$ of $\mathbf{a}_{t+\theta}$ on E_t is called the "*statistical prediction* of $\mathbf{a}_{t+\theta}$ by means of the values $\mathbf{a}_s$ for $s \leqq t$". Show that $\|\mathbf{a}_{t+\theta} - \mathbf{c}_{t,\theta}\|$ (the "error in prediction") is equal to

$$(|(h * h)(0)|^2 - |(h * h)(\theta)|^2)^{1/2}$$

(*Wiener–Kolmogoroff theorem*).

22. (a) Let f be a function in $\mathscr{E}^{(N)}(\mathbf{T})$, identified with a periodic function of class C^N on $\mathbf{R}$. Show that the sequence $(N_2(D^n f))_{0 \leq n \leq N}$ is increasing and that the sequence $(\log N_2(D^n f))_{0 \leq n \leq N}$ is convex (Problem 10). (Use the expression of $N_2(D^n f)$ in terms of the Fourier coefficients of f, and the Cauchy–Schwarz inequality.)
(b) Let $(M_n)_{n \geq 0}$ be an increasing sequence of positive real numbers such that the sequence $(\log M_n)$ is convex and $(\log M_n)/n$ tends to $+\infty$ with n. Put $\tau(r) = \inf_{n \geq 0}(M_n r^{-n})$ for $r > 0$. If $\mu_0 = M_0^{-1}$ and $\mu_n = M_{n-1}/M_n$ for $n \geqq 1$, the sequence $(\mu_n)_{n \geq 1}$ is decreasing; deduce that

$$\tau(r) = \mu_0 \cdot \prod_{n \geq 1,\, \mu_n r > 1} (\mu_n r)^{-1}.$$

Let $m(r)$ denote the number of indices n such that $\mu_n r \geqq e$. Show that

$$m(r) \leqq -\log \tau(r) + \log \mu_0$$

and deduce the inequality

$$\sum_{n=1}^{\infty} \mu_n \leqq 2e^4\left(\int_{e^2}^{\infty} \frac{-\log \tau(r)}{1+r^2}\, dr + \pi \log \mu_0\right).$$

(Split up the range of integration into the intervals $[e^k, e^{k+1}]$, $k \geqq 2$.)

(c) Under the conditions of (b), let $C\{M_n\}$ denote the set of functions $f \in \mathscr{E}(\mathbf{T})$, identified with periodic C^∞-functions on $\mathbf{R}$, such that $N_2(D^n f) \leqq M_n R^n$ for all $n \geqq 0$ and some $R > 0$ (depending on f). For $f \in C\{M_n\}$, put

$$F(\zeta) = \int_0^1 f(t) e^{-2\pi i \zeta t}\, dt$$

(the Fourier–Laplace transform of the function $f\varphi_{[0,1]}$). Show that if $D^n f(0) = 0$ for all n, then

$$\log|F(\zeta)| \leqq \log|\tau(|\zeta|)|$$

for $\mathscr{I}\zeta < 0$. Deduce that if

$$\int_1^{\infty} \frac{\log \tau(r)}{1+r^2}\, dr = -\infty,$$

then $f = 0$. (Use Problem 20.)

(d) Show that if $\sum_{n=1}^{\infty} \mu_n < +\infty$, there exists a function $f \in C\{M_n\}$ which is not identically zero but vanishes at all points of a nonempty open interval. (Use Section **22.18**, Problem 19(e).)

(e) The set $C\{M_n\}$ is said to be *quasi-analytic* if the only function $f \in C\{M_n\}$ such that all the derivatives of f vanish at some $x_0 \in \mathbf{R}$ is the zero function. Show that the following three conditions are equivalent:

(α) $C\{M_n\}$ is quasi-analytic.

(β) $\displaystyle\int_1^{\infty} \frac{\log \tau(r)}{1+r^2}\, dr = -\infty.$

(γ) $\displaystyle\sum_{n=1}^{\infty} (M_n/M_{n+1}) = +\infty.$

(*Denjoy–Carleman theorem*; use (b), (c), and (d).)

20. SOBOLEV SPACES

(22.20.1) Let s be any real number, let λ be Lebesgue measure on $\mathbf{R}^n$, and let μ_s denote the measure $(1+r^2)^s \cdot \lambda$ on $\mathbf{R}^n$. The linear mapping

$$\tilde{f} \mapsto ((1+r^2)^{-s/2} f)^{\sim}$$

is an isometry of the Hilbert space $L^2_{\mathbf{C}}(\mathbf{R}^n)$ onto the Hilbert space $L^2_{\mathbf{C}}(\mathbf{R}^n, \mu_s)$ **(13.14.3)**. Since the functions in $\mathscr{L}^2_{\mathbf{C}}(\mathbf{R}^n)$ may be identified with tempered distributions (22.17.2), and since the function $(1 + r^2)^{-s/2}$ is tempered, it follows that the functions in $\mathscr{L}_{\mathbf{C}}^2(\mathbf{R}^n, \mu_s)$ are also identified with tempered distributions (22.17.3). We denote by $\mathrm{H}^s(\mathbf{R}^n)$ (or simply H^s) the vector space of *Fourier cotransforms* of distributions belonging to $\mathrm{L}_{\mathbf{C}}^2(\mathbf{R}^n, \mu_s)$, endowed with the structure of a (separable) *Hilbert space* obtained by transporting the Hilbert space structure of $\mathrm{L}_{\mathbf{C}}^2(\mathbf{R}^n, \mu_s)$ by means of the bijection $\overline{\mathscr{F}}$; H^s is called the *Sobolev space* of index s. In other words, H^s is the vector space of distributions $\mathrm{T} \in \mathscr{S}'(\mathbf{R}^n)$ such that the distribution $\mathscr{F}\mathrm{T}$ is a locally integrable *function*, and such that the function $(1 + r^2)^s |\mathscr{F}\mathrm{T}|^2$ is integrable; and the scalar product and norm on the Hilbert space H^s are given by the formulas

(22.20.1.1) $$(\mathrm{S} \mid \mathrm{T})_s = \int_{\mathbf{R}^n} (1 + |\xi|^2)^s \mathscr{F}\mathrm{S}(\xi)\overline{\mathscr{F}\mathrm{T}(\xi)}\, d\xi,$$

(22.20.1.2) $$\|\mathrm{T}\|_s^2 = \int_{\mathbf{R}^n} (1 + |\xi|^2)^s |\mathscr{F}\mathrm{T}(\xi)|^2\, d\xi.$$

It is clear that if $\mathrm{T} \in \mathrm{H}^s$, then also $\overline{\mathrm{T}} \in \mathrm{H}^s$ and $\check{\mathrm{T}} \in \mathrm{H}^s$, and we have

(22.20.1.3) $$\|\overline{\mathrm{T}}\|_s = \|\mathrm{T}\|_s, \qquad \|\check{\mathrm{T}}\|_s = \|\mathrm{T}\|_s$$

by virtue of (22.17.5.2). For $s = 0$, Plancherel's theorem shows that $\mathrm{H}^0(\mathbf{R}^n)$ may be canonically identified with $\mathrm{L}_{\mathbf{C}}^2(\mathbf{R}^n)$.

(22.20.2) (i) *The spaces* $\mathscr{S}(\mathbf{R}^n)$ *and* $\mathscr{D}(\mathbf{R}^n)$ *are dense in* $\mathrm{H}^s(\mathbf{R}^n)$, *and the canonical injection* $\mathscr{S} \to \mathrm{H}^s$ *is continuous.*

(ii) *If* $s \geqq t$ *we have* $\mathrm{H}^s \subset \mathrm{H}^t$, *and the canonical injection* $\mathrm{H}^s \to \mathrm{H}^t$ *is a continuous linear mapping of norm* $\leqq 1$.

(iii) *For* $s > \frac{1}{2}n$, *the distributions in* H^s *are bounded continuous functions.*

(iv) *For* $s < -\frac{1}{2}n$, *the space* H^s *contains the Fourier cotransforms of the functions in* $\mathscr{L}_{\mathbf{C}}^{\infty}(\mathbf{R}^n)$.

(i) Since the mapping $f \mapsto (1 + r^2)^{-s/2} f$ is an automorphism of the Fréchet space $\mathscr{S}(\mathbf{R}^n)$ transforming $\mathscr{D}(\mathbf{R}^n)$ into itself (22.16.9), it is enough to prove the assertion when $s = 0$. In that case we know that $\mathscr{D}(\mathbf{R}^n)$ is dense in $\mathrm{L}_{\mathbf{C}}^2(\mathbf{R}^n)$ ((14.11.1) and (17.11.11)); also we have

$$\mathrm{N}_2(f) \leqq q_{0,n}(f) \cdot \int_{\mathbf{R}^n} \frac{dx}{(1 + r^2(x))^n}$$

for all functions $f \in \mathscr{S}(\mathbf{R}^n)$, which shows that the canonical injection $\mathscr{S} \to \mathrm{L}^2$ is continuous.

(ii) The fact that $H^s \subset H^t$ if $s > t$ is a consequence of the inequality $(1 + |\xi|^2)^{s-t} \geqq 1$, and this same inequality shows that $\|T\|_s^2 \leqq \|T\|_t^2$.

(iii) It is enough to show that $L^1_{\mathbf{C}}(\mathbf{R}^n) \supset L^2_{\mathbf{C}}(\mathbf{R}^n, \mu_s)$ when $s > \frac{1}{2}n$. But the function $(1 + r^2)^{-s/2}$ then belongs to $\mathscr{L}^2_{\mathbf{C}}(\mathbf{R}^n)$ (16.24.9.6) and therefore, for each $f \in \mathscr{L}^2_{\mathbf{C}}(\mathbf{R}^n, \mu_s)$, by writing f in the form $((1 + r^2)^{s/2} f) \cdot (1 + r^2)^{-s/2}$, we see that f is the product of two functions in $\mathscr{L}^2_{\mathbf{C}}(\mathbf{R}^n)$ and hence belongs to $\mathscr{L}^1_{\mathbf{C}}(\mathbf{R}^n)$ (13.11.7).

(iv) It is enough to show that $L^\infty_{\mathbf{C}}(\mathbf{R}^n) \subset L^2_{\mathbf{C}}(\mathbf{R}^n, \mu_s)$ for $s < -\frac{1}{2}n$. This follows again from the fact that $(1 + r^2)^{s/2}$ belongs to $\mathscr{L}^2_{\mathbf{C}}(\mathbf{R}^n)$ for $s < -\frac{1}{2}n$ (16.24.9.6), hence so also does $(1 + r^2)^{s/2} f$ for $f \in \mathscr{L}^\infty_{\mathbf{C}}(\mathbf{R}^n)$.

(22.20.3) When s is an *integer* $m \geqq 0$, to say that $(1 + r^2)^m |\mathscr{F}T|^2$ is integrable is equivalent to saying that $r^{2h} |\mathscr{F}T|^2$ is integrable for $0 \leqq h \leqq m$, or again that the function $\xi \mapsto \xi^\nu \mathscr{F}T(\xi)$ belongs to $\mathscr{L}^2_{\mathbf{C}}(\mathbf{R}^n)$ for all multi-indices ν satisfying $|\nu| \leqq m$. By virtue of (22.17.5.5) and Plancherel's theorem, this also means that the distribution T and its derivatives $D^\nu T$, $|\nu| \leqq m$, *may be identified with functions in* $\mathscr{L}^2_{\mathbf{C}}(\mathbf{R}^n)$; for the same reason, the norm $\|T\|_m$ on H^m is *equivalent* to the norm

$$f \mapsto \left(\sum_{|\nu| \leq m} \int_{\mathbf{R}^n} |D^\nu f(x)|^2 \, dx \right)^{1/2}.$$

(22.20.4) Likewise, when $s = -m$, where m is an *integer* $\geqq 0$, we have by hypothesis $\mathscr{F}T = (1 + r^2)^{m/2} g$, where $g \in \mathscr{L}^2_{\mathbf{C}}(\mathbf{R}^n)$. Now we may write $(1 + r^2)^{m/2} = P + h$, where P is a *polynomial* of degree m, and h is a bounded function (by using the asymptotic expansion of $(1 + t)^{m/2}$ in a neighborhood of $+\infty$). Hence an equivalent statement is that $\mathscr{F}T$ may be written as a finite sum $\xi \mapsto \sum_{|\nu| \leq m} \xi^\nu g_\nu(\xi)$, with the g_ν in $\mathscr{L}^2_{\mathbf{C}}(\mathbf{R}^n)$. In view of (22.17.5.6) and Plancherel's theorem, this may also be expressed by saying that $T = \sum_{|\nu| \leq m} D^\nu f_\nu$, *where the* f_ν *belong to* $\mathscr{L}^2_{\mathbf{C}}(\mathbf{R}^n)$.

(22.20.5) By definition, the mapping $T \mapsto ((1 + r^2)^{s/2} \mathscr{F}T)^\sim$ is an isomorphism of the Hilbert space $H^s(\mathbf{R}^n)$ onto $L^2_{\mathbf{C}}(\mathbf{R}^n)$. The elementary properties of Hilbert spaces (6.3.2) therefore show that, for each distribution $S \in H^{-s}(\mathbf{R}^n)$, there exists a unique distribution $l(S) \in H^s(\mathbf{R}^n)$ such that

$$(22.20.5.1) \quad (T \mid l(S))_s = \int_{\mathbf{R}^n} (1 + |\xi|^2)^{s/2} \mathscr{F}T(\xi) \cdot (1 + |\xi|^2)^{-s/2} \overline{\mathscr{F}S(\xi)} \, d\xi,$$

and furthermore l is a linear isometry of H^{-s} onto H^s, or in other words

$$\|l(S)\|_s = \|S\|_{-s}.$$

By means of the isometry l, we may therefore identify H^{-s} with the *dual* of the Banach space H^s (12.15).

(22.20.6) *For each index j such that $1 \leqq j \leqq n$, the derivation D_j is a continuous linear mapping of H^s into H^{s-1}.*

By (22.17.5.5), for each distribution $T \in H^s$, $\mathscr{F}(D_j T)$ is equal to the locally integrable function $\xi \mapsto 2\pi i \xi_j . \mathscr{F}T(\xi)$, and we have

$$\xi_j^2(1 + |\xi|^2)^{s-1} \leqq (1 + |\xi|^2)^s,$$

from which the result follows.

(22.20.7) It follows from (22.20.6) that, for each integer $m \geqq 0$, the distributions in H^s, where $s > m + \frac{1}{2}n$, are *functions of class* C^m. The *intersection* of the spaces H^s for all $s \in \mathbf{R}$, which we denote by $H^\infty(\mathbf{R}^n)$ or just H^∞, is therefore a vector space consisting of C^∞-functions; it may be described as the space of C^∞-functions *all* of whose derivatives belong to $\mathscr{L}_C^2(\mathbf{R}^n)$.

The *union* of the spaces H^s for all $s \in \mathbf{R}$, which we denote by $H^{-\infty}(\mathbf{R}^n)$ or just $H^{-\infty}$, is a space of distributions of *finite order*, which may be characterized as finite sums of derivatives (of any order) of functions in $\mathscr{L}_C^2(\mathbf{R}^n)$ (22.20.4). In particular, the space $\mathscr{E}'(\mathbf{R}^n)$ of distributions with compact support is *contained in* $H^{-\infty}(\mathbf{R}^n)$, by virtue of (17.12.4).

(22.20.8) *For each real number s, each function $g \in \mathscr{S}(\mathbf{R}^n)$ and each distribution $T \in H^s(\mathbf{R}^n)$, the distribution $g \cdot T$ belongs to $H^s(\mathbf{R}^n)$, and the bilinear mapping $(g, T) \mapsto g \cdot T$ of $\mathscr{S}(\mathbf{R}^n) \times H^s(\mathbf{R}^n)$ into $H^s(\mathbf{R}^n)$ is continuous.*

Let $u \in \mathscr{S}(\mathbf{R}^n)$, then we have $gu \in \mathscr{S}(\mathbf{R}^n)$. We shall show that there exist two integers m, t and a constant c_s depending only on s, such that

(22.20.8.1)
$$\|gu\|_s \leqq c_s q_{m,t}(g) \|u\|_s.$$

Since $\mathscr{S}(\mathbf{R}^n)$ is dense in $H^s(\mathbf{R}^n)$ (22.20.2), and $H^s(\mathbf{R}^n)$ is complete, this will establish (5.5.4) that $g \cdot T \in H^s$ for all distributions $T \in H^s$, and that in (22.20.8.1) we may replace u by T and gu by $g \cdot T$, which will complete the proof. Since $\mathscr{F}(gu) = (\mathscr{F}g) * (\mathscr{F}u)$ (22.18.2), we have by definition

(22.20.8.2)
$$\begin{aligned}\|gu\|_s^2 &= \int \left| \int \mathscr{F}g(\xi - \eta) \mathscr{F}u(\eta)\, d\eta \right|^2 (1 + |\xi|^2)^s \, d\xi \\ &\leqq \int \left(\int (1 + |\xi|^2)^{s/2} |\mathscr{F}g(\xi - \eta) \mathscr{F}u(\eta)| \, d\eta \right)^2 d\xi.\end{aligned}$$

Next, there exists a constant $a_s > 0$ such that, for each pair of vectors $\xi, \eta \in \mathbf{R}^n$ we have

(22.20.8.3) $$(1 + |\xi|^2)^s \leqq a_s(1 + |\xi - \eta|^2)^{|s|}(1 + |\eta|^2)^s.$$

For if $s \geqq 0$ this follows directly from **(22.18.3.1)**, with $a_s = 2^s$. If $s < 0$, we have therefore

$$(1 + |\eta|^2)^{-s} \leqq 2^{-s}(1 + |\xi - \eta|^2)^{|s|}(1 + |\xi|^2)^{-s}$$

and **(22.20.8.3)** follows immediately, with $a_s = 2^{-s}$. If we now put

$$g_1(\xi) = (1 + |\xi|^2)^{s/2}|\mathscr{F}g(\xi)|,$$
$$u_1(\xi) = (1 + |\xi|^2)^{s/2}|\mathscr{F}u(\xi)|,$$

it follows from **(22.20.8.2)** and **(22.20.8.3)** that

$$\|gu\|_s^2 \leqq a_{s/2}^2(\mathrm{N}_2(g_1 * u_1))^2 \leqq a_{s/2}^2(\mathrm{N}_1(g_1))^2(\mathrm{N}_1(u_1))^2,$$

by making use of **(14.10.6.1)**. But by definition $\mathrm{N}_2(u_1) = \|u\|_s$; also we have

$$\mathrm{N}_1(g_1) \leqq q_{0,\,n+\frac{1}{2}|s|}(\mathscr{F}g) \cdot \int_{\mathbf{R}^n} \frac{d\xi}{(1 + |\xi|^2)^n},$$

and by reason of the continuity of $\mathscr{F}$ on $\mathscr{S}(\mathbf{R}^n)$, this completes the proof of **(22.20.8.1)**.

(22.20.9) Consider a *regularizing sequence* **(17.1.2)** (g_k) in $\mathbf{R}^n$ defined by $g_k(x) = k^n g(kx)$. For each distribution $\mathrm{T} \in \mathrm{H}^s(\mathbf{R}^n) \subset \mathscr{S}'(\mathbf{R}^n)$, we know that the convolutions $g_k * \mathrm{T}$ are *tempered functions* **(22.18.4)**. Furthermore:

(22.20.10) *For each distribution* $\mathrm{T} \in \mathrm{H}^s(\mathbf{R}^n)$, *the functions* $g_k * \mathrm{T}$ *belong to* $\mathrm{H}^s(\mathbf{R}^n)$. *Moreover, the sequence of linear mappings* $\mathrm{T} \mapsto g_k * \mathrm{T}$ *of* $\mathrm{H}^s(\mathbf{R}^n)$ *into itself is equicontinuous, and converges simply to the identity mapping.*

Put $h = \mathscr{F}g$, which belongs to $\mathscr{S}(\mathbf{R}^n)$. Then the Fourier transform $\mathscr{F}g_k$ is the function $\xi \mapsto h(\xi/k)$, and because $|h(\xi)| \leqq \int g(x)\,dx = 1 = h(0)$, it follows that $|h(\xi/k)| \leqq 1$ for all $\xi \in \mathbf{R}^n$ and all integers $k \geqq 1$, and the sequence $(h(\xi/k))$ tends to 1 for each $\xi \in \mathbf{R}^n$. Since $\mathscr{F}(g_k * \mathrm{T}) = (\mathscr{F}g_k)(\mathscr{F}\mathrm{T})$ **(22.18.4)**, we see that $g_k * \mathrm{T} \in \mathrm{H}^s(\mathbf{R}^n)$ and that $\|g_k * \mathrm{T}\|_s \leqq \|\mathrm{T}\|_s$, thus proving the first two assertions **(12.15.7.1)**. Finally, we have

$$\|g_k * \mathrm{T} - \mathrm{T}\|_s^2 = \int (1 + |\xi|^2)^s |(\mathscr{F}g_k(\xi) - 1)\mathscr{F}\mathrm{T}(\xi)|^2\,d\xi$$

and the convergence of $(g_k * \mathrm{T})$ to T results from the dominated convergence theorem **(13.8.4)**.

PROBLEMS

1. For each p such that $1 \leqq p \leqq +\infty$, let $\mathscr{S}'_p$ (or $\mathscr{S}'_p(\mathbf{R}^n)$) be the vector space of distributions on $\mathbf{R}^n$ which are sums of derivatives (of arbitrary order) of functions in $\mathscr{L}^p_{\mathbf{C}}(\mathbf{R}^n)$. We have therefore $\mathscr{S}'_p \subset \mathscr{S}'$ for all p, and $\mathscr{S}'_2 = \mathrm{H}^{-\infty}$. Show that for T to belong to $\mathscr{S}'_p$ it is necessary and sufficient that, for each function $f \in \mathscr{D}(\mathbf{R}^n)$, the function $\mathrm{T} * f$ should belong to $\mathscr{L}^p_{\mathbf{C}}(\mathbf{R}^n)$. (If this condition is satisfied, observe that if $q = p/(p-1)$ and if B denotes the set of functions $g \in \mathscr{L}^q_{\mathbf{C}}(\mathbf{R}^n)$ such that $\mathrm{N}_q(g) \leqq 1$, the set of functions $\mathrm{T} * g$ with $g \in \mathrm{B} \cap \mathscr{D}(\mathbf{R}^n)$ is bounded in $\mathscr{D}'(\mathbf{R}^n)$. Argue as in Section **17.12**, Problem 6 to show that for each relatively compact open set $\mathrm{U} \subset \mathbf{R}^n$ there exists an integer $m > 0$ such that $\mathrm{T} * f \in \mathscr{L}^p_{\mathbf{C}}(\mathbf{R}^n)$ for all $f \in \mathscr{D}^{(m)}(\mathbf{R}^n; \mathrm{U})$, and complete the proof as in *loc. cit.*).

   Deduce that $\mathscr{S}'_1$ is the space of *summable* distributions on $\mathbf{R}^n$ (Section **17.11**, Problem 1).

2. Show that if

$$\frac{1}{p} + \frac{1}{q} - 1 = \frac{1}{r}$$

   with $p, q, r \in [1, +\infty]$, the convolution product $\mathrm{S} * \mathrm{T}$ of two distributions $\mathrm{S} \in \mathscr{S}'_p$, $\mathrm{T} \in \mathscr{S}'_q$ is defined, and belongs to $\mathscr{S}'_r$.

3. Show that if $1 \leq p \leq 2$, for each distribution $\mathrm{T} \in \mathscr{S}'_p$ the Fourier transform $\mathscr{F}\mathrm{T}$ is the product of a polynomial and a function belonging to $\mathscr{L}^q_{\mathbf{C}}(\mathbf{R}^n)$, where $q = p/(p-1)$.

REFERENCES

VOLUME I

[1] Ahlfors, L., "Complex Analysis," McGraw-Hill, New York, 1953.
[2] Bachmann, H., "Transfinite Zahlen" (Ergebnisse der Math., Neue Folge, Heft 1). Springer, Berlin, 1955.
[3] Bourbaki, N., "Eléments de Mathématique," Livre I, "Théorie des ensembles" (Actual. Scient. Ind., Chaps. I, II, No. 1212; Chap. III, No. 1243). Hermann, Paris, 1954–1956.
[4] Bourbaki, N., "Eléments de Mathématique," Livre II, "Algèbre" (Actual. Scient. Ind., Chap. II, Nos. 1032, 1236, 3rd ed.). Hermann, Paris, 1962.
[5] Bourbaki, N., "Eléments de Mathématique," Livre III, "Topologie générale" (Actual. Scient. Ind., Chaps. I, II, Nos. 858, 1142, 4th ed.; Chap. IX, No. 1045, 2nd ed.; Chap. X, No. 1084, 2nd ed.). Hermann, Paris, 1958–1961.
[6] Bourbaki, N., "Eléments de Mathématique," Livre, V, "Espaces vectoriels topologiques" (Actual. Scient. Ind., Chap. I, II, No. 1189, 2nd ed.; Chaps. III–V, No. 1229). Hermann, Paris, 1953–1955.
[7] Cartan, H., Séminaire de l'Ecole Normale Supérieure, 1951–1952: "Fonctions analytiques et faisceaux analytiques."
[8] Cartan, H., "Théorie Élémentaire des Fonctions Analytiques." Hermann, Paris, 1961.
[9] Coddington, E., and Levinson, N., "Theory of Ordinary Differential Equations." McGraw-Hill, New York, 1955.
[10] Courant, R., and Hilbert, D., "Methoden der mathematischen Physik," Vol. I, 2nd ed. Springer, Berlin, 1931.
[11] Halmos, P., "Finite Dimensional Vector Spaces," 2nd ed. Van Nostrand-Reinhold, Princeton, New Jersey, 1958.
[12] Ince, E., "Ordinary Differential Equations," Dover, New York, 1949.
[13] Jacobson, N., "Lectures in Abstract Algebra," Vol. II, "Linear algebra." Van Nostrand-Reinhold, Princeton, New Jersey, 1953.
[14] Kamke, E., "Differentialgleichungen reeller Funktionen." Akad. Verlag, Leipzig, 1930.
[15] Kelley, J., "General Topology." Van Nostrand-Reinhold, Princeton, New Jersey, 1955.
[16] Landau, E., "Foundations of Analysis." Chelsea, New York, 1951.
[17] Springer, G., "Introduction to Riemann Surfaces." Addison-Wesley, Reading, Massachusetts, 1957.
[18] Weil, A., "Introduction à l'Étude des Variétés Kählériennes" (Actual. Scient. Ind., No. 1267). Hermann, Paris, 1958.
[19] Weyl, H., "Die Idee der Riemannschen Fläche," 3rd ed. Teubner, Stuttgart, 1955.

VOLUME II

[20] Akhiezer, N., "The Classical Moment Problem." Oliver and Boyd, Edinburgh–London, 1965.

[21] Arnold, V. and Avez, A., "Théorie Ergodique des Systèmes Dynamiques." Gauthier-Villars, Paris, 1967.

[22] Bourbaki, N., "Eléments de Mathématique," Livre VI, "Intégration" (Actual. Scient. Ind., Chap. I–IV, No. 1175, 2nd ed., Chap. V, No. 1244, 2nd ed., Chap. VII–VIII, No. 1306). Hermann, Paris, 1963–67.

[23] Bourbaki, N., "Eléments de Mathématique: Théories Spectrales" (Actual. Scient. Ind., Chap. I, II, No. 1332). Hermann, Paris, 1967.

[24] Dixmier, J., "Les Algèbres d'Opérateurs dans l'Espace Hilbertien." Gauthier-Villars, Paris, 1957.

[25] Dixmier, J., "Les C*-Algèbres et leurs Représentations." Gauthier-Villars, Paris, 1964.

[26] Dunford, N. and Schwartz, J., "Linear Operators. Part II: Spectral Theory." Wiley (Interscience), New York, 1963.

[27] Hadwiger, H., "Vorlesungen über Inhalt, Oberfläche und Isoperimetrie." Springer, Berlin, 1957.

[28] Halmos, P., "Lectures on Ergodic Theory." Math. Soc. of Japan, 1956.

[29] Hoffman, K., "Banach Spaces of Analytic Functions." New York, 1962.

[30] Jacobs, K., "Neuere Methoden und Ergebnisse der Ergodentheorie" (Ergebnisse der Math., Neue Folge, Heft 29). Springer, Berlin, 1960.

[31] Kaczmarz, S. and Steinhaus, H., "Theorie der Orthogonalreihen." New York, 1951.

[32] Kato, T., "Perturbation Theory for Linear Operators." Springer, Berlin, 1966.

[33] Montgomery, D. and Zippin, L., "Topological Transformation Groups." Wiley (Interscience), New York, 1955.

[34] Naimark, M., "Normal Rings." P. Nordhoff, Groningen, 1959.

[35] Rickart, C., "General Theory of Banach Algebras." Van Nostrand-Reinhold, New York, 1960.

[36] Weil, A., "Adeles and Algebraic Groups." The Institute for Advanced Study, Princeton, New Jersey, 1961.

VOLUME III

[37] Abraham, R., "Foundations of Mechanics." Benjamin, New York, 1967.

[38] Cartan, H., Séminaire de l'École Normale Supérieure, 1949–50: "Homotopie; espaces fibrés."

[39] Chern, S. S., "Complex Manifolds" (Textos de matematica, No. 5). Univ. do Recife, Brazil, 1959.

[40] Gelfand, I. M. and Shilov, G. E., "Les Distributions," Vols. 1 and 2. Dunod, Paris, 1962.

[41] Gunning, R., "Lectures on Riemann Surfaces." Princeton Univ. Press, Princeton, New Jersey, 1966.

[42] Gunning, R., "Lectures on Vector Bundles over Riemann Surfaces." Princeton Univ. Press, Princeton, New Jersey, 1967.

[43] Hu, S. T., "Homotopy Theory." Academic Press, New York, 1969.

[44] Husemoller, D., "Fiber Bundles." McGraw-Hill, New York, 1966.

[45] Kobayashi, S., and Nomizu, K., "Foundations of Differential Geometry," Vols. 1 and 2. Wiley (Interscience), New York, 1963 and 1969.
[46] Lang, S., "Introduction to Differentiable Manifolds." Wiley (Interscience), New York, 1962.
[47] Porteous, I. R., "Topological Geometry." Van Nostrand-Reinhold, Princeton, New Jersey, 1969.
[48] Schwartz, L., "Théorie des Distributions," New ed. Hermann, Paris, 1966.
[49] Steenrod, N., "The Topology of Fiber Bundles." Princeton Univ. Press, Princeton, New Jersey, 1951.
[50] Sternberg, S., "Lectures on Differential Geometry." Prentice-Hall, Englewood Cliffs, New Jersey, 1964.

VOLUME IV

[51] Abraham, R. and Robbin, J., "Transversal Mappings and Flows." Benjamin, New York, 1967.
[52] Berger, M., "Lectures on Geodesics in Riemannian Geometry." Tata Institute of Fundamental Research, Bombay, 1965.
[53] Carathéodory, C., "Calculus of Variations and Partial Differential Equations of the First Order," Vols. 1 and 2. Holden-Day, San Francisco, 1965.
[54] Cartan, E., "Oeuvres Complètes," Vols. 1_{I} to 3_{II}. Gauthier-Villars, Paris, 1952–1955.
[55] Cartan, E., "Leçons sur la Théorie des Espaces à Connexion Projective." Gauthier-Villars, Paris, 1937.
[56] Cartan, E., "La Théorie des Groupes Finis et Continus et la Géométrie Différentielle traitées par la Méthode du Repère Mobile." Gauthier-Villars, Paris, 1937.
[57] Cartan, E., "Les Systèmes Différentiels Extérieurs et leurs Applications Géométriques." Hermann, Paris, 1945.
[58] Gelfand, I. and Fomin, S., "Calculus of Variations." Prentice Hall, Englewood Cliffs, New Jersey, 1963.
[59] Godbillon, C., "Géométrie Différentielle et Mécanique Analytique." Hermann, Paris, 1969.
[60] Gromoll, D., Klingenberg, W. and Meyer, W., "Riemannsche Geometrie im Grossen," Lecture Notes in Mathematics No. 55. Springer, Berlin, 1968.
[61] Guggenheimer, H., "Differential Geometry." McGraw-Hill, New York, 1963.
[62] Helgason, S., "Differential Geometry and Symmetric Spaces." Academic Press, New York, 1962.
[63] Hermann, R., "Differential Geometry and the Calculus of Variations." Academic Press, New York, 1968.
[64] Hochschild, G., "The Structure of Lie Groups." Holden-Day, San Francisco, 1965.
[65] Klötzler, R., "Mehrdimensionale Variationsrechnung." Birkhäuser, Basel, 1970.
[66] Loos, O., "Symmetric Spaces," Vols. 1 and 2. Benjamin, New York, 1969.
[67] Milnor, J., "Morse Theory," Princeton University Press, Princeton, New Jersey, 1963.
[68] Morrey, C., "Multiple Integrals in the Calculus of Variations." Springer, Berlin, 1966.
[69] Reeb, G., "Sur les Variétés Feuilletées." Hermann, Paris, 1952.
[70] Rund, H., "The Differential Geometry of Finsler Spaces." Springer, Berlin, 1959.
[71] Schirokow, P. and Schirokow, A., "Affine Differentialgeometrie." Teubner, Leipzig, 1962.
[72] Serre, J. P., "Lie Algebras and Lie Groups." Benjamin, New York, 1965.
[73] Wolf, J., "Spaces of Constant Curvature." McGraw-Hill, New York, 1967.

VOLUMES V AND VI

[74] "Algebraic Groups and Discontinuous Subgroups" (Proceedings of Symposia in Pure Mathematics, Vol. IX), American Math. Soc., Providence, 1966.
[75] Bellman, R., "A Brief Introduction to Theta Functions." Holt, Rinehart and Winston, New York, 1961.
[76] Bernat, P. *et al.*, "Représentations des Groupes de Lie Résolubles" (Monographies de la Soc. math. de France, n° 4), Dunod, Paris, 1972.
[77] Borel, A., "Linear Algebraic Groups." Benjamin, New York-Amsterdam, 1969.
[78] Borel, A., "Introduction aux Groupes Arithmétiques." Hermann, Paris, 1969.
[79] Bourbaki, N., "Éléments de Mathématique, Groupes et Algèbres de Lie" (Actual. Scient. Ind., Chap. I, n° 1285, Chap. II–III, n° 1349, Chap. IV–V–VI, n° 1337). Hermann, Paris, 1960–1972.
[80] Carter, R., "Simple Groups of Lie Type." Wiley, New York, 1972.
[81] Chevalley, C., "Classification des Groupes de Lie algébriques," 2 vol.. Séminaire de l'École Normale Supérieure 1956–1958, Paris (Secr. math., 11, Rue P.-Curie).
[82] Conference on Harmonic Analysis (College Park, 1971), "Lecture Notes in Math.," n° 266, Springer, Berlin-Heidelberg-New York, 1972.
[83] Edwards, R., "Fourier Series," 2 vol.. Holt, Rinehart and Winston, New York, 1967.
[84] Gunning, R., "Lectures on Modular Forms." Princeton Univ. Press, 1962.
[85] Hausner, M., Schwartz, J., "Lie Groups, Lie algebras." Gordon Breach, New York, 1968.
[86] Igusa, J., "Thêta Functions." Springer, Berlin-Heidelberg-New York, 1972.
[87] Kahane, J.-P., Salem, R., "Ensembles Parfaits et Séries Trigonométriques." Hermann, Paris, 1963.
[88] Katznelson, Y., "An Introduction to Harmonic Analysis." Wiley, New York, 1968.
[89] Kawata, T., "Fourier Analysis in Probability Theory." Academic Press, New York, 1972.
[90] Meyer, Y., "Trois Problèmes sur les Sommes Trigonométriques," Astérisque, n° 1, 1973.
[91] Miller, W., "Lie Theory and Special Functions." Academic Press, New York, 1968.
[92] Pukansky, L., "Leçons sur les Représentations des Groupes" (Monographies de la Soc. math. de France, n° 2). Dunod, Paris, 1967.
[93] Rudin, W., "Fourier Analysis on Groups." Interscience, New York, 1968.
[94] Serre, J.-P., "Cours d'Arithmétique" (Collection SUP). Presses Univ. de France, Paris, 1970.
[95] Stein, E., Weiss, G., "Introduction to Fourier Analysis on Euclidean Spaces." Princeton Univ. Press, 1971.
[96] Vilenkin, N., "Fonctions Spéciales et Théorie de la Représentation des Groupes." Dunod, Paris, 1969.
[97] Warner, G., "Harmonic Analysis on Semi-simple Lie Groups," Vols. I and II. Springer, Berlin-Heidelberg-New Hork, 1972.
[98] Weil, A., "Basic Number Theory," Springer, Berlin-Heidelberg-New York, 1967.
[99] Zygmund, A., "Trigonometric Series," 2e éd., 2 vol.. Cambridge Univ. Press, 1968.

INDEX

In the following index the first reference number refers to the chapter in which the subject may be found and the second to the section within the chapter.

Q

R

S

T

V

W

Z

Pure and Applied Mathematics

A Series of Monographs and Textbooks

RECENT TITLES

Samuel Eilenberg. Automata, Languages, and Machines: Volumes A and B

Morris Hirsch and Stephen Smale. Differential Equations, Dynamical Systems, and Linear Algebra

Wilhelm Magnus. Noneuclidean Tesselations and Their Groups

François Treves. Basic Linear Partial Differential Equations

William M. Boothby. An Introduction to Differentiable Manifolds and Riemannian Geometry

Brayton Gray. Homotopy Theory: An Introduction to Algebraic Topology

Robert A. Adams. Sobolev Spaces

John J. Benedetto. Spectral Synthesis

D. V. Widder. The Heat Equation

Irving Ezra Segal. Mathematical Cosmology and Extragalactic Astronomy

J. Dieudonné. Treatise on Analysis: Volume II, enlarged and corrected printing; Volume IV; Volume V; Volume VI

Werner Greub, Stephen Halperin, and Ray Vanstone. Connections, Curvature, and Cohomology: Volume III, Cohomology of Principal Bundles and Homogeneous Spaces

I. Martin Isaacs. Character Theory of Finite Groups

James R. Brown. Ergodic Theory and Topological Dynamics

C. Truesdell. A First Course in Rational Continuum Mechanics: Volume 1, General Concepts

George Gratzer. General Lattice Theory

K. D. Stroyan and W. A. J. Luxemburg. Introduction to the Theory of Infinitesimals

B. M. Puttaswamaiah and John D. Dixon. Modular Representations of Finite Groups

Melvyn Berger. Nonlinearity and Functional Analysis: Lectures on Nonlinear Problems in Mathematical Analysis

Charalambos D. Aliprantis and Owen Burkinshaw. Locally Solid Riesz Spaces

In preparation

Jan Mikusinski. The Bochner Integral

Michiel Hazewinkel. Formal Groups and Applications

Thomas Jech. Set Theory

Sigurdur Helgason. Differential Geometry, Lie Groups, and Symmetric Spaces

Carl L. DeVito. Functional Analysis

8 9 0 1 2 3 4 5